Alessandro Soro, Eloisa Vargiu, Giuliano Armano, and Gavino Paddeu (Eds.)

Information Retrieval and Mining in Distributed Environments

Studies in Computational Intelligence, Volume 324

Editor-in-Chief
Prof. Janusz Kacprzyk
Systems Research Institute
Polish Academy of Sciences
ul. Newelska 6
01-447 Warsaw
Poland
E-mail: kacprzyk@ibspan.waw.pl

Further volumes of this series can be found on our homepage: springer.com

Vol. 301. Giuliano Armano, Marco de Gemmis, Giovanni Semeraro, and Eloisa Vargiu (Eds.)
Intelligent Information Access, 2010
ISBN 978-3-642-13999-4

Vol. 302. Bijaya Ketan Panigrahi, Ajith Abraham, and Swagatam Das (Eds.)
Computational Intelligence in Power Engineering, 2010
ISBN 978-3-642-14012-9

Vol. 303. Joachim Diederich, Cengiz Gunay, and James M. Hogan
Recruitment Learning, 2010
ISBN 978-3-642-14027-3

Vol. 304. Anthony Finn and Lakhmi C. Jain (Eds.)
Innovations in Defence Support Systems, 2010
ISBN 978-3-642-14083-9

Vol. 305. Stefania Montani and Lakhmi C. Jain (Eds.)
Successful Case-Based Reasoning Applications-1, 2010
ISBN 978-3-642-14077-8

Vol. 306. Tru Hoang Cao
Conceptual Graphs and Fuzzy Logic, 2010
ISBN 978-3-642-14086-0

Vol. 307. Anupam Shukla, Ritu Tiwari, and Rahul Kala
Towards Hybrid and Adaptive Computing, 2010
ISBN 978-3-642-14343-4

Vol. 308. Roger Nkambou, Jacqueline Bourdeau, and Riichiro Mizoguchi (Eds.)
Advances in Intelligent Tutoring Systems, 2010
ISBN 978-3-642-14362-5

Vol. 309. Isabelle Bichindaritz, Lakhmi C. Jain, Sachin Vaidya, and Ashlesha Jain (Eds.)
Computational Intelligence in Healthcare 4, 2010
ISBN 978-3-642-14463-9

Vol. 310. Dipti Srinivasan and Lakhmi C. Jain (Eds.)
Innovations in Multi-Agent Systems and Applications – 1, 2010
ISBN 978-3-642-14434-9

Vol. 311. Juan D. Velásquez and Lakhmi C. Jain (Eds.)
Advanced Techniques in Web Intelligence, 2010
ISBN 978-3-642-14460-8

Vol. 312. Patricia Melin, Janusz Kacprzyk, and Witold Pedrycz (Eds.)
Soft Computing for Recognition based on Biometrics, 2010
ISBN 978-3-642-15110-1

Vol. 313. Imre J. Rudas, János Fodor, and Janusz Kacprzyk (Eds.)
Computational Intelligence in Engineering, 2010
ISBN 978-3-642-15219-1

Vol. 314. Lorenzo Magnani, Walter Carnielli, and Claudio Pizzi (Eds.)
Model-Based Reasoning in Science and Technology, 2010
ISBN 978-3-642-15222-1

Vol. 315. Mohammad Essaaidi, Michele Malgeri, and Costin Badica (Eds.)
Intelligent Distributed Computing IV, 2010
ISBN 978-3-642-15210-8

Vol. 316. Philipp Wolfrum
Information Routing, Correspondence Finding, and Object Recognition in the Brain, 2010
ISBN 978-3-642-15253-5

Vol. 317. Roger Lee (Ed.)
Computer and Information Science 2010
ISBN 978-3-642-15404-1

Vol. 318. Oscar Castillo, Janusz Kacprzyk, and Witold Pedrycz (Eds.)
Soft Computing for Intelligent Control and Mobile Robotics, 2010
ISBN 978-3-642-15533-8

Vol. 319. Takayuki Ito, Minjie Zhang, Valentin Robu, Shaheen Fatima, Tokuro Matsuo, and Hirofumi Yamaki (Eds.)
Innovations in Agent-Based Complex Automated Negotiations, 2010
ISBN 978-3-642-15611-3

Vol. 320. xxx

Vol. 321. Dimitri Plemenos and Georgios Miaoulis (Eds.)
Intelligent Computer Graphics 2010
ISBN 978-3-642-15689-2

Vol. 322. Bruno Baruque and Emilio Corchado (Eds.)
Fusion Methods for Unsupervised Learning Ensembles, 2010
ISBN 978-3-642-16204-6

Vol. 323. Yingxu Wang, Du Zhang, Witold Kinsner (Eds.)
Advances in Cognitive Informatics, 2010
ISBN 978-3-642-16082-0

Vol. 324. Alessandro Soro, Eloisa Vargiu, Giuliano Armano, and Gavino Paddeu (Eds.)
Information Retrieval and Mining in Distributed Environments, 2010
ISBN 978-3-642-16088-2

Alessandro Soro, Eloisa Vargiu, Giuliano Armano,
and Gavino Paddeu (Eds.)

Information Retrieval and Mining in Distributed Environments

Springer

Alessandro Soro
CRS4, Center of Advanced Studies Research
and Development in Sardinia
Parco Scientifico della Sardegna,
Ed. 1 09010 Loc. Piscinamanna,
Pula, (CA) - Italy
E-mail: asoro@crs4.it

Giuliano Armano
Department of Electrical and
Electronic Engineering
University of Cagliari
Piazza d'Armi
09123 Cagliari - Italy
E-mail: armano@diee.unica.it

Eloisa Vargiu
Department of Electrical and
Electronic Engineering
University of Cagliari
Piazza d'Armi
09123 Cagliari - Italy
E-mail: vargiu@diee.unica.it

Gavino Paddeu
CRS4, Center of Advanced Studies Research
and Development in Sardinia
Parco Scientifico della Sardegna,
Ed. 1 09010 Loc. Piscinamanna,
Pula (CA) - Italy
E-mail: gavino@crs4.it

ISBN 978-3-642-16088-2 e-ISBN 978-3-642-16089-9

DOI 10.1007/978-3-642-16089-9

Studies in Computational Intelligence ISSN 1860-949X

Library of Congress Control Number: 2010936351

Typeset & Cover Design: Scientific Publishing Services Pvt. Ltd., Chennai, India.

Printed on acid-free paper

9 8 7 6 5 4 3 2 1

springer.com

Preface

The Web is increasingly becoming a vehicle of shared, structured, and heterogeneous contents. Thus one goal of next generation information retrieval tools will be to support personalization, context awareness and seamless access to highly variable data and messages coming both from document repositories and ubiquitous sensors and devices.

This book is partly a collection of research contributions from the DART 2009 workshop, held in Milan (Italy) in conjunction with the 2009 IEEE/WIC/ACM International Conference on Web Intelligence (WI 2009) and Intelligent Agent Technology (IAT 2009). Further contributions have been collected and added to the book following a subsequent call for a chapter on the same topics. At DART 2009 practitioners and researchers working on pervasive and intelligent access to web services and distributed information had the opportunity to compare their work and exchange views on such fascinating topics.

Among the several topics addressed, some emerged as the most intriguing. Community oriented tools and techniques form the necessary infrastructure of the Web 2.0. Solutions in this directions are described in Chapters 1-6.

In Chapter 1, *State-of-the-Art in Group Recommendation and New Approaches for Automatic Identification of Groups*, Boratto and Carta present a comprehensive survey on algorithms and systems for group recommendations. Moreover, they propose a novel approach for group recommendation able to adapt to technological constraints (e.g., bandwidth limitations) by automatically identifying groups of users with similar interests, together with a suitable analysis framework and experimental results that support the authors conclusions.

In the following Chapter 2, *Reputation-based Trust Diffusion in Complex Socio-Economic Networks*, Hauke, Pyka, Borschbach, and Heider present a study on the diffusion of reputation-based trust in complex networks. First, they present relevant related work on trust and reputation, as well as their computational adaptation. Then, an outline of complex networks is provided. Finally, they propose a conceptual distributed trust framework, together with a simulation that shows how reputation information can be made available in complex social networks.

In Chapter 3, *From Unstructured Web Knowledge to Plan Descriptions*, Addis and Borrajo present a solution aimed at bridging the gap between automatic extraction of information from the web and automated planning. To this end, they propose an architecture, called PAA (Plan Acquisition Architecture), that performs plan and action acquisition starting from semi-structured information (i.e., web pages). The corresponding system is presented through an example taken from WikiHow, a well-known collaborative project that provides how-to guidelines.

In Chapter 4, *Semantic Desktop: a Common Gate on Local and Distributed Indexed Resources*, Moulin and Lai describe a Web application designed to organize, share and retrieve documents over the Internet with a desktop-like interaction. They consider communities structured as a network of peers without any centralized support. The proposed solution is based on semantic indexing using concepts of domain ontologies automatically downloaded from the network.

In Chapter 5, *An Agent-Oriented Architecture for Researcher Profiling and Association using Semantic Web Technologies*, Adnan, Tahir, Basharat, and de Cesare describe SEMORA, an architecture that combines agent technologies and Semantic Web in order to acquire information about researchers, so as to enable the retrieval and matching of scored profiles. The overall agent architecture is detailed in the papers, together with use cases.

In Chapter 6, *Integrating Peer-to-Peer and Multi-Agent Technologies for the Realization of Content Sharing Applications*, Poggi and Tomaiuolo describe how the well-known multiagent framework JADE can be extended to take advantage of JXTA networking infrastructure and protocols. To this end, they propose RAIS (Remote Assistant for Information Sharing), a peer-to-peer system that provides a set of advanced services for content sharing and retrieval. In particular, RAIS offers a search power comparable with web search engines, but avoids the burden of publishing the information on the web and ensures controlled and dynamic access to the information. In this context, the adoption of agent technologies simplifies the realization of the main features required by the system.

Chapters 7 and 8 are concerned with the exploitation of agent technology applying it to virtual world scenarios.

In the Chapter *Intelligent Advisor Agents in Distributed Environments*, Augello, Pilato, and Gaglio present a decision support system composed of intelligent conversational agents that play the role of advisors explicitly specialized for the government of a virtual town. After a review of knowledge representation models and agent learning, the authors discuss how their intelligent agents work in distributed environments. The chapter ends illustrating a case study in which a real-world town is simulated.

In the Chapter *Agent-based Search and Retrieval in Virtual World Environments*, Eno, Gauch, and Thompson present an intelligent agent crawler designed to collect user-generated content in the Second Life and related virtual worlds. In particular, the authors demonstrate that a crawler able to emulate normal user behavior can successfully collect both static and interactive user-created contents.

In Chapter 9, *Contextual Data Management and Retrieval: a Self-organized Approach*, Castelli and Zambonelli discuss the central topic of context aware information retrieval, presenting a self-organizing agent-based approach to autonomously manage distributed contextual data items into sorts of knowledge networks. Services access contextual information via a knowledge network layer, which encapsulates mechanisms and tools to analyze and self-organize contextual information into sorts. A data model is proposed, meant to represent contextual information, together with a suitable programming interface. Experimental results are provided that show an improvement in efficiency with respect to state of the art approaches.

In the next chapter, *A Relational Approach to Sensor Network Data Mining*, Esposito, Di Mauro, Basile, and Ferilli propose a powerful and expressive description language able to represent the spatio-temporal evolution of a sensor network, together with contextual information. Authors extend a previous framework for mining complex patterns expressed in first-order language. They adopt their framework to discover interesting and human-readable patterns by relating spatio-temporal correlations with contextual ones.

Content based information retrieval is the central topic of Chapters 11-14.

In Chapter 11, *Content-based retrieval of distributed multimedia conversational data*, Pallotta discusses in depth multimedia conversational systems, analyzing several real world implementations and providing a framework for their classification along the following dimensions: conversational content, conversational support, information architecture, indexing and retrieval, and usability. Taking earlier research as the starting point, the author shows how the identification of argumentative structure can improve content based search and retrieval on conversational logs.

In the next Chapter, *Multimodal Aggregation and Recommendation Technologies Applied to Informative Content Distribution and Retrieval*, Messina and Montagnuolo also consider multimedia data, presenting a framework for multimodal information fusion. They propose a definition of semantic affinity for heterogeneous information items and a technique for extracting representative elements. Then, they describe a service platform used for aggregating, indexing, retrieving, and browsing news contents taken from different media sources.

In Chapter 13, *Using a network of scalable ontologies for intelligent indexing and retrieval of visual content*, Badii, Lallah, Zhu, and Crouch present the DREAM framework, whose goal is to support indexing, querying and retrieval of video documents based on content, context and search purpose. The overall architecture and usage scenarios are also provided. Usage studies show a good response in terms of accuracy of classifications.

In the next Chapter, *Integrating Sense Discrimination in a Semantic Information Retrieval System*, Basile, Caputo, and Semeraro propose an information retrieval system that integrates sense discrimination to overcome the problem of word ambiguity. The chapter has a dual goal: (i) to evaluate the effectiveness of an information retrieval system based on Semantic Vectors, and (ii) to describe how they have been integrated into a semantic information retrieval framework to build semantic spaces of words and documents. The authors' main motivation for focusing on the evaluation of disambiguation and discrimination systems is that word ambiguity resolution can improve the performance of information retrieval systems.

Finally, in Chapter 15, *Intelligent Information Processing in Smart Grids and Consumption Dynamics*, Simonov, Zich, and Mussetta describe an industrial application of intelligent information retrieval. The authors describe a distributed environment and discuss the application of data mining and knowledge management techniques to the information available in smart grids, outlining their industrial and commercial potential. The concept of digital energy is introduced here and a system for distributed event delivery is described.

We would like to thank all the authors for their excellent contributions and the reviewers for their careful revision and suggestions for improving them. We are grateful to the Springer-Verlag Team for their assistance during preparation of the manuscripts.

We are also indebted to all the participants and scientific committee members of the three editions of the DART workshop, for their continuous encouragement, support and suggestions.

Cagliari (Italy)
May 2010

Alessandro Soro, Eloisa Vargiu
Giuliano Armano, Gavino Paddeu

Contents

State-of-the-Art in Group Recommendation and New Approaches for Automatic Identification of Groups

Ludovico Boratto and Salvatore Carta

Abstract. Recommender systems are important tools that provide information items to users, by adapting to their characteristics and preferences. Usually items are recommended to individuals, but there are contexts in which people operate in *groups*. To support the recommendation process in social activities, group recommender systems were developed. Since different types of groups exist, group recommendation should adapt to them, managing *heterogeneity* of groups. This chapter will present a survey of the state-of-the-art in group recommendation, focusing on the type of group each system aims to. A new approach for group recommendation is also presented, able to adapt to technological constraints (e.g., bandwidth limitations), by automatically identifying groups of users with similar interests.

1 Introduction

Recommender systems aim to provide information items (web pages, books, movies, music, etc.) that are of potential interest to a user. To predict the items to suggest, the systems use different sources of data, like preferences or characteristics of users.

However, there are contexts and domains where classic recommender systems cannot be used, because people operate in *groups*. Here are some examples of such contexts:

- a system has to provide recommendations to an established group of people who share the same interests and do something together;

Ludovico Boratto · Salvatore Carta
Dipartimento di Matematica e Informatica,
Università di Cagliari,
Via Ospedale 72 - 09124
Cagliari, Italy
e-mail: boratto@sc.unica.it, salvatore@unica.it

Soro et al. (Eds.): Inform. Retrieval and Mining in Distrib. Environments, SCI 324, pp. 1–20.
springerlink.com

- recommendations are provided to an heterogeneous group of people who has a common, specific aim and shares the system on a particular occasion;
- a system tries to recommend items in an environment shared by people who don't have anything in common (e.g., background music in a room);
- when a limitation in the number of available recommendations to be provided is present, individuals with similar preferences have to be grouped.

To manage such cases, group recommendation was introduced. These systems aim to provide recommendations to groups, considering the preferences and the characteristics of more than a user. But what is a *group*? As we can see from the list above, there are at least four different notions of group:

1. **Established group**: a number of persons who explicitly choose to be a part of a group, because of shared, long-term *interests*;
2. **Occasional group**: a number of persons who do something occasionally together, like visiting a museum. Its members have a common *aim* in a particular moment;
3. **Random group**: a number of persons who share an environment in a particular moment, without explicit interests that link them;
4. **Automatically identifed group**: groups that are automatically detected considering the preferences of the users and/or the resources available.

Of course the way a group is formed affects the way it is modeled and how recommendations are predicted.

This chapter will present a survey of the state-of-the-art in group recommendation. A few years ago [29] presented a state-of-the-art survey too, dividing the group recommendation process into four subtasks and describing how each system handles each subtask. Here we will try to describe the existing approaches, focusing on the different notions of group and how the type of group affects the way the system works. Table 1 presents an overview of these systems. Moreover, we will present a new approach, proposed in [8], able to adapt to technological constraints and automatically detect groups of different granularities to fulfill the constraints.

The rest of the chapter is organized as follows: section 2 describes approaches that consider groups with an a priori known structure; section 3 considers systems that automatically identify groups and in 3.2 the new approach cited above is presented; in section 4 we will try to draw some conclusions.

2 Group Recommendation for Groups with an a Priori Known Structure

2.1 Systems That Consider Established Groups

An *established group* is formed by people who share common interests for a long period of time. According to [44] established groups have the property to be *persistent* and users actively *join* the group.

Table 1 Overview of the existing group recommender systems

System	Domain of recommendation	Example of group	Type of group
GRec_OC (Group Recommender for Online Communities) [31]	Books	Online communities that share preferences	1. Established group
Jukola [45]	Music	People attending a party	
PartyVote [53]	Music	People attending a party	
[47]	Movies	Interacting members that share opinions	
I-SPY [51, 50, 52, 49, 9, 22]	Web pages	Communities of like-minded users	
Glue [12]	Web pages	Online communities	
CAPS (Context Aware Proxy based System) [48]	Web pages	Colleagues that browse the web together	
[5]	Documents	Conference committees	
PolyLens [44]	Movies	People who want to see a movie together	2. Occasional group
[14]	Movies	People that share opinions	
[1]	Movies	People that share their *disagreement* with other members	
[18, 19]	Movies	People making decision for a group	
CATS (Collaborative Advisory Travel System) [36, 39, 40, 38, 37]	Travel vacation	Friends planning ski holidays	
INTRIGUE (INteractive TouRist Information GUidE) [3, 2]	Sightseeing destinations	People traveling together	
Travel Decision Forum [27, 26, 28]	Travel vacation	People planning a vacation together	
[33]	Travel vacation	People planning a vacation together	
e-Tourism [23]	Tourist tours	People traveling together	
Pocket RestaurantFinder [34]	Restaurants	People who want to dine together	
FIT (Family Interactive TV System) [25]	TV programs	Family members watching TV together	
[54]	TV programs	Family members watching TV together	
TV4M [56]	TV programs	People watching TV together	
Adaptive Radio [13]	Music	People who share an environment	3. Random group
In-Vehicle Multimedia Recommender [57]	Multimedia items	Passengers traveling together in a vehicle	
Flytrap [17]	Music	People in a public room	
MusicFX [35]	Music	Members of a fitness center	
Let's Browse [32]	Web pages	People that browse the web together	
GAIN (Group Adapted Interaction for News) [46, 11]	News items	People who share an environment	
[10]	Ontology concepts	People that share same interests	4. Automatically identified group
[8]	Movies	People with similar preferences	

As Table 1 shows, group recommender systems that aim to established groups are designed for domains of recommendation like:

- entertainment/cultural items (books, music and movies);
- documents (web pages and conferences documents).

2.1.1 Group Recommender Systems for Entertainment/Cultural Items

GRec_OC (Group Recommender for Online Communities) [31] is a book recommender system for online communities (i.e., people with similar interests that share information). The system aims to improve satisfaction of individual users.

The approach works in two phases. Since the system aims to established groups, the first phase uses a classic Collaborative Filtering (CF) method to build a group profile, by merging the profiles of its members. Each group's nearest neighbors are found and a "candidate recommendation set" is formed by selecting the top-n items. To achieve satisfaction of each member, the second phase evaluates the relevance of the books in the candidate recommendation set for each member. Items not preferred by any member are eliminated and a list of books is recommended to the group.

Jukola [45] and *PartyVote* [53] are two systems able to provide music to an established social group of people attending a party/social event.

The type of group and the context in which the systems are used, make these systems work without any user profiles. In fact, in order to select the music to play, each user is allowed to express preferences (like the selection of a song, album, artist or genre) in a digital musical collection. The rest of the group votes for the available selections and a weight/percentage is associated to each song (i.e., the probability for the song to be played). The song with the highest vote is selected to be played.

The system proposed in [47] aims to produce personality aware group recommendations, i.e., recommendations that consider the personality of its members ("group personality composition") and how conflicts affect the recommendation process.

To measure the behaviors of people in conflicts, each user completes a test and a profile is built computing a measure called *Conflict Mode Weight (CMW)*. Recommendations are calculated using three classic recommendation algorithms, integrated with the CMWs of the group members.

2.1.2 Group Recommender Systems for Documents

I-SPY [51, 50, 52, 49, 9, 22, 16] is a search engine that personalizes the results of a web search, using the preferences of a community of like-minded users.

When a user expresses interest in a search result by clicking on it, I-SPY populates a *hit matrix* that contains relations between the query and the results pages (each community populates its own matrix). Relations in the hit matrix are used to re-rank the search results to improve search accuracy.

Glue [12] is a collaborative retrieval algorithm that monitors the activity of a community of users in a search engine, in order to exploit implicit feedbacks.

A feedback is collected each time a user finds a relevant resource during a search in the system. The algorithm uses the feedback to dynamically strengthen associations between the resource indicated by the user and the keywords used in the search string. Retrieval is based on the feedbacks, so it's not just dependent on the resource's content, making it possible for the system to retrieve even non-textual resources and update its performances dynamically (i.e., the community of users decides which resources are described by which keywords).

CAPS (Context Aware Proxy based System) [48] is an agent that recommends pages and annotates links, based on their popularity among a user's colleagues and the user's profile. The system focuses on two aspects: page enhancement, with symbols that indicate its popularity, and search queries augmentation, with the addition of relevant links for a query. Since the system was designed to enhance the search activity of a user considering the experience of a user's colleagues, a CF approach and a zero-input interface (able to gather implicit information) were used.

The approach proposed in [5] was developed to help a group of conference committees selecting the most suitable items in a large set of candidates.

The approach is based on the *relative preference* of each reviewer, i.e., a rank of the preferred items, with no numeric score given to express the preferences. All the preferences ordering of the reviewers are aggregated through a variable neighborhood search algorithm improved by the authors for the recommendation purpose.

2.2 Systems That Consider Occasional Groups with a Particular Aim

There are lots of contexts in which a group of people is not established but might be interested in getting together for a common aim. This is for example the case of people traveling together: they might not know each other, but they share interest for a common place. In such cases, a group recommender system could be useful, since it would be able to put together the preferences of an heterogeneous group, in order to achieve the common aim. As mentioned in Table 1, group recommender systems that work for occasional groups were developed for the following domains:

- movies;
- tourist destinations;
- TV programs;

Group recommender systems for TV programs consider occasional groups that get togeher for a specific aim (watch TV together) and randomly share an environ-ment (approaches for random groups are described next). Since the approaches focus on the group's aim, this category of systems was placed in this subsection.

2.2.1 Group Recommendation for Movies

PolyLens [44] is a system built to produce recommendations to groups of users who want to see a movie. To produce recommendations for each user of the group a CF algorithm is used. The movies with the highest recommended rates are considered and a "least misery" strategy is used: the recommended rating for a group is the lowest predicted rating for a movie, to ensure that every member is satisfied.

The system proposed in [14] considers interactions among group members, assuming that in a group recommender system ratings are not given just by individuals, but also by subgroups. If a group G is composed of members u_1, u_2 and u_3, ratings might be given by both individuals and subgroups (e.g., $\{u_1,u_2\}$ and $\{u_1,u_3\}$).

The system learns the ratings of a group using a Genetic Algorithm (GA), that uses the ratings of both individuals and subgroups to learn how users interact. For example, if an item is rated by users u_1 and u_2 as 1 and 5 but as a whole they rate the item as 4, it is possible to derive that u_2 plays a more influential role in the group.

The group recommendation methodology used combines an item-based CF algorithm and the GA, to improve the quality of the system.

In [1] an approach to compute group recommendation that introduces *disagreement* between group members as an important aspect to efficiently compute group recommendations is presented. The authors introduce a *consensus function*, which combines *relevance* of the items for a user and *disagreement* between members. After the *consensus function* is built, an algorithm to compute group recommendation (based on the class of Threshold algorithms) is proposed.

The system proposed in [18, 19] presents a group recommendation approach based on Bayesian Networks (BN). The system was developed to help a group of people making decisions that involve the whole group (like seeing a movie) or in situations where individuals must make decisions for the group (like buying a company gift). The system was empirically tested in the movie recommendation domain.

To represent users and their preferences a BN is built. The authors assume that the composition of the groups is a priori known and model the group as a new node in the network that has the group members as parents. A collaborative recommender system is used to predict the votes of the group members. A posteriori probabilities are calculated to combine the predicted votes and build the group recommendation.

2.2.2 Group Recommendation for Tourist Destinations

In [36, 39, 40, 38, 37] a group recommender system called *CATS (Collaborative Advisory Travel System)* is presented. Its aim is to help a group of friends plan and arrange ski holidays. To achieve the objective, users are positioned around a device called "DiamondTouch table-top" [20] and the interactions between them (since they physically share the device) help the development of the recommendations.

To produce the recommendations, the system collects *critiques*, which are feedbacks left by users while browsing the recommended destinations (e.g., a user might specify that he/she is looking for a cheaper hotel, by *critiquing* the price feature).

Interactions with the DiamondTouch device are used to build an individual personal model (IM) and a group user model (GUM). Individual recommendations are built using both the IM and the GUM to maximize satisfaction of the group, whereas group recommendations are based on the critiques contained in the GUM.

INTRIGUE (INteractive TouRist Information GUidE) [3, 2] is a system that recommends sightseeing destinations using the preferences of the group members.

Heterogeneity of a group is considered in several ways. Each group is subdivided into homogeneous subgroups of similar members that fit a stereotype (e.g., children). Recommendations are predicted for each subgroup and an overall preference is built considering some subgroups more influential (e.g., disabled people).

Travel Decision Forum [27, 26, 28] is a system that helps groups of people plan a vacation. Since the system aims to find an agreement between the members of a group, asynchronous communication is possible and, through a web interface, a member can view (and also copy) other members' preferences. Recommendations are made using a simple aggregation (the *median*) of the individual preferences.

In [33] a multiagent system in which agents work on behalf of a group of customers, in order to produce group recommendations, is presented. A formalism, named DCOP (Distributed Constraint Optimization Problem), is proposed to find the best recommendation considering the preferences of the users.

The system works with two types of agents: a user agent (UA), who works on behalf of a user and knows his preferences, and a recommender agent (RA), who works on behalf of suppliers of travel services. An optimization function is proposed to handle the agents' interactions and find the best recommendation.

e-Tourism [23] is a system that plans tourist tours for groups of people. The system considers different aspects, like a group tastes, its demographic classification and places previously visited. A taxonomy-driven recommendation tool called GRSK (Generalist Recommender System Kernel), provides individual recommendations using three techniques: demographic, content-based and preference-based filtering. For each technique group preferences are computed using aggregation, intersection and incremental intersection methods and a list of recommended items is filtered.

Pocket RestaurantFinder [34] is a system that suggests restaurants to groups of people who want to dine together. The system was designed for contexts like conferences, where an occasional group of attendees decides upon a restaurant to visit.

Each user fills a profile with preferences about restaurants, like the price range or the type of cuisine they like (or don't like). Once the group composition is known, the system estimates a user's individual preference for each restaurant and averages those values to build a group preference and produce a list of recommendations.

2.2.3 Group Recommendation for TV Programs

FIT (Family Interactive TV System) [25] is a recommender system that aims to filter TV programs considering the preferences of the viewers.

The only input required by the system is a stereotype user representation (i.e., a class of viewers that would suit the user, like *women*, *businessmen*, *students*, etc.), along with the user preferred watching time. The system automatically updates a profile, by collecting implicit feedbacks from the watching habits of the user.

When someone starts watching TV, the system looks at the probability of each family member to watch TV in that time slot and predicts who there might be watching the TV. Programs are recommended through an algorithm that combines such probabilities and users' preferences.

The system proposed in [54] recommends TV programs to a family.

To protect the privacy of each user and avoid the sharing of information, the system observes the habits of a user and adds contextual information about what is monitored. By observing indicators like the amount of time a TV program has been watched, a user's preferences are exploited and a profile is built.

To estimate the interests of the users in different aspects, the system trains on each family history three Support Vector Machine (SVM) models for program name, genre and viewing history. After the models are trained, recommendation is performed with a Case-Based Reasoning (CBR) technique.

TV4M [56] is a TV programs recommender system for multiple viewers.

To identify who is watching TV, the system provides a login feature. To build a group profile that satisfies most of its members, all the current viewers' profiles are merged, by doing a total distance minimization of the features available (e.g., genre, actor, etc.). According to the built profile, programs are recommended to the group.

2.3 *Systems That Consider Random Groups Who Share an Environment*

A random group is formed by people who share an environment without a specific purpose. Its nature is *heterogeneous* and its members might not share interests.

Group recommender systems that work with random groups calculate the list of predicted items frequently, as people might join or leave the environment. This section will describe group recommender systems that work with random groups. Two main recommendation domains are related to this type of systems:

- multimedia items (e.g., music) broadcast in a shared environment;
- information items (e.g., news or web pages).

2.3.1 Group Recommendation for Broadcast Multimedia Items

Adaptive Radio [13] is a system that broadcasts songs to a group of people who share an environment. The approach tries to improve satisfaction of the users by focusing on *negative preferences*, i.e., it keeps track of which songs a user does not like and avoids playing them. Moreover, the songs similar to the ones rejected by a user are reject too (the system considers two songs similar if they belong to the same album). The highest rated between the remaining songs is automatically played.

In-Vehicle Multimedia Recommender [57] is a system that aims to select multimedia items for a group of people traveling together.

The system aggregates the profiles of the passengers and merges them using a notion of *distance* between the profiles. Once the profiles are merged, a content-based recommender system is used to compare multimedia items and group preferences.

Flytrap [17] is a group recommender system that selects music to be played in a public room. Since people in a room (i.e., the group members) change frequently, the system was designed to predict the song to play considering the preferences of the users present in the room at the moment of the song selection.

A 'virtual DJ' agent is used to automatically decide the song to play. To build a model of the preferences of each user the agent analyzes the MP3 files played by a user in his/her computer and considers the information available about the music (like similar genres, artists, etc.). The song is selected through a voting system in which an agent represents each user in the room and rates the candidate tracks.

MusicFX [35] is a system that recommends music to members of a fitness center.

Since the group structure (i.e., the people in the room) varies continuously, the system gives the users working out in the fitness center the possibility to login. To let users express their preferences about a particular genre, the system has a database of music genres. The music to play is selected considering the preferences of each user in a summation formula.

2.3.2 Group Recommendation for Information Items

Let's Browse [32] is a system that recommends pages to people browsing the web together. Since the group is random (a user might join or leave the group at any time), the system uses an electronic badge to detect the presence of a user.

The system builds a user profile analyzing the words present in his/her homepage. The group is modeled by a linear combination of the individual profiles and the system analyzes the words that occur in the pages browsed by the group.

The system recommends pages that contain keywords present in the user profile.

GAIN (Group Adapted Interaction for News) [46, 11] is a system that selects background information to display in a public shared environment.

The authors assumed that the group of users may be totally unknown, partially or completely known. The group is modeled by splitting it in two subgroups: the *known subgroup* (i.e., people that are certainly near the display for a period of time) and the *unknown subgroup* (i.e., people not recognized by the system). Recommendations are predicted using a statistical dataset built from the group modeling.

3 Group Recommendation with Automatic Group Identification

As shown in Table 1, two group recommender systems automatically detect groups of users. Such an approach is interesting for various reasons: (I) people change their mind frequently, so a user membership in a group might not be long-term, or (II) technological constraints might allow the system to handle only a certain number of groups (or a maximum number of members per group). Group recommender systems that automatically detect groups were developed for the following domains:

- identification of Communities of Interests (groups of similar and previously unrelated people);
- movies recommendation in case of limited bandwidth;

3.1 Group Recommendation with Communities of Interest Identification

The approach proposed in [10] aims to automatically discover Communities of Interest (CoI) (i.e., a group of individuals who share and exchange ideas about a given interest) and produce recommendations for them.

CoI are identified exploiting the preferences expressed by users in personal ontology-based profiles. Each profile measures the interest of a user in concepts of the ontology. The interest expressed by users is used to cluster the concepts.

User profiles are then split into subsets of interests, to link the preferences of each user with a specific cluster of concepts. Hence it is possible to define relations among users at different levels, obtaining a multilayered interest network that allows to find multiple CoI. Recommendations are built using a content-based CF approach.

3.2 Group Recommendation with Automatic Identification of Users' Communities in Case of Bandwidth Limitations

None of the approaches described takes into account the fact that it might be necessary to identify groups of people with similar interests because of technological constraints, like bandwidth limitations.

For example, in multiple access systems with limited transmission capacity like Mobile IPTV or Satellite Systems, it might not be possible to create personalized program schedules for each user. In such cases, the problem relies in identifying groups of related users to fulfill the constraints.

Here we present an approach proposed in [8] to generate group recommendations, able to detect intrinsic communities of users whose preferences are similar. The algorithm takes as input a matrix that associates a set of *users* to a set of *items* through a *rating*. This matrix will be called the *ratings matrix*. Based on ratings expressed by each user in the ratings matrix, the algorithm evaluates the level of similarity between users and generates a network that contains the similarities. A modularity-based Community Detection algorithm proposed by [7] will be run on the network, to find partitions of users in communities. For each community, ratings for all the items will be calculated.

Since the Community Detection algorithm is able to produce a dendrogram, i.e., a tree that contains hierarchical partitions of the users in communities of increasing granularity, experiments were conducted in order to evaluate the quality of the recommendation for the different partitions. Results show that the quality of group recommendations increases linearly with the number of communities created.

The scientific contribution of the recommendation algorithm is the capability to automatically detect intrinsic communities of users who share similar preferences, making it possible for a content provider to explore the trade off between the level of personalization of the recommendation and the number of channels.

3.2.1 Group Recommendation with Automatic Identification of Users Communities

The group recommendation algorithm works in four steps:

Users similarity evaluation

In order to create communities of users, the algorithm takes as input a *ratings matrix* and evaluates through a standard metric (cosine similarity) how similar the preferences of two users are. The result is a weighted network where nodes represent users and a weighted edge represents the similarity value of the users it connects.

Communities detection

To identify intrinsic communities of users, a Community Detection algorithm proposed in [7] is applied to the users similarity network and partitions of different granularities are generated.

Ratings prediction for items rated by enough users of a group

A group's ratings are evaluated by calculating, for each item, the mean of the ratings expressed by the users of the group. In order to predict meaningful ratings, the algorithm calculates a rating only if an item was evaluated by a minimum percentage of users in the group. With this step it is not possible to predict a rating for each item, so another step has been created to predict the remaining ratings.

Ratings prediction for the remaining items

For some of the items, ratings could not be calculated by the previous step. In order to estimate such ratings, similarity between items is evaluated, and the rating of an item is predicted considering the items most similar to it.

The four steps that constitute the algorithm will now be described in detail.

Step 1. Users similarity evaluation

Here it is described how a ratings matrix can be used to evaluate similarity between users. Let v_i be the vector of the ratings expressed by a user i for the items and v_j be the vector of the ratings expressed by a user j for the items. The similarity s_{ij} between users i and j can be measured by the cosine similarity between the vectors:

$$s_{ij} = cos(v_i, v_j) = \frac{v_i \cdot v_j}{\|v_i\| \times \|v_j\|}$$

Similarities can be represented in a network, the *users similarity network*, that links each couple of associated users with a weighted edge.

As highlighted by [24], in networks like the one built, edges have intrinsic weights and no information is given about the real associations between the nodes. Edges are usually affected by noise, which leads to ambiguities in the communities detection. Moreover, the weights of the edges in the network are calculated considering the ratings and it is well known that people have different rating tendencies: some users tend to express their opinion using just the end of the scales, expressing if they loved or hated an item. To eliminate noise from the network and reduce its complexity by removing weak edges, a parameter called *noise* was set in the algorithm. The parameter indicates the weight that will be subtracted by every edge.

Step 2. Communities Detection

This step of the algorithm has the goal to find intrinsic communities of users, accepting as input the weighted users similarity network that was built in the previous step. Another requirement is to produce the intrinsic users communities in a hierarchical structure, in order to deeper understand and exploit its inner partition. Out of all the existing classes of clustering algorithms, complex network analysis [21] was identified as the only class of algorithms fulfilling the requirements. In 2004 an optimization function has been introduced, the modularity [41], that measures for a generic partition of the set of nodes in the network, the number of internal (in each partition) edges respect to the random case. The optimization of this function gives, without a previous assessment of the number and size of the partitions [21], the natural community structure of the network. Moreover it is not necessary to embed the network in a metric space like in the k-means algorithm. A notion of distance or link weight can be introduced but in a pure topological fashion [42].

Recently a very efficient algorithm has been proposed, based on the optimization of the weighted modularity, that is able to easily handle networks with millions of nodes, generating also a dendrogram; a community structure at various network resolutions [7]. Since the algorithm had all the characteristics needed, it was chosen to create the groups of users used by the group recommendation algorithm.

Step 3. Ratings prediction for items rated by enough users of a group

To express a group's preference for an item, the algorithm calculates its rating, considering the ratings expressed by the users of the community for that item.

An average is a single value that is meant to typify a list of values. The most common method to calculate such a value is the arithmetic mean, which also seems an effective way to put together all the ratings expressed by the users in a group. So, for each item i, its rating r_i is expressed as:

$$r_i = \frac{1}{n} \sum_{u=0}^{n} r_u$$

where n is the number of users of the group who expressed a rating for item i and r_u is the rating expressed by each user for that item. In order to calculate meaningful ratings for a group, a rating r_i is considered only if a minimum part of the group has rated the item. This is done through a parameter, called *co-ratings* which expresses the minimum percentage of users who have to rate an item in order to calculate the rating for the group.

Step 4. Ratings prediction for the remaining items

For some of the items, ratings could not be calculated by the previous step. In order to estimate such ratings, a network that contains similarities between items was built. Like the users similarity network presented in 3.2.1, the network is built through the ratings matrix, considering the ratings expressed for each item. Let w_i be the vector of the ratings expressed by all the users for item i and w_j be the vector of the ratings expressed by all the users for item j. The similarity t_{ij} between item i and item j is measured with the cosine similarity and the similarities are represented in a network called *items similarity network*, from which noise was removed through the *noise* parameter presented in 3.2.1.

For each item not rated by the group, a list is produced with its nearest neighbors, i.e., the most similar items already rated by the group, considering the similarities available in the *items similarity network*. Out of this list, the *top* items are selected. Parameter *top* indicates how many similarities the algorithm considers to predict the ratings. An example of how the *top* similar items are selected is shown in Table 2. The algorithm needs to predict a rating for Item 1. The most similar items are shown in the list. For each similar item j, the table indicates the similarity with Item 1 (column t_{1j}) and the rating expressed by the group (column r_j). In the example, the *top* parameter is set to 3 and items with similarity 0.95, 0.88 and 0.71 are selected.

it is now possible to predict the rating of an unrated item by considering both the rating and the similarity of its *top* similar items:

$$\bar{r}_i = \frac{\sum_{j=0}^{n} r_j \cdot t_{ij}}{\sum_{j=0}^{n} t_{ij}}$$

Table 2 Top similar items of an unrated item

Item j	t_{1j}	r_j
Item 2	0.95	3.5
Item 3	0.95	4.2
Item 4	0.88	2.8
Item 5	0.71	2.6
Item 6	0.71	3.9
Item 7	0.71	4.3
Item 8	0.63	1.2
Item 9	0.55	3.2

where n is the number of items selected in the list. Given the example in Table 2, $\bar{r}_1 = 3.55$.

To make meaningful predictions, an evaluation of how "reliable" the predictions are is needed. This is done by calculating the mean of the *top* similarities and by setting a *trust* parameter. The parameter indicates the minimum value the mean of the similarities has to get, in order to be considered reliable and consider the predicted rating. The mean of the similarities in the previous example is 0.85 so, to consider $\bar{r}_1$, the *trust* parameter has to be lower than 0.85.

3.2.2 Algorithm Experimentation

To evaluate the quality of the recommendations, the algorithm was tested using MovieLens[1], a dataset widely used to evaluate CF algorithms. A framework that extracts a subset of ratings from the dataset, predicts group recommendations through the presented algorithm and measures the quality of the predictions in terms of RMSE was built. Details of the algorithm experimentation will now be described.

Experimental methodology and setup

The experimentation was made with the MovieLens dataset, which is composed of 1 million ratings, expressed by 6040 users for 3900 movies. To evaluate the quality of the ratings predicted by the algorithm, around 10% of the ratings was extracted as a probe test set and the rest of the dataset was used as a training set for the algorithm.

The group recommendation algorithm was run with the training set and, for each partition of the users in communities, ratings were predicted. The quality of the predicted ratings was measured through the Root Mean Squared Error (RMSE). The metric compares the probe test set with the ratings predicted: each rating r_i expressed by a user u for an item i is compared with the rating $\bar{r}_i$ predicted for the item i for the group in which user u is. The formula is shown below:

$$RMSE = \sqrt{\frac{\sum_{i=0}^{n}(r_i - \bar{r}_i)^2}{n}}$$

[1] http://www.grouplens.org/

where n is the number of ratings available in the test set. To evaluate the performances of the algorithm, they were compared with the results obtained considering a single group with all the users (predictions are calculated considering all the preferences expressed for an item), and the results obtained using a classic CF algorithm proposed in [15], where recommendations are produced for each user.

Experimental results

To evaluate the algorithm's performances the quality of the recommendations was studied, considering different values of each parameter. The only value that could not be changed was *noise*, because if more than 0.1 was subtracted to the edges of the *users similarities network*, the network would become disconnected.

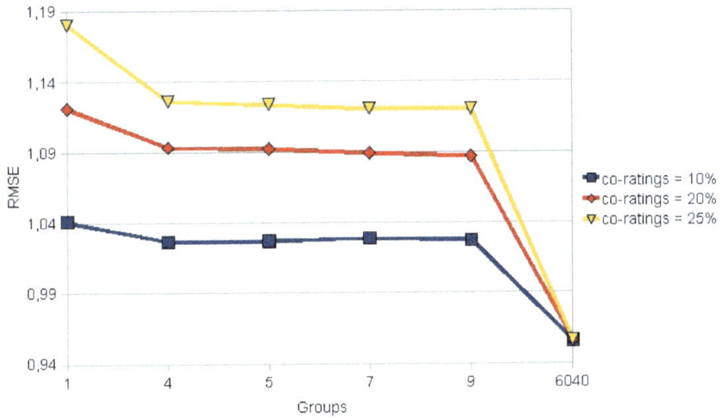

Fig. 1 Algorithm's performances with different co-ratings values

The first experiment conducted evaluated the quality of the recommendations for different values of the *co-ratings* parameter, i.e., the minimum percentage of users who have to rate an item, in order to calculate the rating for the group. Parameter *top* was set to 2 and parameter *trust* was set to 0.0. Fig. 1 shows how RMSE varies with the number of groups, for different values of *co-ratings* (10%, 20% and 25%). It is possible to see that as the number of groups grows, the quality of the recommendations improves, since groups get smaller and the algorithm predicts more precise ratings. To conduct the following experiments, the value of *co-ratings* chosen was 20%. The next experiment conducted was to evaluate the quality of recommendations for different values of the *top* parameter, i.e., the number of similarities considered to select the nearest neighbors of an item. Fig. 2 shows how RMSE varies with the number of groups, for different values of *top* (2 and 3). It is worth noting that the quality of the recommendations improves when parameter *top* is set to 3 (i.e., the top 3 similarities are selected from the list), so this was the value set for the next experiment. The last parameter to evaluate is *trust*, i.e., the minimum value the mean of the similarities has to get when the algorithms predicts a rating considering

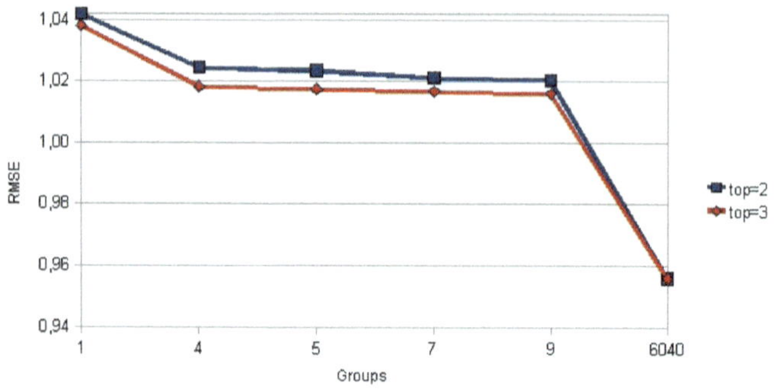

Fig. 2 Algorithm's performances for different values of top

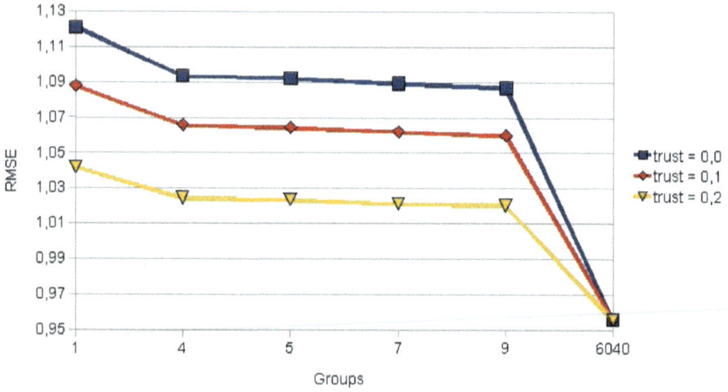

Fig. 3 Algorithm's performances with different trust values

the nearest neighbors of an item. Fig. 3 shows how RMSE varies with the number of groups, for different values of the parameter (0.0, 0.1 and 0.2). In Fig. 3 is shown that the quality of the performances improves for higher values of *trust*, i.e., when the ratings predicted can be considered more "reliable".

4 Conclusions and Discussion

Recommender systems have become important tools that help people making decisions, by adapting to preferences or characteristics of a user and effectively suggesting items that might interest him/her. However, there are contexts in which people operate in groups and in the last years several approaches to produce recommendations for groups of users were developed.

This chapter presented a state-of-the-art survey on group recommendation, focusing on the nature of the group considered by each system. Moreover, a new approach able to adapt to technological constraints (e.g., bandwidth limitations) and produce recommendations for automatically detected groups was presented.

As we can see, nearly all the approaches take for granted the type of group they are aimed to: whether the group is *established*, *occasional* or *random*, its structure is taken "as is". However, there might be contexts in which groups are not available and just two approaches focus on the identification of groups. We believe that the study of algorithms specifically designed for group recommendation, able to model and identify groups, might improve the quality of the recommendation process.

References

1. Amer-Yahia, S., Roy, S.B., Chawla, A., Das, G., Yu, C.: Group recommendation: Semantics and efficiency. PVLDB 2(1), 754–765 (2009)
2. Ardissono, L., Goy, A., Petrone, G., Segnan, M.: A multi-agent infrastructure for developing personalized web-based systems. ACM Trans. Internet Technol. 5(1), 47–69 (2005)
3. Ardissono, L., Goy, A., Petrone, G., Segnan, M., Torasso, P.: Intrigue: Personalized recommendation of tourist attractions for desktop and handset devices. Applied Artificial Intelligence 17(8), 687–714 (2003)
4. Baccigalupo, C., Plaza, E.: A case-based song scheduler for group customised radio. In: Weber and Richter [55], pp. 433–448
5. Baskin, J.P., Krishnamurthi, S.: Preference aggregation in group recommender systems for committee decision-making. In: Bergman, et al. [6], pp. 337–340
6. Bergman, L.D., Tuzhilin, A., Burke, R.D., Felfernig, A., Schmidt-Thieme, L. (eds.): Proceedings of the 2009 ACM Conference on Recommender Systems, RecSys 2009, October 23-25. ACM, New York (2009)
7. Blondel, V.D., Guillaume, J.-L., Lambiotte, R., Lefebvre, E.: Fast unfolding of communities in large networks. J. Stat. Mech. (10), P10008+ (2008)
8. Boratto, L., Carta, S., Chessa, A., Agelli, M., Clemente, M.L.: Group recommendation with automatic identification of users communities. In: Web Intelligence/IAT Workshops, pp. 547–550. IEEE, Los Alamitos (2009)
9. Briggs, P., Smyth, B.: Modeling trust in collaborative web search. In: AICS, Coleraine, NI (2005)
10. Cantador, I., Castells, P., Superior, E.P.: Extracting multilayered semantic communities of interest from ontology-based user profiles: Application to group modelling and hybrid recommendations. In: Computers in Human Behavior, special issue on Advances of Knowledge Management and the Semantic. Elsevier, Amsterdam (2008) (in press)
11. De Carolis, B., Pizzutilo, S.: Providing relevant background information in smart environments. In: Noia and Buccafurri [43], pp. 360–371
12. Carta, S., Alimonda, A., Clemente, M.L., Agelli, M.: Glue: Improving tag-based contents retrieval exploiting implicit user feedback. In: Hoenkamp, E., de Cock, M., Hoste, V. (eds.) Proceedings of the 8th Dutch-Belgian Information Retrieval Workshop (DIR 2008), pp. 29–35 (2008)
13. Chao, D.L., Balthrop, J., Forrest, S.: Adaptive radio: achieving consensus using negative preferences. In: Pendergast, M., Schmidt, K., Mark, G., Ackerman, M. (eds.) GROUP, pp. 120–123. ACM, New York (2005)
14. Chen, Y.-L., Cheng, L.-C., Chuang, C.-N.: A group recommendation system with consideration of interactions among group members. Expert Syst. Appl. 34(3), 2082–2090 (2008)

15. Clemente, M.L.: Experimental results on item-based algorithms for independent domain collaborative filtering. In: AXMEDIS 2008: Proceedings of the 2008 International Conference on Automated solutions for Cross Media Content and Multi-channel Distribution, Washington, DC, USA, pp. 87–92. IEEE Computer Society, Los Alamitos (2008)
16. Coyle, M., Smyth, B.: Explaining search results. In: Kaelbling and Saffiotti [30], pp. 1553–1555
17. Crossen, A., Budzik, J., Hammond, K.J.: Flytrap: intelligent group music recommendation. In: IUI, pp. 184–185 (2002)
18. de Campos, L.M., Fernández-Luna, J.M., Huete, J.F., Rueda-Morales, M.A.: Group recommending: A methodological approach based on bayesian networks. In: ICDE Workshops, pp. 835–844. IEEE Computer Society, Los Alamitos (2007)
19. de Campos, L.M., Fernández-Luna, J.M., Huete, J.F., Rueda-Morales, M.A.: Managing uncertainty in group recommending processes. User Model. User-Adapt. Interact. 19(3), 207–242 (2009)
20. Dietz, P.H., Leigh, D.: Diamondtouch: a multi-user touch technology. In: UIST, pp. 219–226 (2001)
21. Fortunato, S., Castellano, C.: Community structure in graphs. Springer's Encyclopedia of Complexity and System Science (December 2007)
22. Freyne, J., Smyth, B.: Cooperating search communities. In: Wade, V.P., Ashman, H., Smyth, B. (eds.) AH 2006. LNCS, vol. 4018, pp. 101–111. Springer, Heidelberg (2006)
23. Garcia, I., Sebastia, L., Onaindia, E., Guzman, C.: A group recommender system for tourist activities. In: Noia and Buccafurri [43], pp. 26–37
24. Gfeller, D., Chappelier, J.C., Los, D.: Finding instabilities in the community structure of complex networks. Physical Review E 72(5 Pt 2), 056135+ (2005)
25. Goren-Bar, D., Glinansky, O.: Fit-recommend ing tv programs to family members. Computers & Graphics 28(2), 149–156 (2004)
26. Jameson, A.: More than the sum of its members: Challenges for group recommender systems. In: Proceedings of the International Working Conference on Advanced Visual Interfaces, Gallipoli, Italy, pp. 48–54 (2004), `http://dfki.de/~jameson/abs/Jameson04AVI.html`
27. Jameson, A., Baldes, S., Kleinbauer, T.: Enhancing mutual awareness in group recommender systems. In: Mobasher, B., Anand, S.S. (eds.) Proceedings of the IJCAI 2003 Workshop on Intelligent Techniques for Web Personalization. AAAI, Menlo Park (2003), `http://dfki.de/~jameson/abs/JamesonBK03ITWP.html`
28. Jameson, A., Baldes, S., Kleinbauer, T.: Two methods for enhancing mutual awareness in a group recommender system. In: Proceedings of the International Working Conference on Advanced Visual Interfaces, Gallipoli, Italy (2004) (in press)
29. Jameson, A., Smyth, B.: Recommendation to groups. In: Brusilovsky, P., Kobsa, A., Nejdl, W. (eds.) Adaptive Web 2007. LNCS, vol. 4321, pp. 596–627. Springer, Heidelberg (2007)
30. Kaelbling, L.P., Saffiotti, A. (eds.): IJCAI 2005, Proceedings of the Nineteenth International Joint Conference on Artificial Intelligence, Edinburgh, Scotland, UK. Professional Book Center (July 30-August 5, 2005)
31. Kim, J.K., Kim, H.K., Oh, H.Y., Ryu, Y.U.: A group recommendation system for online communities. International Journal of Information Management (2009) (in press, corrected proof)
32. Lieberman, H., Van Dyke, N.W., Vivacqua, A.S.: Let's browse: A collaborative web browsing agent. In: IUI, pp. 65–68 (1999)
33. Lorenzi, F., Santos, F., Ferreira Jr., P.R., Bazzan, A.L.: Optimizing preferences within groups: A case study on travel recommendation. In: Zaverucha, G., da Costa, A.L. (eds.) SBIA 2008. LNCS (LNAI), vol. 5249, pp. 103–112. Springer, Heidelberg (2008)

34. McCarthy, J.F.: Pocket restaurantfinder: A situated recommender system for groups. In: Workshop on Mobile Ad-Hoc Communication at the 2002 ACM Conference on Human Factors in Computer Systems, Minneapolis (2002)
35. McCarthy, J.F., Anagnost, T.D.: Musicfx: an arbiter of group preferences for computer supported collaborative workouts. In: CSCW, p. 348 (2000)
36. McCarthy, K., McGinty, L., Smyth, B.: Case-based group recommendation: Compromising for success. In: Weber and Richter [55], pp. 299–313
37. McCarthy, K., McGinty, L., Smyth, B., Salamó, M.: The needs of the many: A case-based group recommender system. In: Roth-Berghofer, T., Göker, M.H., Güvenir, H.A. (eds.) ECCBR 2006. LNCS (LNAI), vol. 4106, pp. 196–210. Springer, Heidelberg (2006)
38. Mccarthy, K., Mcginty, L., Smyth, B., Salamo, M.: Social interaction in the cats group recommender. In: Brusilovsky, P., Dron, J., Kurhila, J. (eds.) Workshop on the Social Navigation and Community-Based Adaptation Technologies at the 4th International Conference on Adaptive Hypermedia and Adaptive Web-Based Systems (June 2006)
39. McCarthy, K., Salamó, M., Coyle, L., McGinty, L., Smyth, B., Nixon, P.: Cats: A synchronous approach to collaborative group recommendation. In: Sutcliffe, G., Goebel, R. (eds.) FLAIRS Conference, pp. 86–91. AAAI Press, Menlo Park (2006)
40. McCarthy, K., Salamó, M., Coyle, L., McGinty, L., Smyth, B., Nixon, P.: Group recommender systems: a critiquing based approach. In: Paris, C., Sidner, C.L. (eds.) IUI, pp. 267–269. ACM, New York (2006)
41. Newman, M.E., Girvan, M.: Finding and evaluating community structure in networks. Phys. Rev. E Stat. Nonlin. Soft. Matter Phys. 69(2 Pt 2) (February 2004)
42. Newman, M.E.J.: Analysis of weighted networks. Phys. Rev. E 70(5), 56131 (2004)
43. Di Noia, T., Buccafurri, F. (eds.): E-Commerce and Web Technologies. LNCS, vol. 5692. Springer, Heidelberg (2009)
44. O'Connor, M., Cosley, D., Konstan, J.A., Riedl, J.: Polylens: a recommender system for groups of users. In: ECSCW 2001: Proceedings of the seventh Conference on European Conference on Computer Supported Cooperative Work, Norwell, MA, USA, pp. 199–218. Kluwer Academic Publishers, Dordrecht (2001)
45. O'Hara, K., Lipson, M., Jansen, M., Unger, A., Jeffries, H., Macer, P.: Jukola: democratic music choice in a public space. In: DIS 2004: Proceedings of the 5th Conference on Designing Interactive Systems, pp. 145–154. ACM, New York (2004)
46. Pizzutilo, S., De Carolis, B., Cozzolongo, G., Ambruoso, F.: Group modeling in a public space: methods, techniques, experiences. In: AIC 2005: Proceedings of the 5th WSEAS International Conference on Applied Informatics and Communications, Stevens Point, Wisconsin, USA, pp. 175–180. World Scientific and Engineering Academy and Society, WSEAS (2005)
47. Recio-García, J.A., Jiménez-Díaz, G., Sánchez-Ruiz-Granados, A.A., Díaz-Agudo, B.: Personality aware recommendations to groups. In: Bergman et al. [6], pp. 325–328
48. Sharon, T., Lieberman, H., Selker, T.: A zero-input interface for leveraging group experience in web browsing. In: IUI, pp. 290–292. ACM, New York (2003)
49. Smyth, B., Freyne, J., Coyle, M., Briggs, P., Balfe, E.: I-SPY: Anonymous, Community-Based Personalization by Collaborative Web Search. In: Proceedings of the 23rd SGAI International Conference on Innovative Techniques, pp. 367–380. Springer, Cambridge (2003)
50. Smyth, B., Balfe, E.: Anonymous personalization in collaborative web search. Inf. Retr. 9(2), 165–190 (2006)
51. Smyth, B., Balfe, E., Boydell, O., Bradley, K., Briggs, P., Coyle, M., Freyne, J.: A live-user evaluation of collaborative web search. In: Kaelbling and Saffiotti [30], pp. 1419–1424

52. Smyth, B., Balfe, E., Briggs, P., Coyle, M., Freyne, J.: Collaborative web search. In: Gottlob, G., Walsh, T. (eds.) IJCAI, pp. 1417–1419. Morgan Kaufmann, San Francisco (2003)
53. Sprague, D., Wu, F., Tory, M.: Music selection using the partyvote democratic jukebox. In: AVI 2008: Proceedings of the Working Conference on Advanced Visual Interfaces, pp. 433–436. ACM, New York (2008)
54. Vildjiounaite, E., Kyllönen, V., Hannula, T., Alahuhta, P.: Unobtrusive dynamic modelling of tv programme preferences in a finnish household. Multimedia Syst. 15(3), 143–157 (2009)
55. Weber, R., Richter, M.M. (eds.): ICCBR 2007. LNCS (LNAI), vol. 4626. Springer, Heidelberg (2007)
56. Yu, Z., Zhou, X., Hao, Y., Gu, J.: Tv program recommendation for multiple viewers based on user profile merging. User Model. User-Adapt. Interact. 16(1), 63–82 (2006)
57. Zhiwen, Y., Xingshe, Z., Daqing, Z.: An adaptive in-vehicle multimedia recommender for group users. In: 2005 IEEE 61st Vehicular Technology Conference on VTC 2005-Spring, vol. 5, pp. 2800–2804 (2005)

Reputation-Based Trust Diffusion in Complex Socio-Economic Networks

Sascha Hauke, Martin Pyka, Markus Borschbach, and
Dominik Heider

Abstract. Trust and reputation form the foundation of most human interactions, they are ubiquitous in everyday life. Over the past years, attempts have been made to model trust relations computationally, either to assist users or for modeling purposes in multi-agent systems. As a fundamentally social phenomenon, trust forms, operates on and changes social networks, an aspect not investigated in detail so far. In this chapter, we aim to investigate how the nature of social networks, such as their quality of being highly clustered, impacts the spread and thus the availability of data to agents. Furthermore, we will propose an extension to state-of-the-art trust frameworks that leverages the capabilities of information spreading in complex networks by decoupling the provisioning process of reputation information from non-neighboring recommenders.

Sascha Hauke
Institute of Computer Science,
University of Münster & FHDW in Bergisch Gladbach,
Germany
e-mail: `Sascha.Hauke@uni-muenster.de`

Martin Pyka
Department of Psychiatry,
University of Marburg, Germany
e-mail: `Martin.Pyka@uni-marburg.de`

Markus Borschbach
FHDW in Bergisch Gladbach, Germany
e-mail: `Markus.Borschbach@fhdw.de`

Dominik Heider
Department of Bioinformatics,
Center for Medical Biotechnology,
University of Duisburg-Essen, Germany
e-mail: `Dominik.Heider@uni-due.de`

Soro et al. (Eds.): Inform. Retrieval and Mining in Distrib. Environments, SCI 324, pp. 21–40.
springerlink.com

1 Introduction

Until quite recently, the study of trust was firmly in the hands of the social sciences. After all, humans intuitively understand the value of trusting (or distrusting) someone and have become adept at deducing whether or not to trust someone from sublime features. With the advent and ever-increasing popularity of the internet—and particularly the world wide web—many of the actions that are normal social or commercial behavior for human beings, congregating and shopping, for instance, have moved into these cyber-regions. But many of the intuitively graspable markers for evaluating the trustworthiness of someone else are not easily transferable to that new technological domain. Yet, just as in real life, humans establish social networks online, and also just like in real life, these networks—possessing particular structural and dynamic features—can be used to exchange information about others, such as recommendations or gossip.

In the following, we will briefly introduce the concepts of trust and reputation, as well as their computational adaptation (section 2), outline the particularities of complex networks (3), present a conceptual distributed trust framework, building on and extending the state-of-the-art (4) and simulate how reputation information is being made available in complex social networks by application of that framework (5).

2 Trust and Reputation

One of the main pillars of personal relationships is the notion of *trust* in association with the related notion of *reputation*. Both of these concepts are social phenomena and humans—as social entities—are intimately familiar with the way they are applied. Therefore, these concepts have long been a forte of the traditional social sciences, such as psychology [8] or economy [28]. Over the past 15 years, however, the relevance of reputation-based trust has increasingly manifested itself in the various fields of computer science.

2.1 Trust

Trust is highly important in personal interactions and business ventures and has been examined by a multitude of scientist in different disciplines of study. While the positive effects of trust are universally accepted, scholars have been unable to come to a general consensus regarding the meaning of the term trust—it has a plethora of meanings [30, 42], depending on the person asked or literature consulted. Even in the field of computer science, where it is usual to deal in well-defined terms, competing views on trust exist. These views can be categorized into two main classes—cognitive and probabilistic.

On the one hand, the cognitive school, represented mainly by Falcone and Castelfranchi [10, 11], argues that trust is based on an internal mental state of beliefs.

On the other hand, the probabilistic (also: computational or game theoretical) school holds the view that trust can be established by evaluating observable data and deriving a, albeit subjective, probability with which some agent will perform a particular action. This view is put forth, among others, by [1, 22, 43]. By employing observed information from the past to predict behavior in the future, trust establishment thus becomes a data driven, rather than a belief driven, process. By concentrating trust establishment on external observations, as opposed to internal states, it becomes well-suited to computational treatment—given the availability of sufficient amounts of data. Commonly, the probabilistic view of trust follows the definition according to Gambetta [17], as this definition is concise and easily adaptable to computational formalisms.

Definition 1. Trust (or, symmetrically, distrust) is a particular level of the subjective probability with which an agent will perform a particular action, both *before* he can monitor such action (or independently of his capacity ever to be able to monitor it) *and* in a context in which it affects *his own* action.

Thus, trust is not an objective measure of reliability, but rather depends on the trusting party and its expectations regarding the actions of the trusted party. These expectations are formulated *prior* to the actions being implemented—and possibly even without any means of verifying if and how the trusted party acted. Also, trust is *situation dependent*, i.e. trust is given to an agent in a certain context, but withheld from the same agent in another. As an example, you might trust your neighbor to clear the sidewalk in front of his house of snow in the winter, but you might not trust him to look after your children. Furthermore, trust is also associated with interaction, as an action is taken by the trusted agent (or *trustee*)*in a context in which it affects [the trusting agent's (or* trustor's*)] action.*

Sabater and Sierra [37] provide a comprehensive overview of different proposed trust and reputation models. As this article is mainly concerned with the diffusion of trust information through a society of agents, and less with the trust decision making process itself, the processes involved in the latter will be covered in an abbreviated and abstract manner. In particular, cognitive models of trust are of little relevance in the course of this article and will not be inspected further.

Trust, as a social concept, influences more than just the relationship between two agents. It impacts, directly or indirectly, the entire community of participating agents. This impact is the result of the diverse nature of the observed data that forms the core of the probabilistic trust formation process. This data does not only include direct interactions and the resulting experiences between two agents, but also recommendations, external observations of the behavior of agents or qualities of the environment the agent jointly inhabit. Therefore, we will have to consider not just single agents or pairs of agents, but entire societies of agents. Here, a society of agents is represented by the nodes forming a network component.

Trust, as a formal relation, possesses a number of relational properties [1]: **(a)** Trust is *unidirectional*; **(b)** Trust is generally *not transitive*; **(c)** Trust is a *binary* relation.

As further experiences are made, and old experiences are literally forgotten, trust in another agent can change over time. Agents can redeem themselves by improving their performance in interactions with other agents, or lose standing, if they behave in a dissatisfactory or erratic manner. Therefore, trust is a situation and time (t) dependent, unidirectional, intransitive, binary relation.

2.2 Reputation

In order to reliably make a decision of whether or not to trust another agent, that trust has to be based on *reliable* information about the actions expected to be performed by the trusted party. In society, this information is usually procured in two different ways: Either through **(a)** Personal experience derived from prior direct interactions with another entity or via **(b)** Reputability of an entity as reported by other members of the society.

Information garnered from personal experience is easier to evaluate for an agent than information it receives from other agents. The source, the situation and the time at which the information was recorded are known to and trusted by the agent. Although this method can yield the most reliable form of trust information, its reliance on a sufficiently large amount of prior interaction with and *direct* knowledge of the foreign agent hampers its efficiency.

In order to alleviate these, reputation can also be derived via including *recommendations* from third-party (trusted) agents. These third-party agents can supply trust information to an agent in order to give it a broader base upon which to build its reasoning for trusting or distrusting a foreign agent. Abdul-Rahman [1] provide an early model for distributed trust that outlines the use of recommendations in order to derive a trust level based upon Gambetta's [17] definition of trust. Recommendations enable an entity to harness not just its own observations in order to reach a verdict on how much to trust, but also employ those observations made by others.

Reputation, as derived from information received from other agents, is harder to judge for an agent in regards to its relevancy and reliability. Nonetheless, it is important mechanisms in society to assist with making trust decisions. Rasmusson [36] describes reputation as a *social control mechanism* that can yield important information for the trust decision making process. It is assumed that reliable members of a community both (1) identify those members that are malicious and (2) propagate this knowledge throughout the community, thereby making the malicious members known. Rasmusson [35] calls this *soft security*, stating that *in a socially controlled system, it is the participants themselves, the agents, who are responsible for collaboratively maintaining security.*

In this chapter, reputation—in the context of probabilistic trust reasoning—will be defined as follows [1]:

Definition 2. Reputation is an expectation of an agent's behavior based on information about or observations of his past actions.

When attempting to reach a *trust decision*, an agent usually evaluates both its own prior interactions with the other agent, as well as reputation information about the other agent. Usually, trust models use parameterized weighing functions to composite these two aspects of the trust mechanism [22, 31]. These equations typically take the reliability of the information into account as perceived by the deciding agent. How reliable a piece of information is can depend, for instance, on the source of the information or its age. Several trust models designed for the use in multi-agent environments [2, 9, 22, 31, 37] have put forth solutions.

However, when determining whether or not an agent will engage in an action with another agent, it has to determine if the strength of the trust relation between itself and its potential interactor is satisfactory. The agent thus makes a binary trust decision; if both agents in the trust relation make positive trust decisions, some sort of action will be initiated.

Trust frameworks developed over the recent years [1, 21, 29, 38, 41, 43] have mainly been concerned with developing robust trust metrics. Most do not, however, take into account the very structure of the social networks upon which they operate. In particular, these frameworks do not categorically distinguish between the origin of recommendations used in the calculation of trust. Relying only on direct recommedations from trusted neighbors has advantages regarding the reliability of the reputation information. However, this forgoes a potential wealth of additional information. Mui [31] has proposed the establishment of (parallel) recommendation chains between two remote—i.e. non-neighboring—agents. This process, however, suffers from distance effects for long chains and problems when determining the reliability of a particular chain (particularly when determining weighing factors). Furthermore, Mui [31] argues that Bayesian aggregation is unfit for establishing the reputation of remote agents. In all cases, reputation information is distributed through the network, when requested. In the following, we will propose to distinguish between information that should be made available on demand and information that should be published.

A recommendation, i.e. the transmission of reputation information about an agent (the recommendee) from another agent (the recommender) to a third (the recipient) is grouped according to a criterion of direct interaction between recommender and recommendee. If the two have had direct interaction at one point, and the recommendation is based on experience from that interaction, it is considered *hard reputation* if the recipient knows and trusts the recommender.

In far-flung social networks, it cannot be guaranteed that an entity has direct access to hard reputation information, or that the amount of reliable hard reputation is sufficient for the recipient to form a trust decision. In order to supplant or supplement hard reputation in the decision making process, *remote reputation* is introduced in the presented framework. This is based on social phenomena such as gossip. While not based directly on *observable* interaction experiences, it is reasonable for an agent/person to consider such information. Mechanisms such as recommendation referral [31] or the certified reputation (CR) component of the FIRE model [22] are compatible with the conceptual framework outlined in section 4 and can be integrated into trust reasoning.

The rationale behind the introduction of a specific category for remote reputation lies primarily in our desire to communicate reputation information in a peer-to-peer environment that may include bandwidth limited agents or network infrastructure. As outlined in [31], long recommendation chains are of questionable reliability, yet would result in considerable communications overhead. Thus, in the proposed protocol, remote reputation information is considered less reliable and consequently should be accorded less time critical resources.

From a perspective of quality, hard reputation—in the form of direct experiences and recommendations from direct neighbors—should be prioritized when requested over the network, while lower quality information should be made available only when free resources permit. Consequently, we suggest a request/pull model for the distribution of hard information, while resorting to a publish/push model when considering remote information (cf. section 4).

Aside from supplying a broader base of information, these processes are included to further reward good hard reputation, facilitate faster permeation through the network and thereby drive preferential attachment [5] to reliable partners. This does not only serve to stress the benefit of reliable behavior on the entity/node level, but should also reinforce the complex structure of the underlying socio-economic network, as preferential attachment to reliable nodes (and—so to speak—*preferential detachment* from unreliable nodes) is a driving force behind the creation of scale-free structures [5].

3 Complex Social Networks

In modern research on various, diverse subjects—such as social interactions, urban development, ecosystems and e-commerce—complex networks are used for modeling the specifics of those intricate, highly adaptive systems investigated. The notion of web-like structures underlying personal relations, city demographics, predator-prey interactions or individual shopping behavior may have been alien only a couple of years ago, but today, with the ever growing prevalence of the Internet in everyday life, it appears to be almost a matter of course. The world wide web in particular has become a medium that facilitates not just the exchange of information but also the creation of social and economic structures that were previously entirely reliant on personal interaction. This trend has manifested itself in the form of Internet communities such as *Facebook*, *Xing* and their various competitors or imitators, as well as in the popularity of online shopping sites that generate significant revenue.

While the emergence of internet communities serves to illustrate the web-like structures underlying interpersonal relations, they are nonetheless present even beyond the pure technical implementation of computers and protocols. Over the past decade, research in the fields of statistical physics and mathematics has closely investigated the structure and behavior of different real-world systems [12, 33]. These networks exhibit particular and useful properties regarding information diffusion, such as in the proliferation of rumors [32], different sorts of trends [40] or diseases [4, 27, 34]. Various kinds of social dynamics have been modeled based on complex networks (for an overview, cf. [12]), and interesting parallels have been drawn between the spread of infectious diseases and the dissemination of ideas [6].

Two specific structural (static) properties of complex networks are the small-world phenomenon and a scale-free degree distribution. These ubiquitous features have a significant impact on the processes—such as the spread of infectious diseases [4, 27, 34]—occurring in complex systems.

Emergent/dynamic properties present in complex networks include non-smooth creation of a giant component encompassing the majority of nodes in the network. Among others, models of self-organized criticality (SOC) have been proposed to account for this emergent phenomenon, as it resembles the *punctuated equilibrium* of a SOC process [7]. Socio-economic networks typically possess this feature [14, 15, 18, 19].

Furthermore, once the phase transition has occurred, social networks tend to be resilient to deterioration of their giant component [15] – another quality generally observed in complex networks [3, 13]. Thus, once the social network has evolved, the community forming the giant component will maintain connections among its members, even if they are not immediate but indirect.

These qualities of the networks underlying human interactions form the foundation for the proposed reputation-based trust model. Similar approaches have so far not explicitly addressed reputation dissemination in social network structures beyond an agent's direct neighborhood and referral models [1, 21, 29] or required sophisticated server infrastructure for supplying sufficient reputation information [41]. Our approach seeks to harness the intrinsic information-spreading qualities of human socio-economic networks to enable agents to make informed trust decisions.

The ultimate goal of the presented research is the development of a trust framework capable of operation in a purely peer-to-peer environment, thus eliminating the need for expensive server infrastructure (as, for instance, deployed in [41]). It is therefore of particular interest to see whether social networks possess sufficient diffusion capabilities to support *reliable* reputation dissemination. Recent literature [6, 12, 23, 24, 32] strongly suggests this to be the case.

4 Conceptual Trust Framework

Given the previous elaborations of trust and reputation, we will in the following outline a framework for reputation-based trust, designed to function in peer-to-peer environments. The framework is inherently experience based with the intention of providing a foundation of data for the computation of trust. The framework in and by itself is *conceptual* in nature, i.e. implementation independent. A potential implementation only has to (be sufficiently extensible to) fulfill a number of structural criteria outlined in this section. The actual trust metric used in the diffusion simulation is based upon the FIRE model [22].

Beyond the stock mechanisms (direct and witness trust), the proposed framework includes an extension permitting an agent the integration of information originating beyond its direct 0-hop neighborhood. This remote reputation information is actively propagated independently of the on-request trust provisioning process. Through an agent voting scheme, the reliability of such information is assured (by applying an agreement metric). Voting by agents on particular pieces of remote information thus serves as a social filtering mechanism. To the best of our knowledge, this is a novel approach in the communication of trust information using reputation-based trust models.

A simple system will be employed, mapping the experiences made during an interaction to the interval $[-1,1[$. $[-1,0[$ represents negative experiences, 0 serves as the element representing entirely neutral experiences, and $]0,1[$ represents positive experiences. Interaction experiences are generally judged according to numerous sub-experiences. These range from entirely objective aspects (such as technical aspects relating to QoS (Quality of Service), speed of broadband network connections) to totally subjective categories (e.g. personal perception of politeness, taste). In order to include these sub-categories in recommending and trust decision making, instead of aggregating all these different factors into a single rating, multi-dimensional reputation/recommendation/trust vectors are employed to represent trust variables.

Thus, reputation information is communicated throughout the network in the form of multi-dimensional vectors, containing real-valued numbers from the interval $[-1,1[$. Each dimension of the vector represents a different implementation-dependent characteristic of interactions between entities. The semantics of these values may vary according to the scenario for which such an implementation occurred.

An entity x builds an *opinion* of another entity y, $Op_{x,y}$, based upon reputation from prior interactions—both from direct experience and recommendations, qualified by a temporal decaying factor and the reliability of the recommendations—as well as from information received through independently propagated 'remote reputation' information.

Hard Opinion Formation

The opinion x has of y, $Op_{x,y} \in \mathbb{R}^n$, is fundamentally based upon y's reputation. Primarily, reputation information is founded on interactions. The most

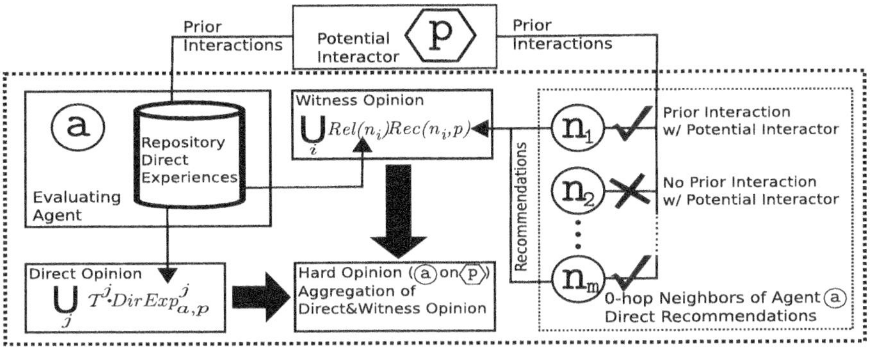

Fig. 1 Hard Opinion formation process, taking into consideration direct experience and witness information (stock mechanism of state-of-the-art trust models)

reliable information, from the subjective perspective of a single agent, is contained within its *own* interaction experiences. *All* prior experiences, no matter the partner, shape an agent's general trusting disposition (i.e. its basic or dispositional trust, cf. [30]). Furthermore, all those interactions made in a particular situation shape an agent's trusting behavior in a comparable situation. However, these factors do not contribute when choosing one potential supplier of a predetermined resource/service from a pool of alternatives, as they do not vary under the given scenario. Pertinent direct interaction thus consists of all the experiences agent x has had with agent y, $DirExp^1_{x,y}, \ldots, DirExp^m_{x,y}$. As experiences are less indicative of expected behavior the older they are, a temporal degradation factor τ^k is introduced, leading to the *direct opinion* x has of y: $Op^{direct}_{x,y} := \tau^1 \cdot DirExp^1_{x,y} \times \ldots \times \tau^m \cdot DirExp^m_{x,y}$.

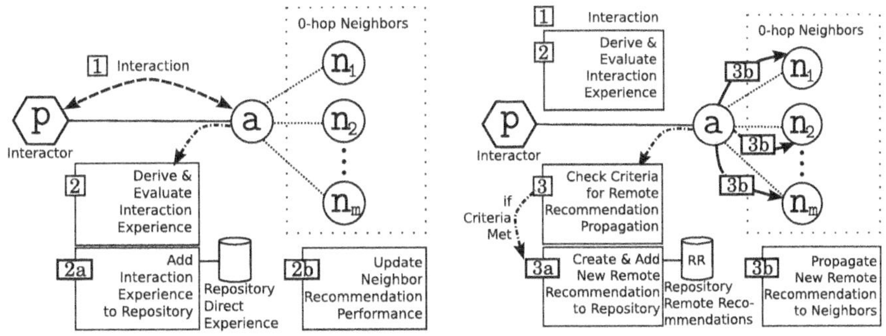

(a) Stock evaluation process (b) Remote recommendation extension

Fig. 2 Interaction evaluation by agent a, including proposed extension for issuing remote recommendations for remote opinion formation

Next in reliability are recommendations by trusted neighboring agents. These recommendations are also based on direct interactions. Specifically, they can be considered to be fundamentally the *transmitted* direct opinion of another agent z on y. Recommendations have to be qualified by factoring in the reliability of the recommender to supply appropriate recommendations – theoretically, direct opinion could be qualified in such a way as well, yet it is probably safe to assume that each agent trusts itself completely. Therefore, $Op_{rec(x,z)} \in \mathbb{R}$ is introduced to represent said recommendation reliability of z as perceived by x; $Op_{rec(x,z)}$ is itself a component of $Op_{x,z}$. Furthermore, the behavior of any entity can change over time, making older information less pertinent to the current reliability of that entity. Therefore, a degradation factor τ is used to account for temporal invalidation of reputation information. Both τ and $Op_{rec(x,z)}$ contribute to the perceived reliability of a recommendation $Rec_{z,y} \subseteq Op_{z,y}$.

When building an opinion $Op_{x,y}$ on y, x uses hard information, i.e. information that can be tied directly to specific interactions, either incurred directly by x or some recommender z, in the form of both it's own prior experience with y, $Op^{direct}_{x,y}$, and any number of recommendations for which it queries known and trusted sources (i.e. it is assumed that $Op_{rec(x,z)}$ is positive). Thus, assuming the availability of recommendations from a number m of recommenders z_1 through z_m, the opinion of x on y based on hard reputation data is computed from: $Op^{hard}_{x,y} := Op^{direct}_{x,y} \times ((Rel_{x,rec(z1,y)} \times Rec_{z1,y}) \times \ldots \times (Rel_{x,rec(zm,y)} \times Rec_{zm,y}))$.

Hard opinion formation can be linked directly to either verifiable direct interactions or attributed to trusted recommenders. It forms the reliable core of information when deriving an overall opinion on another entity. This source of reputation information is both finite and immediately accessible to an agent. Acquiring this sort of information involves only directly neighboring agents, representing only a comparatively small fraction of agents within the social network. Therefore, expenditure of network and agent resources for storing and transmitting hard opinion information is justified, as overall network load is limited.

Remote Opinion Formation

Remote opinion formation [20] is based upon information that is not readily verifiable in any way by an agent, from beyond its immediate neighborhood. As such, remote opinions are less reliable than hard opinions and should be factored accordingly. In spite of this reservation, soft remote opinions offer an additional source of information and enable the presented protocol to leverage the complex structure of its underlying social network further. At the agent level, a remote opinion $Op^{remote}_{x,y}$ represents the aggregate of all reputation information x holds on y, that can not be associated with interaction experiences the way hard opinions can. Rather, the information has been actively propagated through the network, similar to rumors/gossip.

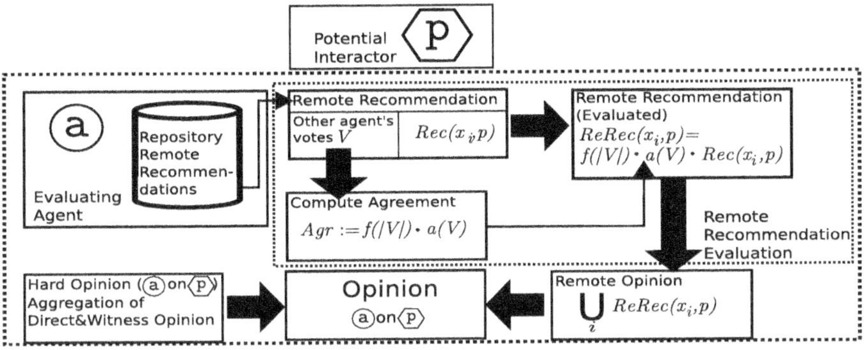

Fig. 3 Remote Opinion formation process and aggregation with Hard Opinion to derive Opinion agent a has of a potential interactor p

The minimum requirement we postulate for such gossip is that its originator, i.e. some remote entity r, is identifiable. This serves both to avoid multiple storage of the same piece of remote reputation information, as well as to afford at least a minimum level of traceability and reliability of the information. Identification can, for example, occur through a digital signature guaranteed by a central identity provider or a certificate authority, or authentication through r's neighbors.

A remote recommendation, i.e. the transmission of remote reputation information, is initiated by entity r as it sees fit, in a broadcast to (all of) its neighboring nodes. In order to avoid overwhelming network traffic, we suggest limiting the occasions on which such a broadcast occurs. After individually assessing the importance of the broadcast, unsolicited recommendation, the receivers of said message decide whether or not to integrate the message into their knowledge base and to send the message along to their respective neighboring nodes, potentially causing dissemination throughout the entire network. In terms of epidemiological spread discussed in [6, 40], these nodes that integrate a particular piece of remote reputation information, which r transmitted with regards to y, become 'infected'.

Entity x, attempting to make a trust decision on entity y, incorporates the remote reputation information it has received indirectly from remote entities r_1 through r_m, i.e. nodes that are not neighbors of x, into its remote opinion of y, $Op^{remote}_{x,y}$. As the unsolicited information that r_1 through r_m have distributed is in principle a recommendation, we designate it $Rec^{remote}_{r1,y}$ through $Rec^{remote}_{rm,y}$. Although x does not know any of the remote issuers that originated these recommendations, it is still able to assess their reliability through its remote opinion on them; in case x holds no opinion on the remote entities at all, a default reliability is assumed. This enables x to calculate the perceived reliability of a remote recommendation. Conceptually, this can be expressed by $Op^{remote}_{x,y} := ((Rel_{x,rec(r1,y)} \times Rec^{remote}_{r1,y}) \times \ldots \times (Rel_{x,rec(rm,y)} \times Rec^{remote}_{rm,y}))$

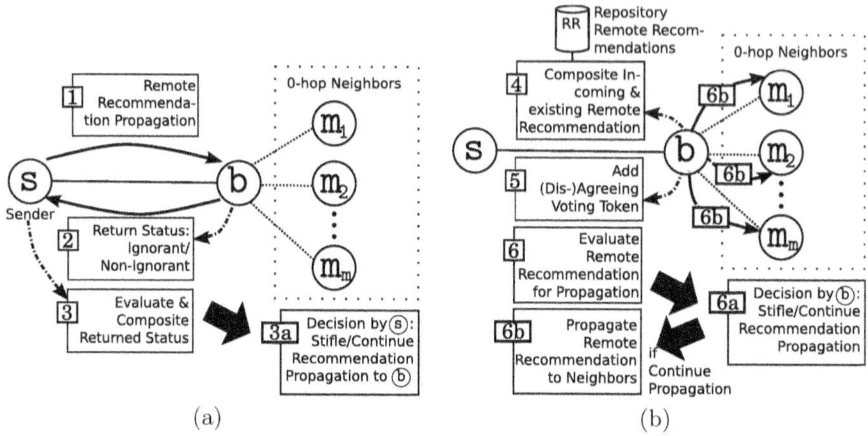

Fig. 4 Remote Recommendation propagation mechanism, involving a sender s, a receiver/propagator b and 0-hop neighbors of b, m_1-m_m

Remote Reputation Propagation

When dealing with remote information in the form of recommendations, an agent has to contend with the problem of uncertainty in regard to recommendation reliability. The recommendation has not been issued by an entity known to the agent, and the agent has to rely on other agents to assess the trustworthiness of the recommender. Furthermore, in case of multi-hop referrals, the assessing agents might also be unknown to the requesting agent. In addition to the uncertainty regarding the quality of the information, an agents request for multi-hop recommendations may cause considerable load on the communication network and create undue waiting periods until the requested recommendation is discovered. In propagating remote reputation through the network independently of hard reputation, we follow a number of goals: **(a)** make remote reputation acquisition time efficient at the time an agent requests information, **(b)** balance the load on the communication network so that remote information will be propagated in periods of low traffic, **(c)** use social filtering mechanisms to increase the quality of the published remote reputation information.

We propose that an agent, upon completing an interaction, can choose to issue remote reputation information. Publishing such an remote recommendation should occur only rarely, so as not to overwhelm the communication network. Therefore, only exceptional information should be communicated, such as: exceptionally good or bad performance of the recommendee, considerable average performance improvement/deterioration of the recommendee or considerable variability in the performance of the recommendee. Furthermore, if relatively recent and similar remote reputation information on the recommendee is already known, the probability of issuing another is reduced according to the newsworthiness of the additional information.

Upon receiving and acknowledging a remote recommendation, the receiving agent has multiple options. After assessing the newsworthiness, relevancy and reliability of the information, it has to decide whether to process or to discard the message. Once the agent chooses to process the information, if it knows the originating agent directly, it can confirm the message's authenticity and augment the information by adding its opinion whether or not the originator is a reliable source for information. Furthermore, any receiving agent can also assess the remote recommendation and add its own rating in according to its opinion of the recommendee, enriching the information content. Potentially, an agent may receive the same original remote recommendation multiple times through different paths of the network. In such a case, it consolidates the information.

Once the remote recommendation has been processed, the agent adds it to its information repository and—if the agent chooses to propagate the message—also to a send queue. If the agent has already sent a remote recommendation onwards and receives the original remote recommendation through another path, it has the option of issuing an update of its own message. Such an update may be issued, given it is sufficiently different under two conditions: **(a)** a considerable increase in the information content of the message and **(b)** the message it already propagated has not yet been consolidated into the newly received remote recommendation.

Thus, the reliability of a remote recommendation can be determined by evaluating the number of agents that have propagated it, the number of agents that have concurred, respectively disagreed, with the recommender's assertion, and the assessment of the recommender's own reliability according to a subset of its neighbors. Based on the reliability of the remote data, the receiving agent may now use the information to supplement its hard opinion on the recommendee.

5 Simulation

For the sake of simulating reputation and trust diffusion in an acquaintance-based recommender network, a trust framework has been developed that is capable of generating complex networks, simulate agent interactions based on the relations modeled through these networks and—potentially—alter the network structure as a consequence of agent-to-agent interactions. The conceptual trust framework outlined in section 4 used the direct and witness trust modules of the FIRE trust model [22], as they are robust and well-tested. For the integration of remote reputation information, a remote opinion component was added to the FIRE model, and given a weighted identical to the witness trust component at 0.5.

Growing the Social Network: In order to evaluate the diffusion of reputation information throughout a society of agents, social networks were grown according to the procedure presented in [25]. The resulting networks ranged from 250 to 4500 agents/vertices in size with a hard cut-off z^* from 5 to 20 connections per node. Parameters γ, r_0 and r_1 were selected in such a way as to result in high clustering coefficients for this paper we chose $C \geq 0.4$) and positive assortativity (e.g. for the 250 agent network, parameter settings of $\gamma = 0.005, r_0 = 0.0001, r_1 = 2$ and $z^* = 5$ yielded a clustering coefficient of $C = 0.52$ and an assortativity coefficient of $r = 0.27$, significantly higher than corresponding random graph, in which almost no clustering would have been present, at $C = 0.02$). The importance of clustering in social networks has been covered extensively in literature [39]. Thus, all networks showed clearly defined communities, that were (for the most part) connected to a giant component containing the large majority of agents.

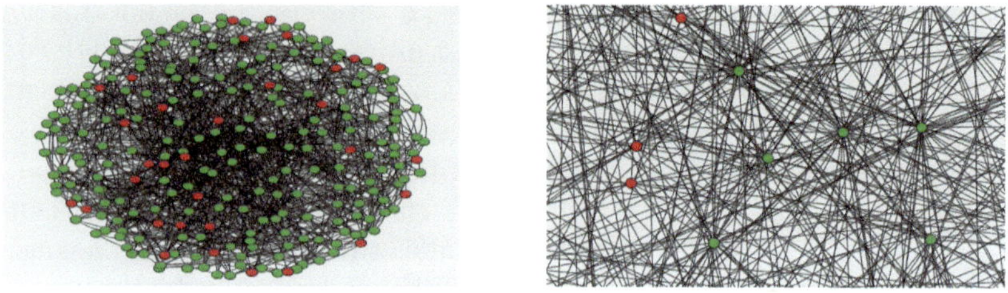

(a) Overview (b) Detail

Fig. 5 Random graph, consisting of 250 nodes. Green = Adopters, Red = Non-Adopters.

These networks, underlying the proposed reputation-based trust framework, simulate human acquaintance relations. They do not show strong signs of preferential attachment, i.e. a power law distribution of node degree. Nonetheless, we consider them an adequate starting point for the simulation, in lieu of real-world data on actual recommender networks and their associated dynamics. Furthermore, friends and acquaintances, from the authors' experience, form the foundation of the reputation gathering process; they represent the first source of information regarding word-of-mouth information.

Additionally, the underlying acquaintance network only serves as a starting point in the generation of a reputation and trust network. Dynamics of reputation and trust formation will affect the structure of the network in such a way, as to make it divergent from the initial network. For the sake of comparison, corresponding random graphs [16] were also created, using the algorithm presented in [23].

Scenario: The primary focus of the presented simulations was on interactions between agents that are part of a particular trust/acquaintance network (thereby forming a society) and agents outside that network. Thus, the recommending agents (recommenders) are either immediately or intermediately connected to the agent requiring the information; ideally, the recommenders will have had prior interaction with the subject of the recommendation (recommendee), although the recommendee itself is not a member of the trust network.

This is akin to the relationship between buyers and sellers, with the seller being, for example, a well-known internet retailer or national retail chain and the buyer being a person. For regular retail goods, such as books, the retailer minimizes risk of loss by relying on established business practices, such as payment before delivery. The buyer has to trust the retailer to fulfill its obligations once the funds have been transferred or credit card information has been provided. This *risk of prior performance* [26] can be mitigated partly through reputation-based trust mechanisms.

In order to compare the behavior of the agents in the trust network, several assumptions were made, namely that the different potential sellers provide essentially the same product at an identical price. This eliminates agent-external differences in the utility of the purchase, due to different levels of functionality in different but similar products, as well as differences in risk. Furthermore, we assume that the sellers are known to the customer agents 'by name'. Thus, an agent in the simulation is aware of all the sellers of a particular product, although it is not necessarily aware of a seller's quality. Furthermore, we assume that agents act truthfully, i.e. do not intentionally misrepresent their opinion when recommending a seller (although the FIRE trust module used in the simulation is capable of compensating for malicious behavior).

For all simulation runs, each seller was assigned a standard service quality from $]0,1[$. Because a seller's quality in providing a resource is thought to be relatively stable with persistent changes occurring only gradually, this value represented the mean of the seller's actual performance in an interaction. Specifically, the actual performance of a seller in an interaction was determined probabilistically via a normal distribution with expectation μ set to the standard service quality, default standard deviation $\sigma = 0.3$ and limited to the interval $[-1,1]$. Additionally, for each agent the initial trust that agent puts in a neighboring agent when evaluating the neighbor as a recommender, a normally distributed value was calculated, with expectation $\mu = 0.5$, standard deviation $\sigma = 0.2$ and a range of $]0,1[$. Assuming that each agent also had prior interaction with its preferred seller, each agent was initialized with a number of direct experiences with that seller, determined as per the method just described. By just requiring a relative advantage of trust in one seller over trust in another one, instead of also requiring an absolute minimum trust value threshold, we seek to eliminate any influence of an agent's internal 'mental' state on its choice.

Remote recommendations were issued if an agent rated an interaction in the top or bottom ten percentile range. The resulting remote opinions constituted a remote opinion component in the trust formation, with the same weight as the witness opinion component .

For the direct and witness trust components of FIRE, parameters were set to the default values presented in [22]. Aside from the parameter settings of the employed trust model, diffusion is dependent on a number of simulation specific conditions: The topology of the social network, the number of interactions taking place in the network per point in time (probability that an agent will interact with a seller) and the initial number of agents preferring a specific seller. If an agent is probabilistically chosen to act at a point in time t, the agent starts to gather reputation information on all potential sellers, evaluate the available data and choose the seller with the highest trust value, according to the methods outlined above. After seller selection, the agent makes a new experience with the selected seller, according to the seller's actual service quality, represented by a random Gaussian value, centered on a seller specific μ. This experience is added to the agent's direct trust knowledge base. Additionally, at each discrete point in time, every agent is 'asked' to state its preferred seller, without interacting.

Results

The trust diffusion process can be divided into three phases: initial, growth and saturation. During the initial phase, little to no increase in the number of agents preferring a particular seller A (the seller with the best standard service quality) over its competitors is observable. If and when the process leaves this phase depends on a number of factors: weighing factors—in particular the temporal decay of experiences—of FIRE, clustering coefficient of the network, the percentage of 'early adopter' seed agents, and the probability of interaction per agent, $p_{interact}$. If $p_{interact} \approx 0.2$ or higher, the initial phase was left consistently and quickly (with 5% adopters, $p_{interact} = 0.2$, FIRE standard parameterization: $t < 40$ for random graphs, $t < 100$ for the corresponding higher clustered acquaintance graphs). At this point, reputation information apparently ceases to be a rare resource and becomes locally available due to the high probability of an agent being active or having an active neighbor. A similar argument holds for the percentage of early adopters; seeding more than 30% leads to the system leaving the initial phase, even under relatively low levels of overall activity ($p_{interact} = 0.03$). When decreasing the temporal decay by increasing parameter λ in the FIRE protocol, an agent's opinion becomes more persistent; this affords the early adopters more time to 'convince' their neighbors before their seeded opinions are invalidated, thereby increasing the amount of information 'in circulation' about the better average behavior of A compared to its competitors. Doubling the half-life of a direct experience (achieved by setting the default for $\lambda = -5/ln(0.5)$ to $-10/ln(0.5)$ for the decay function $exp(-1 * \delta t/\lambda)$), is equivalent in effect to increasing the uniformly distributed adoptors.

Network topology has a considerable impact on the speed of leaving the initial phase. When comparing highly clustered acquaintance network components (clustering coefficient $C > 0.4$) with random graph components (clustering coefficient $C \approx 0.02$) of the same mean degree and number of nodes, diffusion on random graphs was significantly faster (Wilcoxon signed-rank test, $\alpha = 0.01$). This effect was present in all configurations for $p_{interact}$ that resulted in leaving the initial phase, with a magnitude of ≈ 1.5 to ≈ 2. It is however more noticeable when reputation information is scarce (cf. Fig. 6(a)). During the initial phase, reputation propagation and information decay are close to

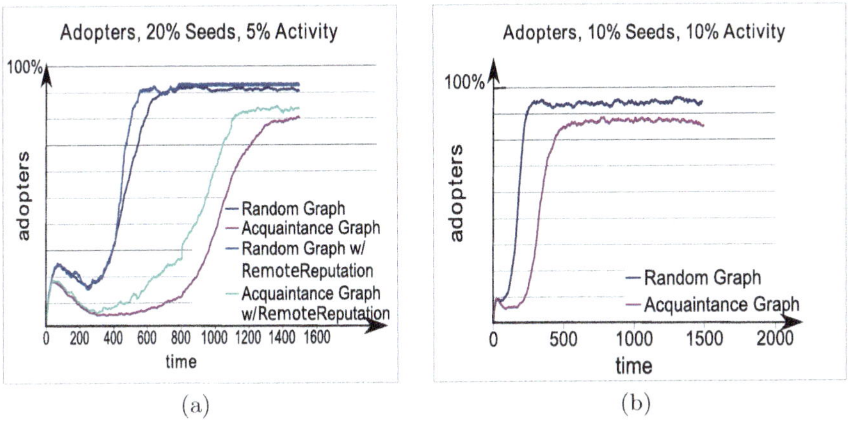

Fig. 6 Growth of adopter population in acquaintance and random graphs (with (a) and without (a, b) remote reputation propagation)

equilibrium. Yet, once enough momentum has been built, the systems enters a phase of rapid growth regarding the number of adopters of seller A.

For the random graphs, this growth phase is stronger, setting in earlier and resulting in higher percentage of adopters at its conclusion than for the more highly clustered graphs (Figures 6(a), 6(b)). The rate of growth is slightly higher for the random graphs as well, although this difference equalizes with increasing $p_{interact}$ and percentage of adopter seed agents. Following the growth phase, the system reaches saturation. Under the given configuration, neither on random nor on highly clustered graphs, did saturation reach 100 per cent of adopters. Final average adopter saturation for random graphs was $\approx 93\%$, for acquaintance graphs $\approx 85\%$. This, however, is at least partly due to the probabilistic nature of the actual service quality provided by the sellers. Tests have shown that reducing the variance of service quality, the saturation increased (data not shown).

When using remote reputation propagation, the performance of the system to choose the objectively better seller A over its competitors improved (cf. Fig. 6(a)). The improvement is more pronounced for acquaintance networks, manifesting itself in faster successful termination of the initial phase, faster growth and higher number of adopters ($4 - 7\%$). Thus, a clear benefit of augmenting a trust model with this mechanism can be witnessed.

6 Conclusion and Future Work

Overall, the simulations indicate, that clustering results in a slower diffusion of trust information. Particularly the existence of clusters that are only weakly connected to the remaining network component are resilient to the adoption of a new seller. The result of this, a higher number of agents preferring other sellers, can be partly overcome by adding a second information diffusion channel through the use of independently propagated remote reputation information, that enabled the system to 'tip' some of the resilient nodes. Reputation information retrieval in a distributed environment has been shown to be a feasible concept, even under relatively scarce initial information (i.e. low $p_{interaction}$ and percentage of seed nodes). The proposed remote recommendation mechanism has been shown to improve the diffusion of reputation information, compared to stock procedures of state-of-the-art trust models.

As the process of reputation dissemination is dependent on the topology of the underlying network, and high clustering appears to inhibit this, the proposed framework should be further expanded to include mechanisms that change and evolve both the network and the trust dynamics adaptively. This may, for instance, be achieved by selecting reliable recommenders as super-agents, whose reliability is spread through the network via rumor spreading mechanisms, thereby driving preferential attachment to these nodes. Furthermore, the parameters of the trust metric could be adaptive to the scarcity of information, slowing the decaying of direct experiences if activity is perceived to be low by the evaluating agent.

Additionally, insights from social sciences, but also from current online communities, regarding the exact nature of human acquaintance, friendship and trust in the emerging—or present—cyber-space we all participate in, be it via networking sites or e-commerce, leave much room for computational adaptation. This can not only serve to better understand human action, but also to assists online users, for instance by offering an automated, distributed (p2p) recommendation network.

References

1. Abdul-Rahman, A., Hailes, S.: A distributed trust model. New Security Paradigms 97 (1997)
2. Abdul-Rahman, A., Hailes, S.: Supporting trust in virtual communities. In: Proceedings of the Hawaii's International Conference on Systems Sciences, Maui Hawaii (2000)
3. Albert, R., Jeong, H., Barabasi, A.L.: Error and attack tolerance of complex networks. Nature 46, 379–381 (2000)
4. Ball, F., Mollison, D., Scalia-Tomba, G.: Epidemics with two levels of mixing. The Annals of Applied Probability 7, 46–89 (1997)
5. Barabasi, A.L., Albert, R.: Emergence of scaling in random networks. Science 286, 509–512 (1999)
6. Bettencourt, L., Cintron-Arias, A., Kaiser, D., Castillo-Chavez, C.: The power of a good idea. Physica A 364, 512–536 (2006)
7. Bianconi, G., Marsili, M., Vega-Redondo, F.: On the non-trivial dynamics of complex networks. Physica A 346, 116–122 (2005)
8. Bromley, D.: Reputation, Image and Impression Management. Wiley & Sons, Chichester (1993)
9. Carbo, J., Molina, J., Davila, J.: Trust management through fuzzy reputation. International Journal of Cooperative Information Systems 12(1), 135–155 (2003)

10. Castelfranchi, C., Falcone, R.: Principles of trust for mas: cognitive anatomy, social importance, and quantification. In: Proceedings of the Third International Conference on Multi-Agent Systems (1998)
11. Castelfranchi, C., Falcone, R.: Trust and Deception in Virtual Societies. In: Social Trust: A Cognitive Approach, pp. 55–90. Kluwer Academic Publishers, Dordrecht (2001)
12. Castellano, C., Fortunato, S., Loreto, V.: Statistical physics of social dynamics. Reviews of Modern Physics 81, 591–646 (2009)
13. Cohen, R., Erez, K., ben Avraham, D., Havlin, S.: Resilience of the internet to random breakdowns. Physical Review Letters 85(21), 4626–4628 (2000)
14. Crane, J.: The epidemic theory of ghettos and neighborhood effects on dropping out and teenage childbearing. American Journal of Sociology 96, 1226–1259 (1991)
15. Ehrhardt, G., Marsili, M., Vega-Redondo, F.: Emergence and resilience of social networks: a general theoretical framework. Annales dconomie et de statistique 86 (2008)
16. Erds, P., Renyi, A.: On the evolution of random graphs. Publications of the Mathematical Institut of the Hungarian Academy of Sciences 5, 17–61 (1960)
17. Gambetta, D.: Can we trust trust? In: Trust: Making and Breaking Cooperative Relations, Basil Blackwell, London (1988)
18. Goyal, S., van der Leijy, M., Moraga-Gonzalez, J.L.: Economics: An emerging small world? Journal of Political Economy 114, 403–412 (2006)
19. Hagedoorn, J.: Inter-firm r&d partnerships: an overview of major trends and patterns since 1960. Research Policy 31, 477–492 (2002)
20. Hauke, S., Pyka, M., Heider, D., Borschbach, M.: A reputation-based trust framework leveraging the dynamics of complex socio-economic networks for information dissemination. Communication of SIWN 7, 53–58 (2009)
21. Huynh, T., Jennings, N., Shadbolt, N.: An integrated trust and reputation model for open multi-agent systems. Autonomous Agents and Multi-Agent Systems 13(2), 119–154 (2006)
22. Huynh, T.D.: Trust and reputation in open multi-agent systems. Ph.D. thesis, University of Southampton (2006)
23. Jackson, M.O., Rogers, B.W.: Search in the formation of large networks: How random are socially generated networks? Game Theory and Information 0503005, EconWPA (2005), `http://ideas.repec.org/p/wpa/wuwpga/0503005.html`
24. Jackson, M.O., Yariv, L.: Diffusion on social networks. Revue d l'Institut d'Economie Publique 16, 3–16 (2005)
25. Jin, E.M., Girvan, M., Newman, M.E.J.: Structure of growing social networks. Phys. Rev. E 64(4), 46132 (2001), doi:10.1103/PhysRevE.64.046132
26. Josang, A., Ismail, R., Boyd, C.: A survey of trust and reputation systems for online service provision. Decision Support Systems 43(2), 618–644 (2007)
27. Keeling, M.J.: The effects of local spatial structure on epidemiological invasions. Proc. R. Soc. Lond B 266, 859–867 (1999)
28. Marimon, R., Nicolini, J., Teles, P.: Competition and reputation. In: Proceedings of the World Conference of the Econometric Society, Seattle (2000)
29. Marsh, S.: Formalising trust as a computational concept. Ph.D. thesis, Department of Computing Science and Mathematics, University of Stirling (1994)
30. McKnight, D., Chervany, N.: The meanings of trust. Tech. rep., University of Minnesota Management Information Systems Research Center (1996)

31. Mui, L., Mohtashemi, M., Halberstadt, A.: A computational model of trust and reputation. In: Proceedings of the 35th Hawaii International Conference on System Science, pp. 280–287 (2002)
32. Nekovee, M., Moreno, Y., Bianconi, G., Marsili, M.: Theory of rumour spreading in complex social networks. Physica A 374 (2007)
33. Newman, M., Barabasi, A.L., Watts, D. (eds.): The structure and dynamics of networks. Princeton University Press, Princeton (2005)
34. Pastor-Satorras, R., Vespignani, A.: Epidemic spreading in scale-free networks. Physical Review Letters 86(14), 3200–3203 (2001)
35. Rasmusson, L.: Socially controlled global agent systems. Master's thesis, Royal Institute of Technology, Dept. of Computer and Systems Science, Stockholm (1996)
36. Rasmusson, L., Janson, S.: Simulated social control for secure internet commerce. In: New Security Paradigms 1996 (1996)
37. Sabater, J., Sierra, C.: Review on computational trust and reputation models. Artificial Intelligence Review 24, 33–60 (2005)
38. Sen, S., Sajja, N.: Robustness of reputation-based trust: Boolean case. In: Proceedings of the First International Joint Conference on Autonomous Agents and Multiagent Systems, Bologna (2002)
39. Watts, D.: Small Worlds. Princeton University Press, Princeton (1999)
40. Watts, D.J.: A simple model of global cascades on random networks. PNAS 99(9), 5766–5771 (2002)
41. Weyl, B.: On interdomain security: Trust establishment in loosely coupled federated environments. Ph.D. thesis, Technische Universität Darmstadt (2007)
42. Williamson, O.: Calculativeness, trust and economic organization. Journal of Law and Economics 36, 453–486 (1993)
43. Yu, B., Singh, M.: Distributed reputation management for electronic commerce. Computational Intelligence 18(4), 535–549 (2002)

From Unstructured Web Knowledge to Plan Descriptions

Andrea Addis and Daniel Borrajo

Abstract. Automated Planning (AP) is an AI field whose goal is to automatically generate sequence of actions that solve problems. One of the main difficulties in its extensive use in real-world application lies in the fact that it requires the careful and error-prone process of defining a declarative domain model. This is usually performed by planning experts who should know about both the domain in hand, and the planning techniques (including sometimes the inners of these techniques or the tools that implement them). In order planning to be widely used, this process should be performed by non-planning experts. On the other hand, in many domains there are plenty of electronic documents (including the Web) that describe processes or plans in a semi-structured way. These descriptions mix natural language and certain templates for that specific domain. One such examples is the *www.WikiHow.com* web site that includes plans in many domains, all plans described through a set of common templates. In this work, we present a suite of tools that automatically extract knowledge from those unstructured descriptions of plans to be used for diverse planning applications.

1 Introduction

This paper tries to build a gap between two fields: automatically extracting information from the Web and AP. As for information extraction, we are assisting a continuous growth in the availability of electronically stored information. In particular, the Web offers a massive amount of data coming from different and

Andrea Addis
Department of Electrical and Electronic Engineering,
University of Cagliari, Italy
e-mail: addis@diee.unica.it

Daniel Borrajo
Department of Computer Science,
University Carlos III of Madrid, Spain
e-mail: dborrajo@ia.uc3m.es

Soro et al. (Eds.): Inform. Retrieval and Mining in Distrib. Environments, SCI 324, pp. 41–59.
springerlink.com

heterogeneous sources. Most of it is in an unstructured format as natural language (blogs, newspapers). However, the new, most overshadowing and noteworthy web information sources are being developed according to the collaborative web paradigm, also known as Web 2.0 [17]. It represents a paradigm shift in the way users approach the web. Users (also called prosumers) are no longer passive consumers of published content, but become involved, implicitly and explicitly, as they cooperate by providing their own content in an *architecture of participation* [6]. Such knowledge repositories are semi-structured mixing some structure with natural language descriptions and predefined ontologies as in Wikipedia,[1] eBay,[2] and Amazon,[3] or IMDb.[4] In Wikipedia, a style template has to be filled in for each category belonging to a hierarchical structure of topics. In commercial sites as eBay, the online auction and shopping website, Amazon, an American-based multinational electronic commerce company website, a predefined list of mandatory attributes within its category is provided for each article. Also a more specialized knowledge base, IMDb, the Internet Movie Database, provides a list of standard attributes such as authors, director, or cast for each stored movie. In the spirit of Wikipedia, WikiHow[5] is a wiki-based web site with an extensive database of how-to guides. They are provided in a standard format (template) consisting of a summary, followed by needed tools (if any), steps to complete the activity, along with tips, warnings, required items, links to related how-to articles, and a section for sources and citations. Nowadays, thanks to advanced publishing tools the semi-structured knowledge base is more common, but not yet dominant. Therefore, it is becoming a primary issue to support applications that require structured knowledge to be able to reason (as in the form of ontologies), in handling with this enormous and widespread amount of web information. To this aim, many automated systems have been developed that are able to retrieve information from the Internet [1, 7], and to select and organize the content deemed relevant for users [3, 4]. Furthermore there has been some work on ontology learning [25, 16] pointing out how it is possible to solve the problem concerning the lack of structure of which the web often suffers. Thus, the structured format of the extracted knowledge is usually in the form of hierarchies of concepts (see for example the DMOZ project[6]) and this can help on developing many different kinds of web-based applications, such as specialized or general purpose search engines, or web directories. Other applications need information in the form of individual actions more than structured hierarchies of concepts, or in the form of plans.

On the other hand, as for AP, making it a widely used technology requires its usage by non-planning experts. Currently, this is quite far from being a reality. So, there is a need for techniques and tools that either allow an interaction with domain experts in their usual language, or automatically (or semi-automatically) acquire knowledge from current sources of plans and actions described in semi-structured

[1] http://www.Wikipedia.org
[2] http://www.eBay.com
[3] http://www.Amazon.com
[4] http://www.IMDb.com
[5] http://www.WikiHow.com
[6] http://www.dmoz.org

or unstructured formats. In the first case, there has been some work on knowledge acquisition tools for planning as GIPO [20], techniques for domain models acquisition [13, 22, 24], or tools that integrate planning and machine learning techniques [26, 11]. In the second case, there has been very little work on building plans from human generated plans or actions models described in semi-structured or unstructured formats, as filling natural language descriptions on templates. Another field that could also benefit from this automatic (or semi-automatic) acquisition of plans is the area of goals/activities/plan recognition [21], where most of its work assumes the existence of plan libraries that are manually coded. Examples are in the health environment [19], helping aging persons to perform their daily activities [18], or assisting a user on performing bureaucratic or tourism related actions [8].

In this chapter, we want to describe some work to deal with the lack of tools to automatically build plans and action models from semi-structured information, and the existence of this kind of knowledge in the Web, as is the case of WikiHow. We believe this is an important step toward a massive usage of planning technology by users in that they can share plans as they are doing now through WikiHow in natural language, and then automatic tools build planning technology on top of those plans. This applies also to other domains as workflow applications, where most big organizations have written processes [14, 5], or hospitals, where many standard procedures are described also in semi-structured formats. Also, we will encourage research in this topic by suggesting potential relevant tools to be built on top of our research for improving the construction of planning and plan recognition tools. This work extends and revises the work by Addis et al.[2] on recovering plans from the web. The main extension consists on adding a more sophisticated tool (i.e., *the Plan Builder*) to translate the article into a final structured plan.

The remainder of the paper is organized as follows: first, we provide an introduction to AP. Subsequently, the proposed architecture is depicted, and the constituting subsystems are separately analyzed. Then, the experiments and their results are presented and evaluated. Finally, we draw some conclusions and outline future research.

2 Automated Planning

AP is the AI field that provides techniques and tools to solve problems (usually combinatorial) that require as output an ordered set of actions. The inputs to planners are: a domain model, that describes, among other components, the available actions in a given domain; and a problem instance, that describes, among other components, the initial state and the goals to be achieved. Both input files are currently represented using a declarative standard language called PDDL (Planning Domain Description Language), that has evolved from the first version in 1998, PDDL1.0, to PDDL3.1 used in the last International Planning Competition.[7] Given the inputs (domain and problem descriptions), planners return an ordered set of actions (usually in the form of a sequence), that is called a plan.

[7] http://ipc.informatik.uni-freiburg.de/

As an example, Figure 1 shows an example of an action definition, taking an image, in a domain that specifies how a set of satellites should take images from space. As it can be seen, the language is based on predicate logic, and also allows numerical computations. In this work, our goal is to generate some declarative representations of the domains. Then, there are currently tools that are able to obtain PDDL models from the kind of output that we generate within this work. Figure 2 shows an example of an output plan in this domain.

```
(:action take_image
  :parameters (?s - satellite ?d - direction ?i - instrument ?m - mode)
  :precondition (and (calibrated ?i)
                     (on_board ?i ?s)
                     (supports ?i ?m)
                     (power_on ?i)
                     (pointing ?s ?d)
                     (power_on ?i)
                     (>= (data_capacity ?s) (data ?d ?m)))
  :effect (and (decrease (data_capacity ?s) (data ?d ?m))
               (have_image ?d ?m)
               (increase (data-stored) (data ?d ?m))))
```

Fig. 1 Example of action definition in PDDL

```
Solution:
    0: (SWITCH_ON INSTRUMENT0 SATELLITE0) [1.000]
    1: (TURN_TO SATELLITE0 GROUNDSTATION2 PHENOMENON6) [1.000]
    2: (CALIBRATE SATELLITE0 INSTRUMENT0 GROUNDSTATION2) [1.000]
    3: (TURN_TO SATELLITE0 PHENOMENON4 GROUNDSTATION2) [1.000]
    4: (TAKE_IMAGE SATELLITE0 PHENOMENON4 INSTRUMENT0 THERMOGRAPH0) [1.000]
    5: (TURN_TO SATELLITE0 STAR5 PHENOMENON4) [1.000]
    6: (TAKE_IMAGE SATELLITE0 STAR5 INSTRUMENT0 THERMOGRAPH0) [1.000]
    7: (TURN_TO SATELLITE0 PHENOMENON6 STAR5) [1.000]
    8: (TAKE_IMAGE SATELLITE0 PHENOMENON6 INSTRUMENT0 THERMOGRAPH0) [1.000]
```

Fig. 2 Example of resulting plan in the Satellite domain

3 From Unstructured Plans to Action and Plan Models

In this section we describe the Plan Acquisition Architecture (PAA) that performs the acquisition of plans and actions from semi-structured information. Then, we will describe the information source that we have used in this paper for showing how it works.

3.1 The PAA

The set of tools we have built are components of a modular structured architecture, which automatically browses some specific category from the ones represented in WikiHow, analyzes individual plans in those web pages, and generates structured representations of the plans described in natural language in those pages. This work

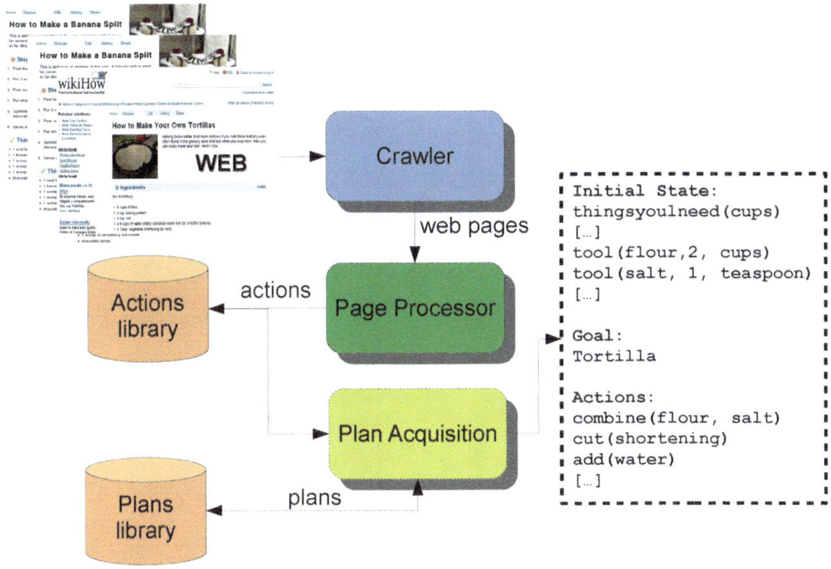

Fig. 3 PAA at a glance

intends to fill the lack of automatic tools to build plans using the semi-structured knowledge, more and more present over the web, which is similar to what is currently done in the Semantic Web, Information Retrieval or Ontology Learning fields.

Figure 3 highlights how the different subsystems, each wrapping different technologies, retrieve and transform the semi-structured information provided by web articles into a semantically overt, declarative structured output. The output contains labeled and recognizable objects (e.g., work tools, ingredients), actions (e.g., cut, peel, fry), and plans (e.g., sequences of instantiated actions), so that they may be reused for planning purposes.

The *Crawler* subsystem is devoted to crawl and store pages and category sections of a web site. The *Page Processor* subsystem is currently the core of PAA. It is aimed at analyzing web pages, storing only the relevant semi-structured contents into an action library after performing an initial pre-processing. In particular (i) the goal, (ii) the initial state (in terms of required tools), (iii) the actions, (iv) tips (to be exploited as heuristic hints), (v) warnings (to be exploited as plan build constraints), and (vi) articles related to the selected page/plan are stored for each input page. The *Plan Acquisition* subsystem processes this information to extract plans. The tools belonging to each subsystem, depicted in Figure 4, will be separately analyzed.

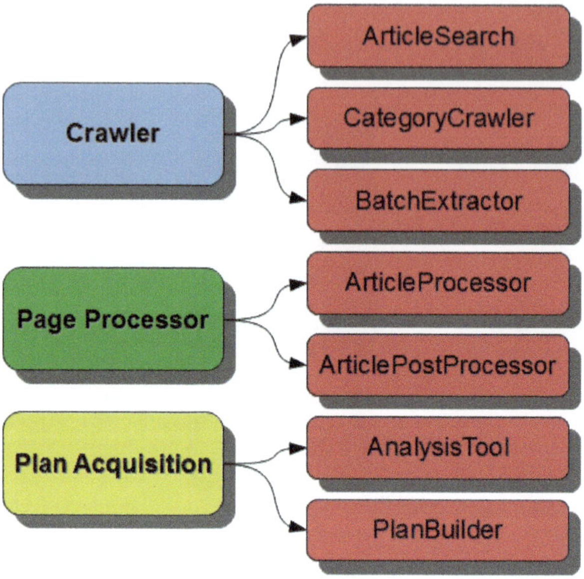

Fig. 4 Subsystems of PAA and corresponding tools

3.2 *WikiHow: A Knowledge Source for Extracting Plans*

WikiHow is a collaborative writing project aimed at building the world's largest and highest quality how-to manual, and it has been used as a benchmark to test the PAA ability on structuring information in the form of plans. WikiHow currently contains more than 70,000 articles written, edited, and maintained primarily by volunteers. Each article contains the necessary tools and describes the sequence of actions required to reach the goal the page is concerned with. As an example, let us take a look at the page *Make Your Own Tortillas* [8], reported in Figure 5, to better understand how the different subsystems parse its different sections. The relevant ground items that can be contained in each WikiHow web page are actions, tools, and related web pages (relatedwps), represented as A, T, and WP respectively. The Page Processor is the subsystem entrusted to process the different sections of the page. The name of the web page is identified by a `<div id=NAME>` HTML tag. Since the name of the how-to web page corresponds to the instructions to achieve something, this will be translated to the goals of the web page. For instance, if we find a how-to web page on "How to make a chocolate cake", then the goal of that AP problem will be translated to something equivalent to how-to-make-a-chocolate-cake. Each section must be associated with a type, being one of the following:

[8] http://www.wikihow.com/Make-Your-Own-Tortillas

Fig. 5 WikiHow sample web page

- **actions**: a sequence of actions, representing the necessary steps to reach the goal. They will form the actions set of the AP domain model. Examples of actions are *combine flour and salt, cut in shortening*;
- **tools**: a set of tools needed to reach the goal with different semantics depending on the selected category. Examples are ingredients for the *cuisine* category or mechanical tools for the *building stuff* category. They will be the objects that will appear in the AP problems, more specifically in the initial state. Examples of tools are: *2 cups of flour*, or *1 spoon of salt*;
- **relatedwps**: other web pages related to the described task. Examples of related pages are *how to make Tortilla de Patatas*, *how to make Flour Tortillas*, *how to make Tacos*. This is usually not used within planning, but they open new possibilities for planning purposes, such as suggesting potentially relevant plans.

Table 1 Equivalence between WikiHow concepts and AP concepts

wikihow	planning
page	plan
title	goal
ingredients	initial state
tools	initial state
steps	instantiated actions
tips	heuristics
warnings	constraints, heuristics
related pages	related plans

In Table 1, we show the equivalence between WikiHow concepts and AP concepts. Within this paper, we covered all aspects except for automatically generating heuristics and constraints from tips and warnings.

Since the Steps, Tips and Warnings sections, that are of type *actions*, the *ThingsYoullNeed* section of type *tools*, and the *RelatedWikiHow* section of type *relatedwps* are suggested by the WikiHow template layout, they occur in almost every page. They are parsed by default by the Page Processor, whereas further sections (e.g., *Ingredients* of type *tools*, usually added by the person that describes a recipe) have to be explicitly declared.

4 The Crawler

The Crawler subsystem is devoted to find an article using natural language or to find all the articles belonging to a specific category. Given the set of natural language queries, and the HowTo categories, the Crawler can perform the following functions:

- ArticleSearch: given a user input stated as a natural language query, it finds all relevant articles, sorted by relevancy rank. Furthermore, it selects the highest ranking page and processes it. As an example, if the user enters the query *Make Tortilla*, pages like:

  - http://www.wikihow.com/Make-Your-Own-Tortillas
  - http://www.wikihow.com/Make-Tortilla-de-Patatas
  - http://www.wikihow.com/Make-Tortilla-Pizzas
  - http://www.wikihow.com/Make-Tortilla-Snacks
  - ...

  are suggested as relevant, and the first one (e.g., Make Your Own Tortillas) is automatically parsed by the ArticleProcessor (described later on)
- CategoryCrawler: finds all the articles belonging to a specific category. For instance, if the user enters the category recipes:

  http://www.wikihow.com/Category:Recipes

the Crawler will find the 3144 recipes that currently belong to its 167 subcategories.[9]

- BatchExtractor: applies the page processing to all pages belonging to a category. It stores results in a file representing the category or in a database. Currently it can handle JDBC-compliant databases.

5 The Page Processor

The Page Processor subsystem is devoted to deal with a web page extracting the contents deemed important for AP purposes; its tools are devised to facilitate the recognition of sections, together with the contents type. Currently, the Page Processor includes the ArticleProcessor and the ArticlePostProcessor tools. The aim of the ArticleProcessor is to process a given HowTo article in order to extract the useful information in terms of semistructured relations. It also keeps some information about the page structure and the raw contents for further potential processing purposes. On the other hand the ArticlePostProcessor, that integrates semantic tools, builds an augmented plan in the form of predicates containing unstructured components. Exploiting the entire knowledge base, this information will be used to build plans.

5.1 The ArticleProcessor

Given as input a web page, the ArticleProcessor returns a tuple $< a,t,r >$ where a is a sequence of actions, t is a set of tools, and r is a list of related web pages. Each tuple can be seen as an augmented plan with information on its actions, a, initial state t, and related plans r. This processing phase tries to remove all noisy information while avoiding to loose the relevant one required for further processing. The ArticleProcessor embeds an HTML parser devoted to cope with several errors, such as the ones related to the `<div>` closure tag, incoherences with the `id` attribute declarations, changes on the main structure of the page, or bad formatted HTML code.

This subsystem incorporates the current standard tools for processing natural language, such as stemming procedures, which remove inflectional and derivational suffixes to conflate word variants into the same stem or root, or stopwording procedures which remove words with a low information content (e.g., propositions, articles, common adverbs) from the text. The semantic analysis is performed by using WordNet,[10] a lexical database considered the most important resource available to researchers in computational linguistics, text analysis, and related areas. Its design is inspired by current psycholinguistic and computational theories of human lexical memory [10].

The raw contents of a sentence is also preserved to permit further tools to reparse it. As for the sections of type *actions*, the action itself of each sentence is recognized by identifying the verb or the corresponding compound. Furthermore, a

[9] Values on September 10, 2009.

[10] http://Wordnet.Princeton.edu/

set of parameters related to the action are stored and separated from the redundant part of the sentence. More specifically, both actions, tools and relatedwps can have related parameters. Next, we define the most relevant parameters of each type of information.

The parameters related to actions are:

- **action**: the main action, represented by a verb or a compound
- **components**: the components of the action
- **components-st**: the stopwording+stemming of the *components* field
- **plus**: sentences related to the action considered redundant
- **plus-st**: the stopwording+stemming of the *plus* field
- **raw**: the raw contents of the sentence

As an example of actions parsing, given two of the input sentences in the Tortillas Web page "Combine flour, salt, and baking powder in a large medium large bowl." and "Cut in shortening until lumps are gone.", the output of the parser would be:

ACTION:combine; COMPONENTS:flour; PLUS:salt, amp baking powder in a large medium large bowl; COMPONENTS-ST:flour; PLUS-ST:bowl larg bake powder amp medium salt; RAW:combine flour, salt, amp baking powder in a large medium large bowl. and

ACTION:cut; COMPONENTS:in shortening; PLUS:until lumps are gone; COMPONENTS-ST:shorten; PLUS-ST:lump gone until; RAW: cut in shortening until lumps are gone.

As for the elements of type tools, the information concerning *Quantity*, *Unit of measure* (e.g., units, grams, centimeters, cups, spoons) and *name of the ingredients* is stored. So their parameters are:

- **quantity**: the quantity/measure of the tool
- **type**: the unit of measure of the tool
- **tool**: the name of the tool
- **tool-st**: the stopwording+stemming of the *tool* field

As an example, given the sentence "$< b > 2 < /b >$ cups of flour", taken from the ingredients section of the *How To Make Your Tortillas* web page, the parser would generate: *QUANTITY:2; TYPE:cups; TOOL:of flour; TOOL-ST:flour; RAW:* $< b > 2 < /b >$ *cups of flour.*

In the case of the relatedwps, only the name and the HTTP URL are stored. They serve as indexes in our plan data base for accessing other plans. As an example of related web pages for the article Make Your Own Tortillas, it would generate the following two relations:

- URL:http://www.wikihow.com/Make-Your-Own-Tortillas;NAME:Make Your Own Tortillas; and
- URL:http://www.wikihow.com/Make-Tortilla-de-Patatas; NAME:Make Tortilla de Patatas; (in Figure 6)

Fig. 6 WikiHow sample of a related web page

5.2 *The ArticlePostProcessor*

The ArticlePostProcessor rounds off progressively the collected information in order to build the plans structure in terms of steps (i.e., actions) and initial state (i.e., tools). This kind of representation is still not good enough for planning purposes, because it is expressed in a propositional representation, but it will be necessary during the further steps to build the plans corresponding to the articles in predicate logic. This will be possible taking into account the entire knowledge base.

Thus, given a web page, the ArticlePostProcessor builds its corresponding augmented plan. For each action and tool, the ArticlePostProcessor uses the information retrieved by the ArticleProcessor, encompassing in particular the semantical annotation, in order to define the end result. The plan representation contains information about

- The goal represented by the name of the web page
- The initial state in the form of needed tools represented as a tuple <name of the section, quantity/measure, unit of measure, name of the tool>
- The actions to reach the goal represented as a tuple <name of the section, ordinal number of the action, action name, action tools (if any)>

As an example, the following is an extracted plan for the web page How To Make Your Own Tortillas:

- **Goal:** make tortilla
- **Initial state:**
  - tool(ingredients,2,cup,flour):
  - tool(ingredients,1,tsp,salt):
  - tool(ingredients,1,cup,water):
  - ...
- **Plan:**
  - action(steps,1,combine,{flour,salt});
  - action(steps,2,cut,{shorten});
  - action(steps,3,make,{indentation});
  - action(steps,4,add,{water});
  - action(steps,5,work,{mixture});
  - ...

6 The Plan Acquisition

The Plan Acquisition subsystem includes tools that allow to create plans from web pages and to build new plans. The tools belonging to the Plan Acquisition subsystem are: (i) the PlanBuilder which aim is to build plans in predicate logic, and (ii) the AnalysisTool that embodies a suite of statistical tools.

6.1 The Plan Builder

Starting from the ArticlePostProcessor output we need to express the contents of the article in predicate logic (close to PDDL). The idea is to reduce the article to a list of two kinds of information:

- **input-tool**: *predicate that describes the tool properties*
- **action**: *action description with action properties*

These two predicates (in this first step, actions are represented as predicates) are sufficient to express all the information we need to build a plan for each article. However, we need to solve some problems due to the unstructured representation of the components given as output from the ArticlePostProcessor. The first problem is that sometimes tools cited in an action are not defined in the *tools* section. For instance, in an action as "put water", usually the water has not been defined as a tool (ingredient). Sometimes, also the opposite is true: a tool defined in the *tools* list, is not used within the actions. In these two cases, the tool is added/removed to/from the tools list.

Another problem, also due to the use of natural language to define the how-to pages, consists on tools being defined (written) in different ways, specially when analyzing different articles. For instance, a "boiled egg" in the recipes context has been found as *boiled egg* or also *eggs boiled* as well as *water hard boiled eggs*. To solve this problem, a partial string matching algorithm has been used. Considering the stemming of the words composing the name of the tool, the algorithm acts so:

- a list of known tools is generated from the most common to the least, associating an identifier to them
- when a new tool has been found, if it "fuzzy" matches with an already indexed one, it is labeled with the same identifier. A specific matching threshold can be specified as parameter.

The result is a table as Table 2. This table shows, for example, that both *olive virgin high quality oil*, and *pure olive virgin oil* are recognized as *olive virgin oil* and their identifier (for relational purposes) is 87.

Table 2 Example of index of tools

tool-id	tool-name	tool-original-name
...	...	...
87	olive virgin oil	pure olive virgin oil
87	olive virgin oil	olive quality extra virgin oil
87	olive virgin oil	olive virgin high quality oil
87	olive virgin oil	olive best virgin oil
...	...	...
...	...	...
175	fine chopped onion	chopped fine onion
175	fine chopped onion	white chopped fine onion
175	fine chopped onion	chopped fine large onion
175	fine chopped onion	chopped fine medium size onion
...	...	...

A third problem relates to different ways of specifying quantities, so we have normalized them, reducing all of them to a double number. For instance, $\frac{1}{3}$ is stored as 0.33. When all information to build the predicates has been collected, predicates will be:

- **Input tool**: *tool(type, article_id, tool_id, quantity_normalized, quantity_type)*
- **Action**: *action_name(article_id, action_step, tool_id)*

where *type* is the type of the tool if known (ingredient, instrument, etc.), *article_id* is the id of the article (i.e., the id of the recipe), *tool_id* is the standard id of the tool, *quantity_normalized* is the *quantity* of the tool expressed in units of type *quantity_type* that belongs to a predefined enumerated set (i.e., spoon, grams, teaspoon, glasses, etc.), *action_name* is the name of the action (i.e., combine, cut, put, etc.), and *action_step* is the ordinal position of the action in the actions list. If the action requires more than one parameter, we define several action predicates with the same action_step (i.e., `combine(water, flour)` → `combine(water), combine(flour)`).

Tables 3, and 4 show examples of the generated output. They can be computed for the plans corresponding to all the articles in WikiHow or all the ones belonging to a subcategory (i.e., recipes that use onions, or recipes for making tortillas).

Table 3 A fragment of the output file: Actions.txt

action_name(article_id,step_id,tool_id)
...
use(2903,3,56)
get(2911,1,103)
spread(2911,2,103)
put(2906,16,128)
...

Table 4 A fragment of the output file: Tools.txt

input(type, article_id, ingredient_id, quantity, quantity_type)
...
input(ingredient,10,634,1,cup)
input(tool,5,651,1,teaspoon)
input(ingredient,5,978,1,units)
...

6.2 *The AnalysisTool*

The AnalysisTool contains data mining and statistical algorithms. Statistics extracted by this tool could be useful to understand which component or action are most likely to be used or applied in a specific context. This could be used to build new plans, or understand which are the most common subsequences of actions. Experiments have been performed exploiting actions, goal, tools frequency tables, and goal → action and goal → tool correlation tables. If we express the correlation between X and Y as $C(X,Y) = F$, where F is the value of the frequency of how many times the object X appears in the context Y, an example of correlation between goal components and actions performed in the context of the recipes category is:

- C(cake, pour) = 19.12934%
- C(sandwich, put) = 19.01585%
- C(cream, add) = 17.94737%
- C(cake, bake) = 14.81189%

This highlights that, for instance, it's likely (with probability above 19%) to perform the action *put* when the goal is to make a *sandwich*. It will be also useful to analyze particular subsequences of plans in specific contexts (e.g., what are the most likely actions to be performed on an onion in recipes needing oil?).

7 Results

PAA has been developed in Java using the version 1.6.0.11 of the Sun Java Development Kit. NetBeans 6.5.1[11] has been used as IDE. For the experimental phase, a GUI and a Glassfish WebService integrating the architecture have been deployed. Most of the libraries and the distributed computing capabilities rely on the X.MAS framework [3].

We have applied PAA to three WikiHow categories. In Table 5 we show some data on the analysis we have performed over different categories within WikiHow. For each category, we show the number of different plans (pages) parsed, the number of subcategories found, and the number of extracted different actions.

Table 5 Some data on performed analysis

Recipes: *http://www.wikihow.com/Category:Recipes*	
3144	recipes parsed/acquired plans
167	sub-categories found
24185	different individual actions
Sports: *http://www.wikihow.com/Category:Team-Sports*	
979	recipes parsed/acquired plans
22	sub-categories found
6576	different individual actions

The error on unrecognized actions (meaning that the current version of PAA could not be able to semantically parse some sentences and recognize their structure) is about 2%. In general, it is difficult to assess PAA performance, mainly due to the fact that there is no known gold standard; that is correct representations in terms of plans of those pages, so that we could automatically compare against. Hence, with regard to the acquired plans, we did some ad-hoc analysis by manually inspecting some output plans.

In order to perform a first measure of the accuracy of the system the matching between the article and the output of the *ArticlePostProcessor* was analyzed for a set of 40 random taken articles (566 single actions). The accuracy has been calculated taking into account the total number of recognized actions: an action is recognized if its name and the tool match the corresponding row of the article. If an action is missed or it is not well translated, this is considered as a non-match.

Then the accuracy is calculated as $Accuracy = \frac{(matching\ actions)}{(total\ actions)}$. The error can be defined as $Error = (1 - Accuracy) = \frac{non-matching\ actions}{total\ actions}$. An example of measurement took on the WikiHow article "*Make your own tortillas*" is shown on table 6.

This first analysis shows that the system performed rather well on plan extraction (above 68% of accuracy), considering the complexity of the semantic analysis tasks and the need to handle many outliers. In fact, parsing an HTML page, even if automatically generated from a php engine, is not trivial due to *code injection* during

[11] http://www.Netbeans.org

Table 6 Measuring the accuracy of the system, an example with the article "make your own tortillas"

Article row	Action(s)	Results and Notes
1) Mix flour, salt, & baking powder in a large medium/large bowl.	mix(flour, salt)	2 correct, 1 mistake (mix baking powder was missed)
2) Cut in shortening/lard until lumps are gone.	cut(shortening)[12]	1 correct
3) Make a hole in the center of the dry ingredients.	make(hole)	1 correct
4) Add water, about a half a cup at a time, and work mixture into a dough. The dough should be slightly sticky but not hard. You can add slightly more water or flour if needed.	add(water), work(mixture)	2 correct
5) Cover and set aside for 10 minutes.	cover(set), set(aside)	2 mistakes (set and aside are not components)
6) Make the dough into balls about the size of eggs.	make(dough)	1 correct
7) Using a rolling pin, roll each dough ball into about a 6 inch circle.	roll(dough)	1 correct
8) Heat griddle or skillet on medium heat without grease.	heat(griddle)	1 correct
9) Cook tortilla 1/2 to 1 minute (if it starts to bubble, that's long enough).	cook(tortilla)	1 correct
10) Flip tortilla to the other side and cook for a few seconds.	flip(tortilla), cook(few)	1 correct, 1 mistake (few is not a component)
11) Continue until all your dough is cooked.	continue(until)	1 mistake (until is not a component)
12) Then you can eat!	eat()	1 correct

the compiling of the predefined structure, and to the addition of different sections depending on the context (e.g., *ingredients* in recipes, or *work tools* in machinery). Besides, sentences are structured in different ways and filled with different kinds of contents more than "simple" steps. For instance, some people add a lot of non descriptive text (e.g., from a step for the "Make Vodka" how-to: *Column stills produce purer alcohol because they essentially redistill the alcohol in a single pass*). Others add playful goals with facetious suggestions (e.g., from the "Microwave a Peep" how-to: *Don't hurt yourself with the fork. You will be a slave to the Peeps if you eat it at all*). Moreover, somebody slangs or adds actions not related to the goal (e.g., *Stir and enjoy!*). The preprocessor has to handle all this, other than trying to manage compound forms, exploiting redundant descriptions of the action in the sentence and attempting to discover tools not explicitly cited.

To test the validity of the subsystems composing the architecture, the system has been tested by five selected users. The users monitored the behavior of the system through a web service that integrates the tools over a two-week period. By conducting regular interviews with each user to estimate her/his satisfaction we could verify the correctness of the process. All users stated their satisfaction with the system that succeeded in translating most of the articles into understandable plans. They also allowed to spot most of the processing problems.

Clearly, it is impossible to perform perfectly well. However, we reached our goal to have a good tradeoff between information retrieved from a web page and information lost during the filtering process. Also, we obtained a reasonably good tradeoff between semantical comprehension and introduction of errors. Thus, even if the integrated technologies that compose our tools are subject to future improvements, they already gave us a base on which to work and play on collected information in the next future.

8 Conclusions and Future Work

In this chapter, we described a work aimed at bridging the gap between the need of tools for automatically building plans and action models from semi-structured information and the existence of this kind of knowledge in the Web (e.g., WikiHow). We believe this work can encourage further research in this topic that can greatly help the massive application of planning to many real-world human related tasks. Experimental results show that the tools performed well on extracting plans, thus establishing a preliminary base on which to work.

As for future work, the *analysis of common subsequences* on multiple plans in the same domain will be performed. We will base this analysis on previous work on planning as macro-operators [12], n-gram analysis of natural language tools [9], or association rules learning [15]. This can be further used for performing case-based planning by recovering subsequences that include some specific object. For instance, subsequences of recipes that use a particular ingredient or tool in general. Also, it could be used by tools that help on inputing recipes on the WikiHow by suggesting previous plans subsequences for the ingredients. Furthermore, we plan to include a Planner subsystem that contains tools necessary to exploit the collected knowledge base (e.g., the plans library) *to build new plans*. Other uses of this library will be aimed at performing *plan recognition* from data coming from sensors and at matching such data against the plans recovered from the web, or at *acquiring complete action models* with preconditions and effects from the input plans [23]. We will also exploit different knowledge bases as *eHow*[13] an online knowledge resource offering step-by-step instructions on *how to do just about everything*. eHow content is created by both professional experts and amateur members and covers a wide variety of topics organized into a hierarchy of categories; or *HowToDoThings*[14] another hierarchically organized knowledge base of how-to

[13] http://www.eHow.com/

[14] http://www.howtodothings.com/

manuals, in order to make our architecture even more general and independent from the particular web site.

References

1. Addis, A., Armano, G., Vargiu, E.: WIKI.MAS: A multiagent information retrieval system for classifying Wikipedia contents. Communications of SIWN 3, 83–87 (2008)
2. Addis, A., Armano, G., Borrajo, D.: Recovering Plans from the Web. In: Proceedings of SPARK, Scheduling and Planning Applications woRKshop, ICAPS 2009 (2009)
3. Addis, A., Armano, G., Vargiu, E.: From a generic multiagent architecture to multiagent information retrieval systems. In: AT2AI-6, Sixth International Workshop, From Agent Theory to Agent Implementation, pp. 3–9 (2008)
4. Birukov, A., Blanzieri, E., Giorgini, P.: Implicit: an agent-based recommendation system for web search. In: AAMAS 2005: Proceedings of the fourth International Joint Conference on Autonomous Agents and Multiagent Systems, pp. 618–624. ACM Press, New York (2005)
5. Rodríguez-Moreno, M.D., Borrajo, D., Cesta, A., Oddi, A.: Integrating Planning and Scheduling in Workflow Domains. Expert Systems with Applications 33, 389–406 (2007)
6. Burdman, J.: Collaborative Web Development: Strategies and Best Practices for Web Teams. Addison-Wesley Longman Ltd, Amsterdam (1999)
7. Camacho, D., Aler, R., Borrajo, D., Molina, J.: A multi-agent architecture for intelligent gathering systems. AI Communications, The European Journal on Artificial Intelligence 18(1), 1–19 (2005)
8. Castillo, L., Armengol, E., Onaindía, E., Sebastiá, L., González-Boticario, J., Rodriguez, A., Fernández, S., Arias, J.D., Borrajo, D.: SAMAP: An user-oriented adaptive system for planning tourist visits. Expert Systems with Applications 34(34), 1318–1332 (2008)
9. Dunning, T.: Statistical identification of language. Technical report (1994)
10. Fellbaum, C.: WordNet An Electronic Lexical Database. The MIT Press, Cambridge (1998)
11. Fernández, S., Borrajo, D., Fuentetaja, R., Arias, J.D., Veloso, M.: PLTOOL. A KE tool for planning and learning. Knowledge Engineering Review Journal 22(2), 153–184 (2007)
12. Fikes, R.E., Hart, P.E., Nilsson, N.J.: Learning and executing generalized robot plans. Artificial Intelligence 3, 251–288 (1972)
13. Gil, Y.: A domain-independent framework for effective experimentation in planning. In: Proceedings of the Eighth International Workshop (ML 1991), pp. 13–17 (1991)
14. Hoffmann, J., Weber, I., Kraft, F.M.: Planning@SAP: An Application in Business Process Management. In: Proceedings of SPARK, Scheduling and Planning Applications woRKshop, ICAPS 2009 (2009)
15. Kavsek, B., Lavrac, N., Jovanoski, V.: APRIORI-SD: Adapting Association Rule Learning to Subgroup Discovery. In: Berthold, M.R., Lenz, H.-J., Bradley, E., Kruse, R., Borgelt, C. (eds.) IDA 2003. LNCS, vol. 2810, pp. 230–241. Springer, Heidelberg (2003)
16. Manzano-Macho, D., Gmez-Prez, A., Borrajo, D.: Unsupervised and domain independent ontology learning. combining heterogeneous sources of evidence. In: Proceedings of the Sixth International Conference on Language Resources and Evaluation, LREC 2008 (2008)
17. O'Reilly, T.: What is Web 2.0, Design Patterns and Business Models for the Next Generation of Software. O'Reilly, Sebastopol (2005)

18. Pollack, M., Brown, L., Colbry, D., McCarthy, C., Orosz, C., Peintner, B., Ramakrishnan, S., Tsamardinos, I.: Autominder: An intelligent cognitive orthotic system for people with memory impairment. Robotics and Autonomous Systems 44, 273–282 (2003)
19. Sánchez, D., Tentori, M., Favela, J.: Activity recognition for the smart hospital. IEEE Intelligent Systems 23(2), 50–57 (2008)
20. Simpson, R.M., Kitchin, D.E., McCluskey, T.L.: Planning domain definition using gipo. Knowl. Eng. Rev. 22(2), 117–134 (2007)
21. Tapia, E.M., Intille, S.S., Larson, K.: Activity Recognition in the Home Using Simple and Ubiquitous Sensors. In: Ferscha, A., Mattern, F. (eds.) PERVASIVE 2004. LNCS, vol. 3001, pp. 158–175. Springer, Heidelberg (2004)
22. Wang, X., Veloso, M.M.: Learning planning knowledge by observation and practice. In: Proceedings of the ARPA Planning Workshop, pp. 285–294 (1994)
23. Yang, Q., Wu, K., Jiang, Y.: Learning action models from plan examples using weighted max-sat. Artificial Intelligence 171(2-3), 107–143 (2007)
24. Yang, H.C.: A general framework for automatically creating games for learning. In: Proceedings of the fifth IEEE International Conference on Advanced Learning Technologies (ICALT 2005), pp. 28–29 (2005)
25. Zhou, L.: Ontology learning: state of the art and open issues. Inf. Technol. and Management 8(3), 241–252 (2007)
26. Zimmerman, T., Kambhampati, S.: Learning-Assisted Automated Planning: Looking Back, Taking Stock, Going Forward. AI Magazine 24, 73–96 (Summer 2003)

Semantic Desktop: A Common Gate on Local and Distributed Indexed Resources

Claude Moulin and Cristian Lai

Abstract. The social Web is characterized by the communication between loose community members and the sharing of resources that have to be managed by adapted indexing and querying systems. In this chapter we present a solution helping users that want to belong to some community to organize in a common way the resources retrieved from the community and their own resources. We only consider communities structured as a network of peers without any centralized support. The implemented system front-end is built as a web application similar to a traditional desktop operating system. It gives interfaces to several tools dealing with resources stored in the memory. A specific one, independent of any other applications is in charge of all the operations concerning the indexing, the publication of resources in the peer to peer network and the retrieving of resources in the local memory and from the network. Our solution is based on semantic indexing using concepts of domain ontologies automatically downloaded from the network. We show the way we have solved the main issues occurring in this research context.

1 Introduction

One of the main features of the Web 2.0, generally called the social Web, is the communication between loose community members and the sharing of resources. We actually focus on the communication of significant resources inside communities having a cultural goal in a specific domain. Community members manage their own resources organized in a personal memory (Abel et al., 2006) and want

---

Claude Moulin
Compiègne University of Technology,
Heudiasyc CNRS, France
e-mail: claude.moulin@utc.fr

Cristian Lai
CRS4, Research Center for Advanced Studies in Sardinia,
Italy
e-mail: clai@crs4.it

Soro et al. (Eds.): Inform. Retrieval and Mining in Distrib. Environments, SCI 324, pp. 61–76.
springerlink.com

to share some of them with the community they belong to. They would like both to retrieve interesting documents from the community and publish some documents that are already stored in their memory.

We only consider loose communities that are structured as a peer to peer (P2P) network allowing publishing and searching for documents. They don't need centralized services and each member only requires an easy to use application for managing the access to the network and thus the community.

We consider that each member owns a memory that has a private part containing personal resources and a public part containing the documents that have been shared with the community. Each member desires to have a unique system for managing both parts of the memory.

Documents contained in the memory may have different types (text, audio, video, archive, etc.) and different storing formats. We cannot consider that the title could be the only way to identify a resource, as in ordinary file sharing systems (Mougul, 1986), because generally these resources are created locally and their meaning is not universally known. Due to the diversity of types and formats, they must be manually indexed. Only some of them could be automatically analyzed. However, external information that cannot be found inside a document has generally to be added in order to better describe the content and its use.

In our study, the community nature is not relevant because the solution we propose to the issues is generic. However, we stress on applications regarding the e-learning domain and most of the examples take part. The challenge is (i) the development of a unique indexing system for supporting the management of resources in the public part of the user's memory and also in its private part, and (ii) the development of an application encapsulating in a transparent manner the functionalities that the user requires for managing the resources. In a P2P network they are managed by a distributed index, i.e. that each peer owns a part of the index in a transparent manner. The indexing is Boolean (Salton et al., 1982) and the keys allow both publishing and searching for resources. Thus, the local index used for managing the private resources must have the same characteristics as those of the distributed index in order to insure the compatibility of the indexing.

We mainly consider a semantic indexing of the resources. It allows reasoning based on ontologies that may enlarge the results of a search or may propose close documents complementary of the research. In our solution, all the keys used for representing a document in the index represent its semantic description and are written in a language based on RDF. Ontologies used for indexing have to be inserted in the network by expert members and used by all the members. We can consider the index as a knowledge base augmented by any user that wants to publish a resource. Sometimes, it is necessary to add some new strings as a value of ontology attributes. We also deal with the possibility to tag a resource with any additional keyword thanks to the system ontology we have developed.

We developed different applications as part of a Semantic Desktop (Sauermann et al., 2005, 2006). They are integrated within a web application. Users rely on a set of tools like in a traditional computer desktop. The architecture backend consists in a set of web services managing the resources and giving access to the P2P network. A unifying web user interface gives a common access to the services and allows an easy communication between them.

In this chapter, we describe the indexing system and in particular the service that manages the ontologies and simplifies the building of indexing keys. We detail the user interface of the system and also show the process that a user must follow in order to publish or retrieve documents.

2 Related Works

Long time ago Vannevar Bush, the director of Office of Scientific Research and Development of United states wrote an article in The Atlantic Monthly Journal, titled "As we may think" (Bush, 1945), describing the idea of an hypertextual machine called Memex, a sort of a electronic desktop equipped with a microfilm storage memory, allowing to save pages of books and documents, play them and to associate each other, in order to make knowledge more accessible. The essay predicted many kinds of technology invented after its publication.

Progresses in Semantic Web, peer to peer, natural language processing, is leading to new forms of collaboration and social semantic information spaces.

Even Tim Berners Lee did not really envision the World Wide Web as a hypertext delivery tool, but as a tool to make people collaborate (Decker, 2006).

2.1 Semantic Desktop

In (Sauerman et al, 2005) the idea of Semantic Desktop is defined as: *A Semantic Desktop is a device in which an individual stores all her digital information like documents, multimedia and messages. These are interpreted as Semantic Web resources, each is identified by a Uniform Resource Identifier (URI) and all data is accessible and queryable as RDF graph. Resources from the web can be stored and authored content can be shared with others. Ontologies allow the user to express personal mental models and form the semantic glue interconnecting information and systems. Applications respect this and store, read and communicate via ontologies and Semantic Web protocols. The Semantic Desktop is an enlarged supplement to the user's memory.*

Many research projects are attempting to merge the focal parts of Semantic Web into desktop computing, P2P and Social Networking.

The Gnowsis project (Sauermann, 2003) (Sauermann et al, 2006) deals with the details of integrating desktop data sources into a unified RDF graph, also addressing the problem of identifying resources with URIs. The main idea was to enhance existing desktop applications and the desktop operating system with Semantic Web features. Whenever a user writes a document, reads e-mails, or browses the web, a terminology addressing the same people, projects, places, and organizations is involved. It is connected by the interests and the tasks of the user. They only propose Gnowsis to be combined with web 2.0 philosophy and semantic web technology as useful basis for future semantic desktops; they didn't design Gnowsis as an integrated web system for large use among internet communities.

Other projects focus on the issue of data integration, aggregating data obtained from the web. On the web services world the SECO project (Harth, 2004) aims at integrating web sources via an infrastructure that lets agents uniformly access data that is potentially scattered across the web. Using a crawler, it collects the RDF data available in files. RDF repositories are used as sources for data. Integration tasks over the various data sources, such as object consolidation and schema mapping, are carried out using a reasoning engine and are encapsulated in mediators to which software agents can pose queries using a remote query interface. SECO includes a user interface component that emits HTML, which allows for human users to browse the integrated data set. To create the user interface are considered portion of the whole represented data set and used to generate the final page. The structure of the pages of the site is created using a query, and transforming the results of the query to HTML, giving three fundamental operations: a list view, keyword search functionality, and a page containing all available statements of an instance.

2.2 Distributed Systems

Nowadays there are more tools than ever to help harness unused computing power in the hundreds of PC being used by users. Traditionally, there have been three categories of "distributed" computing: *Cluster computing*, similar machines (generally servers of similar power and configuration) are joined to form a virtual machine. Linux clusters are good examples; *Peer to peer*, many desktop computers are linked to aggregate processing power. The distinguishing characteristic is the machine itself, which almost exclusively is a low-power client PC. Often, the link is via the Internet; *Distributed computing*, increasingly known as grid computing, this approach connects a wide variety of computer types and computing resources, such as storage area networks, to create vast "virtual" reservoirs of computers serving geographically widely separated users.

The traditional client-server internet model is beginning to give some ground to P2P networking, where all network participants are approximately equal. The primary advantage of P2P networks is that large numbers of people share the burden of providing computing resources (processor time and disk space), administration effort, creativity and legal liability. It's relatively easy to create community of users in such an environment and it's harder for opponents of a P2P service to bring it down.

Distributed Hash Tables (DHT) are permanently considered as a key technology in P2P applications as a consequence of their robustness and scalability. Several projects (Chawathe et al, 2005) (Druschel and Rowstron 2001), as well as popular file sharing applications [1] make use of DHTs in order to distribute the data over a large number of peers, that contribute storage to a community of users. In the last years the research and the development in the P2P field has been

[1] Vuze - Java BitTorrent Client.http://azureus.sourceforge.net/
eMule - http://www.emule-project.net

considerable. Napster, Gnutella2, Edonkey2K, Bit Torrent, Kademlia (Mayamounkov and Maziers, 2002) are only a few examples of consolidated protocols/architectures. The use of a such shaped infrastructure in general is justified due to the most relevant features of P2P systems, such as resistance to the censorship, decentralization, scalability, security, etc. To distribute data among thousand or million of peer not only means to have a huge amount of information, but also means to make confidence to a robust system free of restriction from a central authority; enabling virtual communities to self-organize and introduce incentives as a resource sharing and cooperation, arguing that what is missing from today's peer-to-peer systems should be seen both as a goal and a means for self-organized virtual communities to be built and fostered.

Between 2001 and 2002 was born almost simultaneously several architectures for DHT, such as CAN (Ratnasamy et al, 2001), Chord (Stoica et al, 2001), Pastry (Rowstron and Druschel, 2001), Kademlia (Mayamounkov and Maziers, 2002), etc. unfortunately too few of them are originated a real and good implementation supported of stable communities of developers.

2.3 Distributed Index

In SA Net (Chatree and Taieb, 2004) an agent-based system achieves its semantic richness through the use of explicit ontologies to represent resources. SA Net further enhances the DHT based resource distribution scheme by using the unique identifier assigned to each ontology as a key to locate the overlay node responsible for maintaining the resource index associated with the underlying ontology. In other words, the ontology-based hashing scheme, utilizes ontologies, instead of resource names, as the hash input to generate the key necessary to distribute the resource among overlay nodes. In our approach we give the same responsibility to all nodes of the network. We have a slightly different meaning of semantic indexing. We do not directly attach resources to ontologies but create keys whose content refer to ontologies.

The SCORE project (Sheth et al, 2002) provides classification and terminological basis for contextual reasoning on metadata. Metadata are divided into two types: syntactic metadata and semantic metadata. Syntactic metadata describe non-contextual information about content, e.g. language, length, date, audio bit-rate, format, etc. Such metadata offer no insight about the content. Semantic metadata describe domain-specific information about content. We consider that meta-data are not enough to describe resources and allow some reasoning upon them. We don’t intend to extract or use metadata from different structured information sources.

The PAGE (Della Valle et al, 2006) (Put And Get Everywhere) project consists in a peer to peer infrastructure for distributed RDF storing and retrieval. It starts from YARS (Hart and Decker, 2005), a solution that defines an optimized index structure for fast retrieval of RDF statements. PAGE implements YARS index structure that indexes resources using RDF encoding called quad (spoc), where s is the subject, p the predicate, o the object and c is the context. The index is achieved creating keys becoming from the combination of quad elements. We adopted a similar approach but moving toward OWL data model, considering a semantic description as a conjunction of compositions of triples. We don't index RDF statements. We build keys based on RDF statements.

RDFGrowth algorithm (Tummarello et al, 2004) introduces a scenario of a peer to peer network where each peer has a local RDF database and is interested in growing its internal knowledge by discovering and importing it from others peers in a group. It is important to consider this kind of approach in order to define a mechanism of queries based on SPARQL formalism. This kind of queries requires RDF knowledge base. The problem should be to distribute a centralized knowledge base on different nodes in order to satisfy a query by accessing only one node.

Our scenarios of use remain similar: browsing several ontologies, a user can index or search for resources. In background, the system builds the indexing keys. The types of allowed requests determine the types of indexing keys and routing algorithms. In a centralized case a compound query is an investigation on a knowledge base (looking for the triples which suit the query in RDF bases). In our case we have to face issues of distributed knowledge bases. A direct interrogation of the overall knowledge base is impossible.

2.4 Progress Beyond State of the Art

Our work aims at demonstrating the benefit of a semantic indexing engine exploited through a set of tools available for a community of users located in different places.

To create a community of users means to tackle the topic of distributed systems. A first requirement is to have systems independent from any central point of aggregation. Among distributed systems, P2P architecture brings most advantages, among them decentralization, scalability, fault tolerance. It is mandatory to have an efficient distributed data structure to efficiently store and retrieve elements from a huge amount of information; the evident efficiency of DHTs relies in the number of messages exchanged to route a query to its destination. The order of magnitude of this number is O(log(N)), where N is the total number of nodes. In this work the low level layer concerning the P2P applications is built on FreePastry[2], the open-source implementation of Pastry, whose significance is guaranties by the support provided by the community of users regularly improving and amending; its features allow for adapting the network to the specific needs.

We don't require to deduce new metadata from different structured information, but simply to create an index whose content refer to ontologies. However some reasoning elements have to be taken into account.

From a user point of view it is necessary to access a set of tools providing an intuitive interaction. The activities around Gnowsis started with the enhancing of desktop applications with the features of Semantic Web, but only standalone applications have been considered. Considering the web 2.0 approach and using

[2] FreePastry - `http://www.freepastry.org/`

semantic web technologies we have created an integrated web system, similar to a common desktop, for a large use among internet communities. There are certainly several advantages in designing a web application (and using a desktop metaphor) rather than designing a standalone application. The first one is that the same UI is presented regardless of platform. Cross-platform GUIs are an old problem. Qt, GTK, wxWindows, Java AWT, Java Swing, XUL, they all suffer from the same problem: the resulting GUI doesn't look native on every platform. Even if it's possible to get a toolkit that looks native on every platform, often it's necessary somehow to code the application to feel native on each platform. Secondly, a web application can be easily upgraded, allowing all users to run the last version at the same time. The main advantage is the simplified access to data, everywhere and at any time. People can consult their resource even with a simple display. The requirement of a network access is not a constraint because nowadays there are good possibilities to be always connected.

As experienced in the SECO project, many RDF compliant heterogeneous data sources are queried and depicted to the user via ad-hoc web user interfaces generation. Our purpose is not to aggregate data obtained generally from the web but from a community allowing to limit the context of data and to ensure a better adequacy between research intention and results.

3 Semantic Indexing

We generally consider two kinds of models for indexing resources: boolean (Salton et al., 1982) and vectorial. In the Boolean model, the index of documents is an inverse file which associates to each keyword of the indexing system the set of documents that contain it. A user's query is a logical expression combining keywords and Boolean operators (NOT, AND OR). The system answers with the list of documents that satisfy the logical expression. The keywords proposed by the user are supposed contained in the index.

In the vectorial model a document is represented by a vector whose dimensions are associated to the keywords occurring in the document and the coordinates correspond to the weight attached to the keywords in the document thanks to a specific calculus. A request is also a vector of the same nature. The system answers a request with the list of documents which present a similarity with the request thanks to a specific measure based on the vectors coordinates.

In centralized indexing both models are available. However, in the case of P2P networks, the index must be distributed among the peers and the numbers of queries sent to the index when searching for resources should be minimized, because they are time consuming. For respecting this constraint, the model of a distributed index is necessarily Boolean.

In traditional file sharing systems, titles of resources, like songs or films are used both for indexing and discovering resources. It is enough because all needed information shared by users is contained these elements. Such a title can be considered as a unique universal identifier of a resource. Each title is associated with the reference to the computers that physically own the corresponding resource.

Our solution also allows a file sharing. The index can be seen as a table associating values to keys. A value is the lists of elements that satisfy the key. According to the type of keys, elements included in a list can be the URLs of document associated with the key or specific textual documents.

For indexing a specific resource, the system provides keys containing all the needed elements that describe its contents and that allow the further discovery of the resource. For example, we can express in a key that associated resources are concerning the concept of grammar in the domain of the theory of languages, and that this resource is a difficult exercise. A key can also mean that associated resources are archives containing a grammar file of a domain specific language and the java binary code of a parser for this grammar.

A key can be seen as a conjunction of simpler elements, each expressing a se-mantic assertion about resources. This way of understanding a description leads to a semantic indexing of the resources i.e. an indexing based on a shared representation of the knowledge that is the subject of the community. A common index of resources semantically identified is distributed among the set of nodes of the P2P community network. The data structure that has been considered suitable for that is a Distributed Hash Table (DHT) (Stoica et al., 2001). Each node of the network contains a part of the whole data structure. The publication, or indexing, is the operation of insertion of a resource inside the DHT. In this operation a new index entry is inserted in the DHT. The discovery is the operation which allows to find some resources in the network that correspond to a research key.

We chose to build keys from ontology elements. Several ontologies are generally necessary for representing either objective or subjective information about a resource. Each key represents a fragment of a knowledge base, updated each time a new resource is published. From a formalization point of view, we considered ontologies represented in OWL (Bechhofer et al., 2004). The theoretical structure represented in the index should be a RDF knowledge base if it was deployed. For brevity reason, we present all the examples using the N3 formalism (Berners-Lee, 2006).

3.1 Expressiveness

A key is a string whose content is based on a RDF description of a resource. Sometimes it is possible to find existing ontologies for expressing all the intention of the user describing a resource. The first example given in the previous section requires two ontologies. The first one is concerning the theoretical domain (theory of language) and contains the concept of Grammar. The namespace of this ontology and the nature of the concepts it contains are enough to be sure it represents the right knowledge domain. The second one is concerning the domain of learning, and contains the concept of Exercise and a way to express the difficulty level of this exercise. More information about this ontology can be found in (Ghebghoub et al., 2009). Using this ontology, we have to say that the resource is a learning object whose education category has very difficult as difficulty level. If we denote the ontology with *lom*, the corresponding description fragment looks like: [a lom:LearningObject lom:hasLomEducational [lom:hasDifficulty lom:very_difficult]. We observe that the description contains blank nodes whose identifiers are useless and the N3 language is interesting for that. Another characteristic is to use individuals of enumerated concepts as values. From a practical point of view, the user needs a specific application whose graphical interface presents a simple navigation through ontologies and allows a path selection inside an oriented graph that automatically generates indexing keys.

Sometimes, the intention of a user is simpler. For example, it should be possible to express that a resource is about Automaton. If the user can find an ontology containing a concept representing this notion, then this ontology should be used for indexing. It is a particular example of indexing on a Concept. We can also conceive to index on a relation or an individual of an ontology. If no ontology can be found with such a concept, the only solution is to tag with a keyword.

Our system allows both simpler intentions. However, it was necessary to represent them according to the same indexing model. We have created a system ontology that allows such descriptions. Denoting by *syst* this ontology and the domain ontology by *lt*, it is possible to describe a document concerning the concept of Automaton with: [a syst:Document] syst:hasInterest lt:Automaton. Formaly, the logical level of this representation is OWL Full. However, it is not directly used for reasoning and thus doesn't present any logical issue.

The system ontology is also useful for associated keywords and for characterizing the ontologies used for indexing and that have to be published in the network. It contains the concept of Ontology, sub-concept of Document for that purpose. Such an ontology is published in the network under the description: [a syst:Ontology]. Thus, our solution lets open the choice of the ontologies that could be useful for the descriptions. The Indexing tool in our platform first looks for all the ontologies used in the network for indexing and can load them in the navigation sub system.

For indexing a document that would happen to be an ontology concerning the Cinema, but not dedicated to the indexing, it is first necessary to find a specific ontology containing a concept denoting knowledge representations and where the document to be indexed could be an instance. The domain of this file is defined by the hasDomain attribute of the system ontology. The following description is an example: [a ont:KnowledgeRepresentation] syst:hasDomain "Cinema".

3.2 Reasoning

In Boolean index, as DHT, keys used for retrieving resources must be equals to keys used for publishing. However resource publishing and retrieving contexts are different. The other issue to take into account is the number of queries that have to be launched through the network in order to minimize the access time.

The solution we propose is to foresee during the publication of a resource different reasonable situations, i.e. different queries to which the resource can respond positively. A resource is then published with an expansion of different keys generated from the one created by the resource provider. At this stage the consumed time is not penalizing. Here are some examples of key expansion.

Generally, a description corresponds to a conjunction of assertions (key: A and B); the resource must be retrieved when keys A, B, (A and B), and (B and A) are proposed. For a conjunction of more than two assertions we must consider all the combinations. A lexicographic ordering approach reduces (A and B), and (B and A) to only one form.

A second expansion type is based on logical inferences that can be done on ontologies. For example, a document concerning deterministic automata is also about automata. We consider that when a resource is indexed on a concept B, subconcept of A, it also has to be indexed on A. We restrict the transitivity of subsumption for avoiding the meaningless indexing on a too general concept.

A third case is concerning the generalization of the last element of a composition of relations used in assertions. In the following example the difficulty of the document has been expressed: [a lom:LearningObject lom:hasLomEducational [lom:hasDifficulty lom:very_difficult]. The difficult level has been expressed with a specific value. We consider that it is a particular case where the difficulty is expressed. We also create the key: [a lom:LearningObject lom:hasLomEducational [lom:hasDifficulty], meaning that the difficulty level has been expressed.

3.3 Community Related Issues

The life of a community is not only due to the indexing, publication and exchange of documents. Other issues concern the interests, the ontologies, the tools used by the members and the information that can be relevant for the community. We call "Community resources" those relating the life of the community. They must be accessible to all members and represent the daily life of the community. The complete decentralized situation of the communities we consider seems to bring a contradiction with the daily life. Where and how diffusing information about events? The main example of community resources is the community wiki, where users can update information regarding the community in a collaborative way. People can describe special events, add notes on tools, etc.

As we are considering communities having a specific nature, it seems completely logical that also for the community resources no official site should be required. We already considered ontologies used for indexing. These specific community resources are stored directly in the index and respond to the following key involving the system ontology: [a syst:Ontology].The system ontology is extended with the definition of all the documents involved in the system. A specific query gives access to the wiki initial pages.

Pages of a wiki, for instance, are generally organized as hypertexts and contain references to other pages. In our case, these pages have to be stored in the index of the network and then can be retrieved thanks to an analogous request. We have substituted the usual static reference links with semantic links, representing keys used for indexing.

4 Implementation

We have developed a platform in order to simplify the indexing task to the user. Even if the semantic indexing is the core of our study, the notions of ontology, concept, etc. are generally ignored. The platform is designed as a set of integrated elements whose aim is to manage electronic resources of different types (texts, images, video, audio) in shared and personal memories. On the network side the technology chosen is built on Pastry, a generic, scalable and efficient substrate for P2P applications. We use FreePastry (Rowstron and Druschel., 2001), the open-source implementation of Pastry. Its features allow for adapting the network to the specific needs.

The global conception of the system is a Web application. The Web design of the front-end guaranties an anywhere and anytime access to the system. The user interface is developed mixing web technologies, HTML, JavaScript, CSS, and animated with AJAX techniques. We have based all this features on the ExtJs library[3].

4.1 Three Layered Architecture

The following picture shows how the system is structured. In the border of the system we can see a community of nodes distributed within a P2P network; the resources are assumed to be available in a user personal archive called Personal

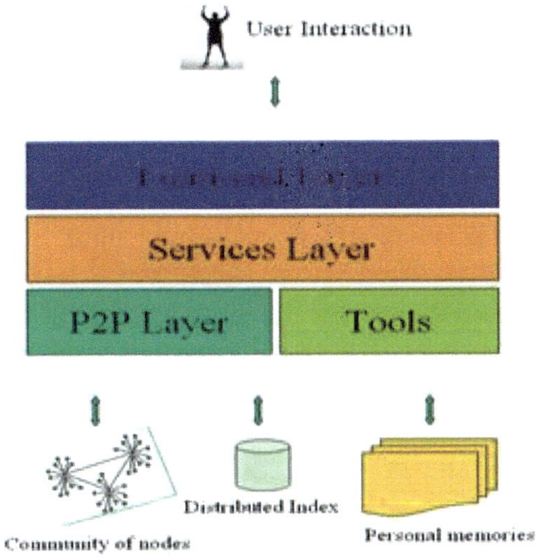

Fig. 1 The architecture

[3] ExtJs - `http://www.extjs.com/`

Memory. The user accesses the system and interacts with it thanks to a graphical interface provided by the Front-end Layer. An extensible set of services provides the necessary support for accessing the system.

The P2P Layer constitutes the infrastructure of the system. Each user is normally associated with a node (a Peer). Such layer gives access to the DHT scattered among a community of nodes. A node is the atomic element of the network. A user system joining the community is located through a node that contains a portion of the distributed index as part of the whole information.

The Service Layer provides an integration substrate exporting the features of the node and the personal memory. The capabilities of the platform are based on the SOA paradigm and each capacity is implemented as a web service, exploiting the features of the REST (Fielding, 2000) technology. Such a layer is mainly designed to interconnect the first layer and required back-end tools with the front-end. In this way, it is easy to implement a new capability for the system as a web service.

The Front-end Layer plays the role of graphical user interface. Each node is associated with one interface, but there are no constraints to connect to a specific one. We have designed a Web application, so as to provide a cross-platform solution with no requirements of specific installation. Such interface is designed as a set of views on tools and looks like a Semantic Desktop.

4.2 Tools

The set of tools available via the web interface is extensible so as to improve the desktop with new applications for different purposes. Fig. 2 gives an idea of the user interface. The "Indexing Tool", "Indexing Pool Service", "Notes", "Local Resources Manager", "Retrieval Service" are currently available.

The "Indexing Tool" is used to browse the ontologies downloaded from the network and selected by the user. Navigating through the ontologies the users made some selections that allow the indexing tool to generate the indexing keys. This point is important because users aren't generally aware of the existence of ontologies. They only know that they have to select elements for indexing. Till now, we simplified the navigation system by presenting only the possible paths that are available at each moment for indexing. With an ontology is associated an entry point representing the concept that can denote a document. It is the starting point, and then users can attach a concept representing the subject of the document by choosing it in the list of presented concepts. The other way is to follow a path because our system only shows the possible relations at any moment.

The "Indexing Pool Service" gives access to the repository of resources intended to be published (or simply stored in case of personal memory). The "Notes" is an application that allows for creating personal notes. The "Local Resources Manager", allows for selecting resources from personal memory. Resources that have to be published must be first uploaded using the Local Resources Manager functionalities. The "Retrieval Service" tool allows queries submissions and is in charge of retrieving the corresponding resources. Then it can display the results.

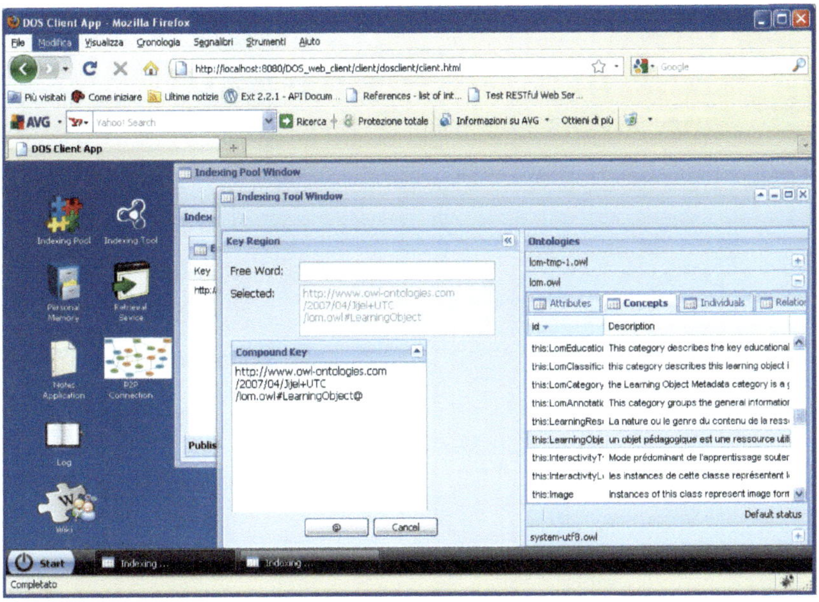

Fig. 2 The web user interface

The Indexing Tool and the Indexing Pool Service have to be used together for creating a key and preparing the publication of a resource. A simple drag and drop between the respective interfaces enables the communication between the tools. Users use the Indexing Tool in order to build resource descriptions. Thus, the user can create a description based on one or more ontologies.

From the Indexing Pool Service, users select the resource they want to index and associate the descriptions built by the indexing tool. The result is a pair containing a file and a description. In backend, a (key, value) pair, called entry is generated. The key identifies the resource and is unique in the whole network. The same key can identify multiple resources. The value is the resource itself. It can be a link to the resource (if the resource is available as an HTTP url) or the content of the resource itself.

The pair may be used either to store the resource in the local memory, or to publish the resource in the network. Fig 3 shows the logical association between a file and a description.

To retrieve a resource semantically described and stored within the system, it is necessary to use the Indexing Tool in order to create a key. The process is similar to the publishing one. As soon as the key is completed, it is submitted to the system via the Retrieval Service that launches a request in the personal and/or the shared memory. Resources belonging to the different memories are retrieved in the same way and shown by the Retrieval Service.

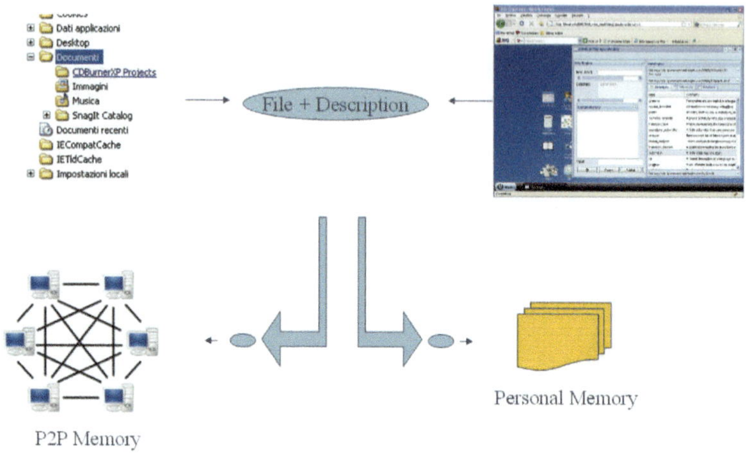

Fig. 3 The web user interface

5 Conclusion

In this work we have presented an approach for managing resources belonging to a personal repository or exchanged with a distributed environment storing the resources of a community. This approach is completely compatible with the definition of Semantic Desktop. The community structure is assumed to be a network of peers. The resources are in every case semantically indexed via domain specific ontologies downloaded from the network. The semantic information strictly related to a resource may also represent a point of view of the user on the resource.

The semantic index, which can be considered as distributed RDF knowledge base is inserted in a Distributed Hash Table whose structure guaranties an efficient management of the resources. Our solution can support the building of online communities of users that want to easily share digital resources. We consider that the ontologies required by the indexing can be provided by some experts of the community the user belongs to.

The user has to navigate the suitable ontologies in order to understand their different concepts and even if this operation is time consuming it is nevertheless useful because a semantic indexing has some advantages regarding a keyword indexing. Ontologies allow some reasoning and we have shown how the structure of the index requires to take into account this characteristics for improving the queries.

Considering the web 2.0 approach and using semantic web technologies we have created an integrated web system, similar to a common desktop, for a large use among internet communities. The web front-end guaranties the anywhere and anytime access to the resources. The system is still under development and the user interfaces designed for the browsing of ontologies are not still satisfactory. Issues of navigating the ontology in order to annotate and search documents will be investigated with respect to usability. The system is extensible and new services could be easily added in order to provide new functionalities that could benefit of the common indexing sub-system.

References

Abel, M.-H., Moulin, C., Lenne, D., Lai, C.: Integrating distributed repositories in learning organizational memories. WSEAS Transactions on Advances in Engineering Education 3(6), 579–585 (2006)

Bechhofer, S., Van Harmelen, F., Hendler, J., Horrocks, I., McGuinness, D.L., Patel-Schneider, P.F., Stein, L.A.: OWL Web Ontology Language (2004), http://www.w3.org/TR/2004/REC-owl-ref-20040210/

Berners-Lee, T.: A readable language for data on the web. n3 formalism (2006), http://www.w3.org/designissues/notation3.html

Bush, V.: As we think. The Atlantic Montly 176(1), 101–108 (1945)

Chatree, S., Taieb, Z.: Semantic Driven Hashing (SDH): An Ontology-Based Search Scheme for the Semantic Aware Network (SA Net). In: Fourth International Conference on Peer-to-Peer Computing (P2P 2004), pp. 270–271 (2004)

Chawathe, Y., LaMarca, A., Ramabhadhran, S., Ratnasamy, S., Hellerstein, J., Shenker, S.: A Case Study in Building Layered DHT Applications. IRS-TR-05-001 (2005)

Decker, S.: The social semantic desktop: Next generation collaboration infrastructure. Information Services and Use 26(2), 139–144 (2006), ISSN 0167-5265 (Print) 1875-8789 (Online)

Della Valle, E., Turati, A., Ghigni, A.: PAGE: A Distributed Infrastructure for Fostering RDF-Based Interoperability. In: Eliassen, F., Montresor, A. (eds.) DAIS 2006. LNCS, vol. 4025, pp. 347–353. Springer, Heidelberg (2006)

Druschel, P., Rowstron, A.: Storage Management and Caching in PAST, a Large-scale, Persistent Peer-to-peer Storage Utility. In: Proceedings of the 18th ACM Symposium on Operating Systems Principles (SOSP 2001), Lake Louise, AB, Canada (2001)

Ghebghoub, O., Abel, M.-H., Moulin, C., Leblanc, A.: A LOM ontology put into practice, Ontology Theory, Management and Design: Advanced Tools and Models (2009)

Fielding, R.T.: Architectural styles and the design of network-based software architectures. PhD Thesis, University of California, Irvine (2000)

Harth, A.: SECO: Mediation Services for Semantic Web Data. IEEE Intelligent Systems, 66–71 (2004)

Harth, A., Decker, S.: Optimized index structures for querying RDF from the web. In: LA-WEB (2005)

Mayamounkov, P., Maziers, D.: Kademlia: A peer-to-peer information system based on the xor metric. In: Proceedings of the 1st International Workshop on Peer-to-Peer Systems (2002)

Mougul, J.: Representing Information About Files. PhD Thesis. Computer science department. Stanford University, California, USA (1986)

Ratnasamy, S., Francis, P., Handley, M., Karp, R., Schenker, S.: A scalable content-addressable network. In: Proceedings of the 2001 conference on Applications, technologies, architectures, and protocols for computer communications, pp. 161–172. ACM, San Diego (2001)

Rowstron, A., Druschel, P.: Pastry: Scalable, decentralized object location and routing for large-scale peer-to-peer systems. In: Guerraoui, R. (ed.) Middleware 2001. LNCS, vol. 2218, pp. 329–350. Springer, Heidelberg (2001)

Salton, G., Fox, E.A., Wu, H.: Extended Boolean information retrieval. Technical Report, Cornell University (1982)

Sauermann, L.: The gnowsis - using semantic web technologies to build a semantic desktop. Diploma thesis, Technical University of Vienna (2003)

Sauermann, L., Bernardi, A., Dengel, A.: Overview and outlook on the semantic desktop. In: Dennis, Sauermann, L. (eds.) Proceedings of the 1st Workshop on The Semantic Desktop at the ISWC 2005 Conference, pp. 1–18 (2005)

Sauermann, L., Grimnes, A., Kiesel, M., Fluit, C., Maus, H., Heim, D., Nadeem, D., Horak, B., Dengel, A.: Semantic Desktop 2.0: The Gnowsis Experience. In: Cruz, I., et al. (eds.) ISWC 2006. LNCS, vol. 4273, pp. 887–900. Springer, Heidelberg (2006)

Sheth, A., Bertram, C., Avant, D., Hammond, D., Kochut, K., Warke, Y.: Semantic Content Management for Enterprises and the Web. IEEE Internet Computing, 80–87 (2002)

Stoica, I., Morris, R., Karger, D., Kaashoek, M.F., Balakrishnan, H.: Chord: A scalable peer-to-peer lookup service for internet applications. In: Proceedings of the ACM SIGCOMM 2001 Conference, San Diego, California (2001)

Tummarello, G., Morbidoni, C., Petersson, J., Puliti, P., Piazza, F.: RDFGrowth, a P2P annotation exchange algorithm for scalable Semantic Web applications. In: Proceedings of the MobiQuitous 2004 Workshop on Peer-to-Peer Knowledge Management (P2PKM 2004). CEUR Workshop Proceedings 108 CEUR-WS, Boston, MA, USA (2004)

An Agent-Oriented Architecture for Researcher Profiling and Association Using Semantic Web Technologies

Sadaf Adnan, Amal Tahir, Amna Basharat, and Sergio de Cesare

Abstract. Collaboration within the international scientific community has steadily increased over the years especially in the presence of complex interdisciplinary problems being investigated. At the same time the amount of research artifacts produced by the research community has grown exponentially making it difficult for individual researchers to filter and search through such information. In the presence of a vast amount of research information the problem of identifying potential project partners or collaborators with specific profiles can become extremely difficult. This paper presents a semantic multi-agent architecture (called SemoRA) aimed at tackling such a problem. The architecture combines agent and Semantic Web technologies in order to develop a framework capable of efficiently acquiring researcher information, making sense of it and giving meaning to it. The architecture ultimately enables the retrieval and matching of scored profiles aimed at enhancing collaborations among researchers – collaborations that can transcend both institutional and national boundaries.

Keywords: Agent, Semantic Web, Ontology, Researcher Profiling, Knowledge Representation.

Sadaf Adnan · Amal Tahir · Amna Basharat
National University of Computer & Emerging Sciences
Computer Science Department
AK Brohi Road, H-11/4
Islamabad, Pakistan
e-mail: sad_totaalus@yahoo.com, amal_tahir1@yahoo.com, amna.basharat@nu.edu.pk

Sergio de Cesare
Brunel University
School of Information Systems, Computing and Mathematics
Uxbridge, Middlesex, U.K.
e-mail: sergio.decesare@brunel.ac.uk

Soro et al. (Eds.): Inform. Retrieval and Mining in Distrib. Environments, SCI 324, pp. 77–91.
springerlink.com

1 Introduction

In recent years, within the international scientific community, the number of multidisciplinary research projects has steadily grown. This rise is primarily due to the growing complexity and scale of the problems undergoing investigation. In the presence of such problems, solutions cannot be sought only within the confines of one academic discipline. Consequently it becomes necessary to bring together researchers from different scientific backgrounds as well as possessing diverse knowledge, skills and expertise. The scale of multidisciplinary research often transcends both institutional and national boundaries. Nowadays it is very common to have research projects, which, in terms of their people and groups, are geographically dispersed both nationally and internationally.

At the same time information technology has steadily advanced making it easier to publish and obtain information. This has led to an exponential rise in the amount of information accessible by individuals and organizations including academic institutions and research groups [16, 17]. From the perspective of a researcher this growth of information implies having to spend more time searching and filtering among the numerous sources available in order to find research work and people that satisfy various requirements (e.g., gathering relevant scientific literature and identifying potential research collaborators with specific skill sets and expertise).

Several research studies have shown that researcher profiling is a problem characterized by many challenging issues including: the automatic extraction of research profiles from distributed sources (homepages, indexing authorities, etc.) [14]; the consistency and completeness of information; and the resolution of ambiguities [9]. These problems are aggravated by the predominantly syntactic and unstructured way in which such information is organized [14], thus minimizing the interoperability between heterogeneous knowledge and information sources.

The research presented in this paper proposes and develops a distributed architecture aimed at profiling scientific researchers, their projects and research deliverables. The proposed architecture combines agent and Semantic Web technologies. On the one hand semantic technologies help to model and structure information resources in a more meaningful and machine-processable manner. On the other hand agents have proved to be effective in dynamic and open environments due mainly to their autonomous and problem-solving capabilities [18, 19]. These capabilities can be effectively utilized in 'social network' environments where multi-agent based autonomous collaboration can help enhance the network effect [22] within research communities [20]. Therefore this paper aims to introduce a synergy between agents and the Semantic Web to facilitate collaboration and associations between distributed researchers and research communities [9].

Studies have shown that the utilization of intelligent agent power facilitated by a semantic-based framework can further enhance Web intelligence [21] and thus help in effective knowledge virtualization. Extensive research, including that of Bryson et al. [30], Blake, et al. [31], Gibbins, et al. [34], Chen et al, [32,33] and many others, has attempted to bring together agents and the Semantic Web. Agents are seen as powerful tools that can be used in combination with semantic mark-up languages such as the Resource Description Framework (RDF), the Web Ontology Language (OWL) [4] and Semantic Web service ontologies (such as OWL-S).

Thus agents were deemed as an appropriate design metaphor to provide an architecture for semantic profiling and association of researchers in academic social networks. The aim is to provide comprehensive and efficient services to research networks in which researchers are not only interested in searching for information (such as publications, co-authors and conferences from existing indexing authorities such as CiteSeer and Google Scholar) [11,12], but are also interested in establishing collaborations with other researchers through semantic (meaningfully structured) [13] research profiles.

The rest of the paper is organized as follows. Section 2 describes the system architecture of SemoRA and gives an overview of all the components. Section 3 gives the detailed design and implementation details of the architectural components. Section 4 provides a Distributed Knowledge Retrieval scenario explaining the researcher's associations. Section 5 provides a discussion and Section 6 concludes the paper and provides direction for future work.

2 System Architecture of SemoRA

Figure 1 shows the conceptual model of the Semantic Agent-Oriented Architecture for Research Profiling and Association (SemoRA). The distributed repository of ontological knowledge models is a means for storing the vocabulary for agents to understand and process knowledge with. These models are designed to be specific to a knowledge domain which, for this paper, is the Research Community. The ontological model would map information such as researcher profiles, conference papers, research centers, etc. Thus the purpose of this modeling is to allow data interoperability and knowledge discovery through implicit/explicit reasoning (described later). The architecture is designed with the aim to automatically share, integrate and link the information in the knowledge base on the basis of the profiles of the researcher. The purpose of combining Semantic Web and agent-oriented frameworks is to understand, discover and manipulate researchers' profiles (available from digital libraries and distributed sources) so as to establish quality associations among researchers and to facilitate research collaborations among them.

The vision is to link research communities, which are like disjoint entities possessing their own data, research works, documents and research profiles. To achieve this objective SemoRA was designed with the following requirements in mind:

1. To ensure efficient retrieval and linking processes;
2. To achieve a transparent retrieval mechanism (through a distributed SPARQL querying engine being used by agents);
3. To develop autonomous subject specific information retrieval, storage and mapping of structured metadata based on users' requirements and
4. To enable reasoning and verification of researchers' information.

Collaboration is made possible by communities of agents cooperating in an intelligent manner so as to retrieve information, minimize the distances of the different researchers and research centers. Through semantic interlinking of data, knowledge from the research communities would be made easily retrievable using reasoning engines and inferencing mechanisms. The roles of the agents are described below.

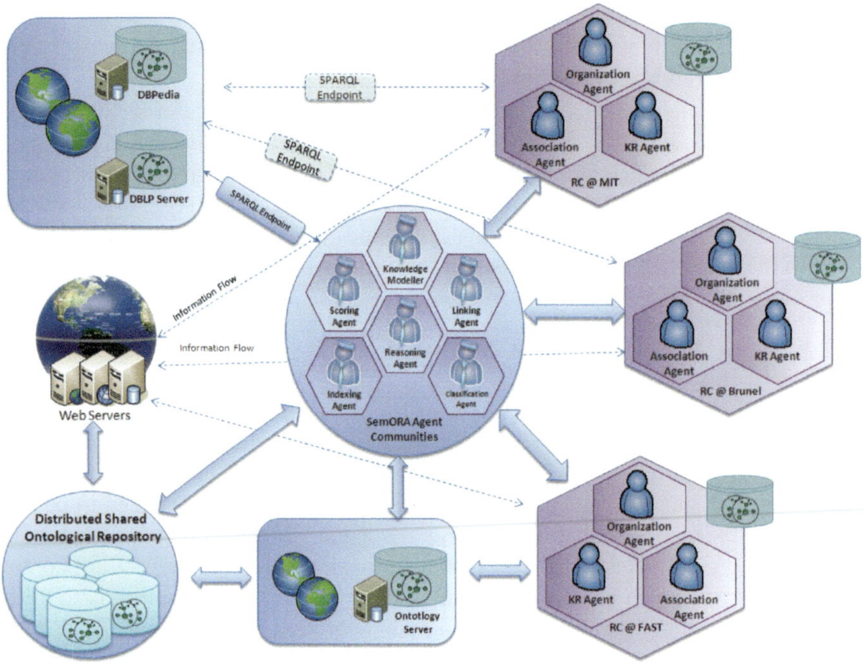

Fig. 1 Conceptual Model for Semantic Agent Oriented Architecture for Researcher profiling and Association (SemoRA)

Agent Communities and their Roles

The following describes the key software agents of SemoRA and their roles:

Knowledge Publishing Agent – The RDF schemas and data (individuals) that are created are passed to this agent in order to be published in the ontology. No other agent has write access to the ontological repository. This agent also checks the consistency and correctness of the information that is added.

Ontological Modeling Agent – This consists of two further agents: the Ontology Modeling Agent and the Schema Modeling Agent. The two agents model RDF schema (classes and properties) and data (individuals) using the Protégé OWL API. The agents first create the Protégé OWL model and then send it to the Publishing Agent that adds the ontology to the Knowledge Base.

Knowledge Retrieval Agent (KR Agent) and the Query Agent - The Query Agent and KR Agent are responsible for query generation and execution. These agents use the inputs provided via the interface to make and execute a SPARQL query on the Knowledge Base. The final results are passed on to the Organization and Acquisition Agent.

Organization and Acquisition Agent - This Agent organizes the search results provided by the Query Agent into a readable format, which is then presented to the User.

Scoring Agent - The Scoring Agent determines the researcher's expertise level in a certain field. The Scoring Agent evaluates the scores of a researcher on the basis of his publications and qualifications. After the scores are calculated they are published in the Knowledge Base.

Linking Agent – The Linking Agent connects the researcher profiles with one another. The links made here are retrieved from DBLP and are then published onto the ontology, improving the ease and efficiency of searching.

Reasoning Agent – The Reasoning Agent reasons and classifies the individuals and classes in the ontology. This can be both DL Based and Rule Based Reasoning. This agent is invoked every time the publishing agent publishes any data in the ontology; the role of the Reasoning Agent is to validate and publish a well-reasoned knowledge.

3 Detailed Design and Implementation

In the following sub-sections, the detailed design and implementation of core functional competencies of the agents, including the Knowledge Profiling, Knowledge Retrieval and Knowledge Linking Agents, are discussed.

Ontological Modeling Agents

Ontological Modelling Agents mainly focus on the conceptualization and representation of researchers' domain concepts. For knowledge representation and ontology implementation the Protégé-ontology editor [1] and the Protégé OWL API [2] were used. Similarly the Jena API [3] (an open source Semantic Web framework which provides mechanisms for retrieval and storage of RDF graphs [4]) was used for knowledge manipulation, querying and access. During knowledge representation the focal concept was the 'Researcher' class as the domain selected is that of scientific research communities. This class is further categorized into various classes according to the domain definitions. A key feature of using the Semantic Web is the possibility of creating in OWL defined and specialized classes based on Description Logic (DL). This allows for automated classification and reasoning (in addition to consistency checking), which proves vital in knowledge representation, publishing, retrieval and maintenance. Figure 2 shows an extract of the knowledge model as represented in Protégé.

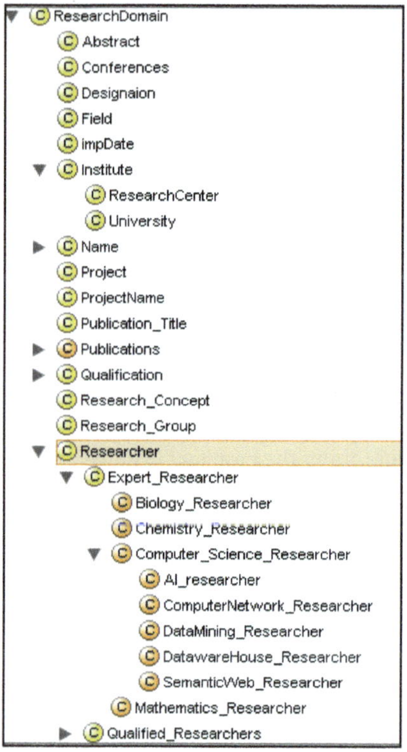

Fig. 2 OWL Classes in Protégé

Knowledge Profiling Agents

Knowledge Profiling consists mainly of Knowledge Representation, Knowledge Publishing and then applying reasoning and classification on the knowledge.

Knowledge Publishing Agent

Publishing knowledge into the ontologies can be done either by manually inserting (entering) information using Protégé, or by using an automatic form-based approach in the application. The form based approach uses the Protégé OWL API [2] and Jena API [3] to access and update the ontology. The publishing agent mechanism is explained in Figure 3. The Publishing Agent includes first the mapping and the arranging of parameters that have either been received by the user or retrieved from any existing data repository (e.g., DBLP) and then creating (instantiating) an RDF schema or individual on the basis of these parameters. The individuals are then passed to the Knowledge Builder so that the information passed can be stored into the Knowledge base.

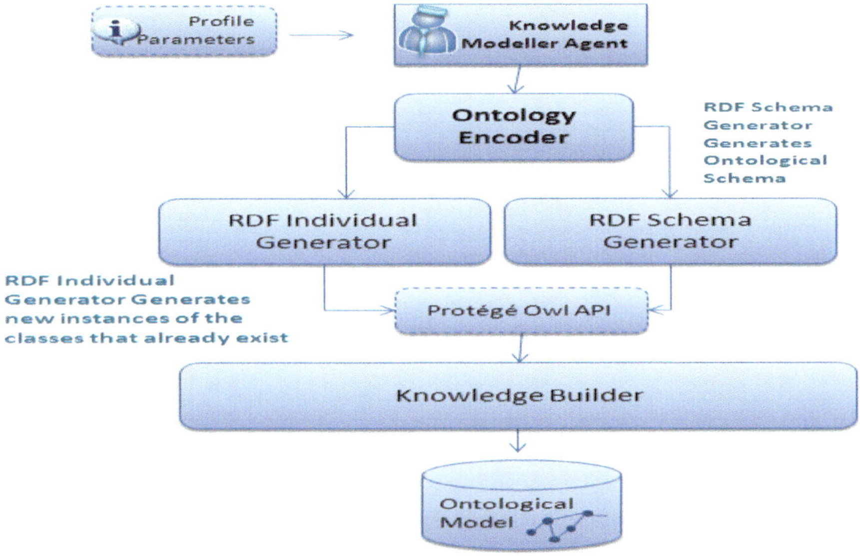

Fig. 3 Knowledge Publishing Mechanism

Knowledge Retrieval Agent

Efficient knowledge retrieval from the ontologies is the main focus of this agent-based architecture. Knowledge Retrieval Agents are primarily responsible for this task. The process of Knowledge Retrieval includes the use of the Query or Knowledge Retrieval (KR) Agents by receiving the search parameters either directly from the user or from another agent. A Query Generator then creates a query on the basis of those parameters and finally the execution of the query is carried out by a Knowledge Acquisition module, which includes the query processing, and execution mechanism. The results are returned in a structured form; RDF triples or simple lists of data items that may then be passed on to the Organization and Acquisition Agents for further processing. Figure 4 demonstrates how the Knowledge Retrieval Agent works and also highlights the main concepts behind the design and implementation.

The Query Generator further includes two sub-modules that first parse the parameters provided by the user and then map the parameter onto a query structure provided by the Query Modeller Module. The connection of the ontology and the application was implemented using the Jena API [3] through which knowledge from the ontological knowledge base can be retrieved.

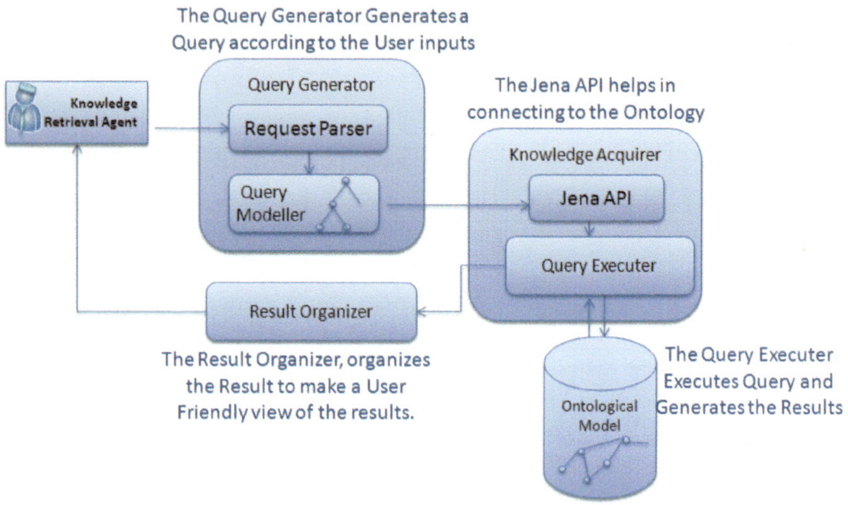

Fig. 4 Knowledge Retrieval Mechanism

Knowledge Linking and Association Agent

Knowledge Linking deals with access, retrieval and linking of data from distributed knowledge repositories. It is the task of the Linking Agent to provide a virtualization of heterogeneous data resources through federated queries. The federated queries require certain wrappers over the normal SPARQL query, for accessing the data from the remote data sources [5].

To perform efficient linking of knowledge sources, various indexing repositories, such as Google Scholar [6], CiteSeer, Digital Bibliography Library Project (DBLP) [7], were reviewed. After a detailed analysis DBLP was selected as a source of data for the framework as DBLP is described fully in RDF format and possesses its own SPARQL endpoint facilitating knowledge retrieval through federated queries. Although Google Scholar and CiteSeer also provide efficient services of retrieving information about a researcher or his/her publications, the retrieval is however mainly done through keyword-based searching, thus these sources are incapable of providing semantic links to the data. Similarly access to these repositories is not allowed and thus relevant knowledge retrieval incurs enormous overheads.

In order to retrieve and discover relevant data for knowledge modeling from DBLP through a SPARQL end-point, the architecture in Figure 5 was proposed for the Association and Linking Agent.

The linking agent has two main features: 1) To establish the links between researchers and publish the associations; and 2) To retrieve the distributed knowledge through federated queries. The connection to DBLP is established when an author is queried (from the DBLP) for his/her publications; the co-authors' records and links between the co-authors are created when the information is published into the researcher ontology. The architecture is designed to publish the information (publications) of co-authors into its ontology as new researchers and by gathering parameters such as interests, qualifications, etc. from their FOAF profiles. Their personal information is finally published into the ontology.

Once the links between co-authors are created and published, the association is created by retrieving the researchers' publication information from the DBLP repository (for example, conference, subject area via the Dublin Core schema). Appropriate expertise scores are created and direct links and possible associations are established between the researchers according to their expertise and current research interest areas.

All the information provided by DBLP is retrieved through a DBLP SPARQL endpoint. SPARQL, as a querying protocol, and its endpoints play a significant

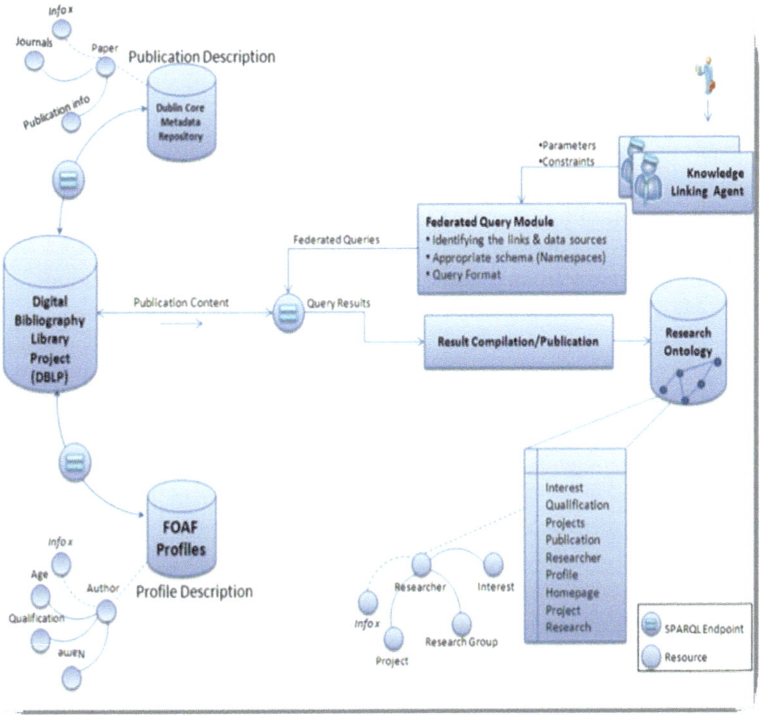

Fig. 5 Knowledge Linking Agent's Mechanism

role in this architecture helping to achieve knowledge virtualization through federated queries. As mentioned earlier, the essence of federated queries is to retrieve knowledge from various heterogeneous data resources enabling a transparent querying mechanism across platforms. Table 1 shows a sample of the SPARQL query implemented.

Table 1: Sample Federated SPARQL Query

```
SELECT ?coauthor ?birthdate
FROM NAMED < http://dblp.l3s.de/d2r/ >
FROM NAMED <http://www.dbpedia.org>
WHERE {
GRAPH < http://dblp.l3s.de/d2r/ > {
?paper dc:creator < http://dblp.l3s.de/d2r/resource/authors/Sergio_de_Cesare >.
?paper dc:creator ?coauthor.
?coauthor foaf:name ?name. }
GRAPH <http://www.dbpedia.org> {
?person foaf:name ?name.
?person dbpedia:birth ?birthdate. }
}
```

4 Validation of the Architecture

Distributed knowledge discovery and information retrieval are key functionalities of the architecture. The high-level architecture of SemoRA was evaluated by using an implementation of a distributed knowledge discovery and information retrieval scenario as shown in Figure 6.

The top-level functional goals of the demonstration application were as follows:

- The multi-agent based application uses agents modeled with SemoRA to generate associations in a specific research area or expertise.
- The agents extract, filter and store information automatically from the Semantic Web using other agents.
- Cognitive sharing of knowledge and different kinds of abilities are demonstrated to gain efficiency in the task of problem solving.
- The agents are simulated to reuse each other's capabilities, behaviors and offer affordances to each other.
- The Behavior API and Interaction Protocol API of JADE [26,27] are used to model the Agent's actions as well as for Inter-Agent Communication.

The scenario shown in Figure 6 is modeled according to the requirements of a user that needs to discover associations that another researcher has with colleagues in a certain field. This information is first checked by the Ontological Knowledge Base to see whether the required profile already exists; then through a SPARQL query the association is retrieved. However, if the required profile does not exist in the

Knowledge Base then the information should be passed on to the Linking Agent. The Linking Agent queries the DBLP SPARQL endpoint and information about the researcher's publications, co-authors, subject areas and so on should be retrieved. All this information is then passed to the Publishing Agent responsible for publishing the links and the information in the Ontological Knowledge Base. At this point the information is sent to the Reasoning Agent and inferences are drawn. Meanwhile the Linking Agent would query all of the researcher's co-authors, retrieve their publication information and pass it on to the Publishing Agent.

When researchers' information is published an acknowledgement is sent to invoke the Scoring Agent that calculates the scores and sends the score to the Publishing Agent. This method continues until the scores of all co-authors and publications of the researchers have been published.

The advantage of the agent-based implementation is that the application can achieve parallelism, functional load distribution and balancing as the Linking, Publishing and Scoring Agents can perform their tasks simultaneously enhancing the performance of the application (this however remains to be empirically tested and is currently beyond the scope of this paper). Figure 6 depicts the sequence of steps involved in the implementation of the above scenario.

5 Discussion

The problem of researcher profiling has been researched previously. Examples include Arnet Miner [15] and OntoWeb [16]. However, the efficient retrieval and linking of knowledge remains a significant issue within the research community. The main focus of this work was to identify associations between researchers on the basis of co-authorship as most of the researcher profiles consist of crawled publication data from various digital libraries [15].

Agents proved to be a suitable design metaphor for implementing the core functional competencies of SemoRA. The recent promotion of the Semantic Web (Berners-Lee, 2001) as a vision of the evolution of the World Wide Web from a publishing medium to that of a general services provider, shares many ideals with the vision of agent researchers as presented above and were validated with the design and implementation of SemoRA Agents.

With current Semantic Web standards, agents and the Semantic Web have come to be considered almost inseparable. The intention of this research was to utilize the power of agents blended with semantics to enable the user to find and establish subject and concept specific associations. For this purpose, ontologies have proven to be a useful means of knowledge representation since they make Web content understandable by machines (in general) or software agents (more specifically). As a result, the information retrieval process, autonomous information discovery and its validation provide results of greater value.

The core functional competencies of SemoRA have been fully tested by creating datasets for various research centers. Some subject specific research datasets have also been used.

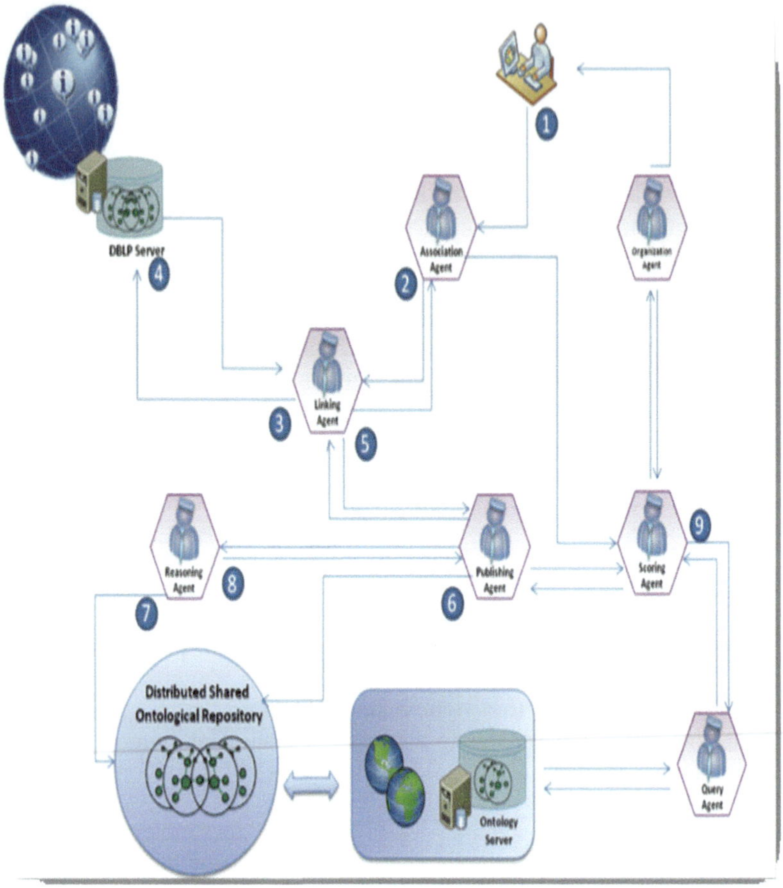

Fig. 6 Distributed Knowledge Retrieval Scenario through Multi-Agent Collaboration

6 Conclusion and Future Research

This paper presented a comprehensive approach to develop a semantic agent-based architecture for researcher profiling and association to facilitate collaboration among distributed research and academic networks. The architecture supports well-defined mechanisms for knowledge publishing, integration, sharing and retrieval from distributed knowledge sources for establishing researchers' associations in specialized areas of expertise. Through its autonomous and transparent mechanisms, supported by well-coordinated agent architecture, the framework is able to link varying research communities on a single platform for collaboration in research. The use of ontology enables agents to understand the domain knowledge and hence making them capable of determining the appropriate associations for the researchers.

With the core architecture in place, a basic framework has been provided for efficient integration and use with existing and future researcher repositories. Although the framework is currently targeting the research communities, its core framework is generic enough to accommodate expansion; it could in the future have a knowledge base that not only covers the research domain but also covers other related content available on the web.

References

[1] Stanford Center for Biomedical Informatics Research, Protégé (2009), http://protege.stanford.edu/ (accessed: 29th December 2008)

[2] SMI Stanford Medical Informatics Groups, Protégé OWL API (2007c) http://protege.stanford.edu/plugins/owl/api/index.html [accessed August 24, 2008]

[3] JENA, HPLabs and Open Source Community, Jena – A Semantic Web Framework for Java (2006), http://jena.sourceforge.net/ (accessed: August 20, 2008)

[4] W3C, W3C: Resource Description Framework RDF (2004), http://www.w3.org/RDF/ (accessed: December 20, 2008)

[5] Kache, H., Han, W., Markl, V., Raman, V., Ewen, S.: POP/FED: Progressive Query Optimization for Federated Queries in DB2. In: VLDB ACM, Seoul, Korea, September 12-15 (2006), http://www.vldb.org/conf/2006/p1175-kache.pdf (accessed January 20, 2009)

[6] About Google Scholar, http://scholar.google.com/intl/en/scholar/about.html (accessed January 9, 2009)

[7] D2R Server publishing the DBLP Bibliography Database, generated by D2R server, http://www4.wiwiss.fu-berlin.de/dblp/ (accessed January 2, 2009)

[8] Aleman-Meza, B., Hakimpour, F., Arpinar, I.B., Sheth, A.P.: SwetoDblp Ontology of Computer Science Publications. Web Semantics: Science, Services and Agents on the World Wide Web 5(3), 151–155 (2007), http://lsdis.cs.uga.edu/projects/semdis/swetodblp/, http://swat.cse.lehigh.edu/resources/onto/dblp.owl (accessed January 15, 2009)

[9] Tang, J., Zhang, D., Yao, L.: Social Network Extraction of Academic Researchers. In: Proceedings of 2007 IEEE International Conference on Data Mining (ICDM 2007), pp. 292–301 (2007)

[10] Yao, L., Tang, J., Li, J.: A Unified Approach to Researcher Profiling. In: Proceedings of the IEEE/WIC/ACM International Conference on Web Intelligence, pp. 359–366 (2007)

[11] Noruzi, A.: Google Scholar: The New Generation of Citation Indexes. Libri 55, 170–180 (2005), http://www.librijournal.org/pdf/2005-4pp170-180.pdf (accessed January 9, 2009)

[12] http://citeseer.ist.psu.edu/

[13] Robal, T., Kalja, A.: Applying User Profile Ontology for Mining Web Site Adaptation Recommendations. In: CEUR Workshop Proceedings ADBIS Research Communications (2007)

[14] Li, J.-Z., Tang, J., Zhang, J., Luo, Q., Liu, Y., Hong, M.: EOS: expertise oriented search using social networks. In: Proceedings of WWW 2007, pp. 1271–1272 (2007)

[15] Tang, J., Zhang, J., Yao, L., Li, J., Zhang, L., Su, Z.: ArnetMiner: Extraction and Mining of Academic Social Networks. In: Proceedings of the Fourteenth ACM SIGKDD International Conference on Knowledge Discovery and Data Mining (SIGKDD 2008), pp. 990–998 (2008)

[16] Oberle, D., Spyns, P.: The Knowledge Portal 'OntoWeb. In: Collection of Handbook on Ontologies, pp. 499–516 (2004)

[17] Dorresteijn, J.: The World Web is growing a billion pages per day (July 29, 2008), http://thenextweb.com/2008/07/29/the-world-wide-web-grows-a-billion-pages-per-day/ (accessed: October 31, 2008)

[18] Dove, M.: eScience - Harnessing the power of the internet for environmental research (October 2002), http://www.allhands.org.uk/2008/conference/USB/NERC%20eScience%20Glossy.pdf (accessed: October 31, 2008)

[19] Dickinson, I., Wooldridge, M.: Towards Practical Reasoning Agents for the Semantic Web. In: Rosenschein, J.S., Sandholm, T., Wooldridge, M., Yakoo, M. (eds.) Proceedings of the Second International Joint Conference on Autonomous Agents and Multiagent Systems, AAMAS 2003, vol. 2, pp. 827–834 (2003)

[20] Bradshaw, J.M.: Software Agents. MIT Press, Cambridge (1997)

[21] Nwana, H.S.: Software agents: An overview. Knowledge Engineering Review 11(3), 205–244 (1996)

[22] Xia, Y., Xia, M.: Agents-based Intelligent Retrieval Framework for the Semantic Web. In: Wireless Communications, Networking and Mobile Computing, September 21-25, pp. 5357–5360 (2007)

[23] Hendler, J., Golbeck, J.: Metcalfe's Law, Web 2.0, and the Semantic Web. In: Web Semantics: Science, Services and Agents on the World Wide Web, vol. 6(1), pp. 14–20

[24] Hendler, J.: IEEE, Agents and the Semantic Web. IEEE Intelligent Systems Journal (March/April 2001)

[25] Bell, D., De Cesare, S., Iacovelli, N., Lycett, M., Merico, A.: A framework for deriving Semantic Web services. Information Systems Frontiers 9(1), 69–84 (2007)

[26] Laclavík, M., Balogh, Z., Babík, M., Hluchý, L.: Agentowl: Semantic knowledge model and agent architecture. Computing and Informatics 25(5), 421–439 (2006)

[27] JADE: JADE (Java Agent DEvelopment Framework) (2004a), http://jade.cselt.it/ (accessed: July 20, 2007)

[28] JADE: JADE Documentation (2004b), http://jade.cselt.it/ (accessed: July 20, 2007)

[29] Gruber, T.R.: A translation approach to portable ontology specifications. Knowledge Acquisition 5(2), 199–220 (1993)

[30] Bryson, J.J., Martin, D.L., McIlraith, S.A., Stein, L.A.: Toward behavioral intelligence in the Semantic Web. Computer 35(11), 48–54 (2002)

[31] Blake, M.B., Parsons, S., Payne, T.R.: The synergy of electronic commerce, agents, and Semantic Web services. Knowledge Engineering Review 19(2), 175–180 (2004)

[32] Chen, L., Shadbolt, N.R., Goble, C.A.: A Semantic Web-based approach to knowledge management for grid applications. IEEE Transactions on Knowledge and Data Engineering 19(2), 283–295 (2007)

[33] Chen, H., Finin, T., Joshi, A., Kagal, L., Perich, F., Chakraborty, D.: Intelligent agents meet the Semantic Web in smart spaces. IEEE Internet Computing 8(6), 69–79 (2004)

[34] Gibbins, N., Harris, S., Shadbolt, N.: Agent-based Semantic Web Services. In: Web Semantics, vol. 1(2), pp. 141–154 (2004)

Integrating Peer-to-Peer and Multi-agent Technologies for the Realization of Content Sharing Applications

Agostino Poggi and Michele Tomaiuolo

Abstract. The combination of peer-to-peer networking and multi-agent systems seems be a perfect solution for the realization of applications that broaden on the Internet. In fact, while peer-to-peer networking infrastructures and protocols provide the suitable discovery and communication services necessary for developing effective and reliable applications, multi-agent systems allow to realize autonomous, social, reactive and proactive peers that make the development of intelligent and flexible application possible. This paper presents how JADE, one of the most known software framework for the development of multi-agent systems, has been extended to take advantage of the JXTA networking infrastructure and protocols, and describes a system, called RAIS, that has been realized thanks to such extended version of the JADE software framework and that provides a set of advanced services for content sharing and retrieval.

1 Introduction

The combination of the distributed capabilities of peer-to-peer networks with multi-agent systems appears to be very promising since it will allow the transparent access to large-scale distributed resources while maintaining high-availability, fault tolerance and low maintenance application deployment through self-organization. Moreover, multi-agent systems can be considered an appropriate framework for the realization of peer-to-peer applications, because they have always been thought as networks of equal peers providing some properties, i.e., autonomy, social ability, reactivity and pro-activeness, that might be very useful for the realization of the new generations of peeer-to-peer applications [1].

This paper presents how JADE, one of the most known software framework for the development of multi-agent systems, has been extended to take advantage of

Agostino Poggi · Michele Tomaiuolo
Dipartimento di Ingegneria dell'Informazione
Università degli Studi di Parma,
Viale U.P. Usberti 181A, 43100 Parma, Italy
e-mail: {Agostino.Poggi,Michele.Tomaiuolo}@unipr.it

Soro et al. (Eds.): Inform. Retrieval and Mining in Distrib. Environments, SCI 324, pp. 93–107.
springerlink.com

the JXTA networking infrastructure and protocols, and describes a system, called RAIS, that has been realized thanks to such extended version of the JADE software framework and that provides a set of advanced services for content sharing and retrieval. The next section discusses relevant works in the field of integration of P2P and MAS. Section three introduces JADE. Section four introduces how JXTA technology was used for improving JADE discovery and communication services. Section five describes RAIS. Finally, section six concludes the paper discussing about the experimentation of the RAIS system and sketching some future research directions.

2 Related Work

Peer-to-peer technologies have been already used in the development of multi-agents systems. RETSINA [2,3] is a multi-agent infrastructure that uses the Gnutella network and some DHT based protocols for extending the discovery services. DIAS [4] is a distributed system for the management of digital libraries based on a network of middle-agents that interacts thanks to some peer-to-peer routing algorithms. Anthill [5] is a mobile agent system that can be used for the realization of peer-to-peer applications by providing a JXTA [6] based network infrastructure where agents can move to perform the required tasks. A-peer [7] is a multi-agent-based P2P system where agents rely on hierarchically arranged advertising elements to find the services they need from other agents.. Bertolini and his colleagues [8] use the JXTA infrastructure as a communication environment for realizing distributed multi-agent systems. Gorodetsky and his colleagues [9] present a peer-to-peer agent platform implementing basic mandatory components of the peer-to-peer agent platform functional architecture [10]. Therefore, a multi-agent system is based on structured Chord network implemented as application layer set up on top of a peer-to-peer network.

Moreover, multi-agents systems have been already used for improving the typical peer-to-peer applications, i.e., the decentralized sharing of computer resources [11,12]. Zhang and his colleagues [13] propose a mediator-free multi-agent information retrieval system that provides some context-sensitive searching algorithms based on the various topologies of peer-to-peer networks. ACP2P (Agent Community based Peer-to-Peer) [14] is an information retrieval system that uses agent communities to manage and look up information related to users. In such a system an agent works as a delegate of its user and searches for information that the user wants by communicating with other agents. Kungas and Matskin [15] describe a multi agent system for distributed composition of semantic web services, where agent and service discovery is facilitated through the use of a peer-to-peer infrastructure. Zhang [16] proposes a peer-to-peer multi-agent system that supports the execution of e-commerce tasks by facilitating a dynamic partner selection and enabling the use of heterogeneous agents.

3 JADE

JADE (Java Agent DEvelopment framework) [17,18] is a software framework designed to aid the development of agent applications in compliance with the FIPA specifications [19] for interoperable intelligent multi-agent systems. The purpose of JADE is to simplify development while ensuring standard compliance through a comprehensive set of system services and agents. JADE is an active open source project, and the framework together with documentation and examples can be downloaded from JADE Home Page [20].

JADE is fully developed in Java and is based on the following driving principles:

- Interoperability: JADE is compliant with the FIPA specifications. As a consequence, JADE agents can interoperate with other agents, provided that they comply with the same standard.
- Uniformity and portability: JADE provides a homogeneous set of APIs that are independent from the underlying network and Java version (edition, configuration and profile). More in details, the JADE run-time provides the same APIs both for the J2EE, J2SE and J2ME environment. In theory, application developers could decide the Java run-time environment at deploy-time.
- Ease of use: the complexity of the middleware is hidden behind a simple and intuitive set of APIs.
- Pay-as-you-go philosophy: programmers do not need to use all the features provided by the middleware. Features that are not used do not require programmers to know anything about them, neither they add a computational overhead.

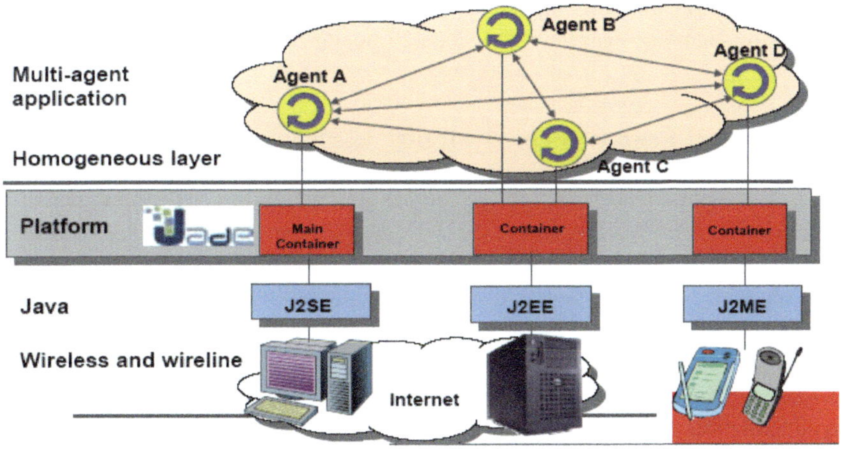

Fig. 1 Architecture of a JADE multi-agent system

JADE includes: i) the libraries (i.e., the Java classes) required to develop the application specific agents, ii) the implementation of the two management agents that a FIPA compliant agent platform must provide, i.e., the AMS (Agent Management System) agent and the DF (Directory Facilitator) agent, and iii) the run-time environment that provides the basic services and that must be active on the device before agents can be executed. Each instance of the JADE run-time is called container (since it "contains" agents). The set of all containers is called platform and it provides a homogeneous layer that hides to agents (and to application developers) the complexity and the diversity of the underlying technologies (hardware, operating systems, types of network, JVM, etc.). Figure 1 draws the architecture of a JADE multi-agent system deployed on a set of heterogeneous computing nodes.

JADE is extremely versatile and therefore it both fits the constraints of environments with limited resources and has already been integrated into complex architectures such as .NET or J2EE [21] where JADE becomes a service to execute multi-party proactive applications. The JADE run-time memory footprint, in a MIDP1.0 environment, is around 100 KB, but can be further reduced until 50 KB using the ROMizing technique [22], i.e., compiling JADE together with the JVM. The limited memory footprint allows installing JADE on all mobile phones provided that they are Java-enabled. Analyses and a benchmarks of scalability and performance of the JADE Message Transport System are reported by different works [23,24].

Moreover, JADE supports mobility of code and of execution state. That is, an agent can stop running on a host, migrate on a different remote host (without the need to have the agent code already installed on that host), and restart its execution from the point it was interrupted. In particular, JADE implements a form of not-so-weak mobility because the stack and the program counter cannot be saved in Java. This functionality allows, for example, distributing computational load at runtime by moving agents to less loaded machines without any impact on the application.

4 Jade and Peer-to-Peer Systems

JADE is based on a peer-to-peer communication architecture. The intelligence, the initiative, the information, the resources and the control can be fully distributed across mobile terminals as well as computers connected to the fixed network. The environment evolves dynamically together with peers – that in JADE are called agents – that appear and disappear in the system according to the needs and the requirements of the application domain. Communication between the peers, regardless of whether they are running in a wireless or wired network is completely symmetric with each peer being able to play both initiator and responder roles.

Nevertheless, JADE does not exploit some important features of modern peer-to-peer networks, in particular:

1. the possibility of building a completely distributed, global index of resources and services, without relying on any centralized entity, and
2. the possibility of building an "overlay network", hiding differences in lower level technologies and their related communication problems.

In fact, in a JADE multi-agent platform, the cooperation among agents is possible thanks to the presence of a yellow pages service that allows them to reciprocally individuate offered services. However this often limits the search inside a single platform. Solutions are possible, which allow the consultation of other yellow pages services, but they necessitate the a priori knowledge of the address of the remote platforms where services are hosted or listed. An alternative solution is represented by a yellow pages service leaning on a peer-to-peer network thanks to which each network device is able to individuate in a dynamic way services and resources of other network device. JADE followed the first solution and then only provides the possibility of federating different agent platforms through a hierarchical organization of the platform directory facilitators on the basis of a priori knowledge of the agent platforms addresses. This solution is adequate for small federations, but the cost of the distributed search and the recurring problems emerged at the level of connection among remote platforms grow with the cardinality and geographical extension of the interconnected infrastructure, with the number of connected platforms.

4.1 FIPA and Peer-to-Peer Systems

FIPA has acknowledged the growing importance of the peer-to-peer technologies and techniques and so it has released some specifications for the interoperability of FIPA platforms with peer-to-peer networks taking advantage of JXTA techniques [25, 26].

JXTA technology [6] is a set of open, general-purpose protocols that allows any connected device on the network (from cell phones to laptops and servers) to communicate and collaborate in a peer-to-peer fashion. The project was originally started by Sun Microsystems, but its development was kept open from the very beginning. JXTA comprises six protocols allowing the discovery, organization, monitoring and communication between peers. These protocols are all implemented on the basis of an underlying messaging layer, which binds the JXTA protocols to different network transports.

JXTA peers can form peer groups, which are virtual networks where any peer can seamlessly interact with other peers and resources, whether they are connected directly or through intermediate proxies. JXTA defines a communication language which is much more abstract than any other peer-to-peer protocol, allowing to use the network for a great variety of services and devices. A great advantage of JXTA derives from the use of XML language to represent, through structured documents, named advertisements, the resources available in the network. XML adapts without particular problems to any transport mean and it is already an affirmed standard, with good support in very different environments, to structure generic data in a form easily analyzable by both humans and machines.

Finally, JXTA is one of the most used technologies to improve connectivity on a global scale. In fact, JXTA does not suppose a direct connection is available between all couples of peers. Peers can use the Peer Endpoint Protocol to discover available routes for sending a message to a destination peer. Particular peers, called routers, are in charge of responding to such queries providing route information, i.e. a list of gateways connecting the sender to the intended receiver. A gateway acts as a communication relay, where messages can be stored and later collected by their intended recipient, overcoming problems related to limited connectivity.

In particular, FIPA defined a set of new components and protocols for the implementation of a DF-like service on a JXTA network. These include:

- Generic Discovery Service: a local directory facilitator, taking part in the peer-to-peer network and implementing the Agent Discovery Service specifications to discover agents and services deployed on remote FIPA platforms working together in a peer-to-peer network.
- Agent Peer Group: a child of the JXTA Net Peer Group that must be joined by each distributed discovery service.
- Generic Discovery Advertisements: to handle agent or service descriptions, for example FIPA df-agent-descriptions.
- Generic Discovery Protocol: to enable the interaction of discovery services on different agent platforms. It's a request/response protocol to discover advertisements, based on two simple messages, one for queries and one for responses.

4.2 Extending JADE with JXTA

Starting from such specification, we extended the JADE software framework for providing a global index of resources and services that does not rely on any centralized entity and to enable the communication among remote agents through an "overlay network" that hides the differences between the lower level communication technologies and reduces the increases the reliability of such a communication.

The global index of resources and services in a federation of JADE platforms has been realized by enabling a JADE directory facilitator to take advantage of JXTA discovery service and by connecting each JADE platform to the Agent Peer Group, defined by the FIPA specifications, as well as to other peer groups that are shared by the agent platforms involved in a specific application. The realized directory facilitator implements a JXTA-based ADS (Agent Discovery Service), which has been developed in the respect of relevant FIPA specifications to implement a GDS (Generic Discovery Service). In particular, the Agent Discovery

Service uses the Generic Discovery Protocol to advertise and discover agents and their services: the agent descriptions are wrapped in Generic Discovery Advertisements that, of course, are spanned over a whole peer group and so all the directory facilitators belonging to such peer group are able to access to such agent descriptions. Figure 2 shows how an ADS agent interacts with JXTA discovery service for finding agents of other JADE platform connected through a JXTA network.

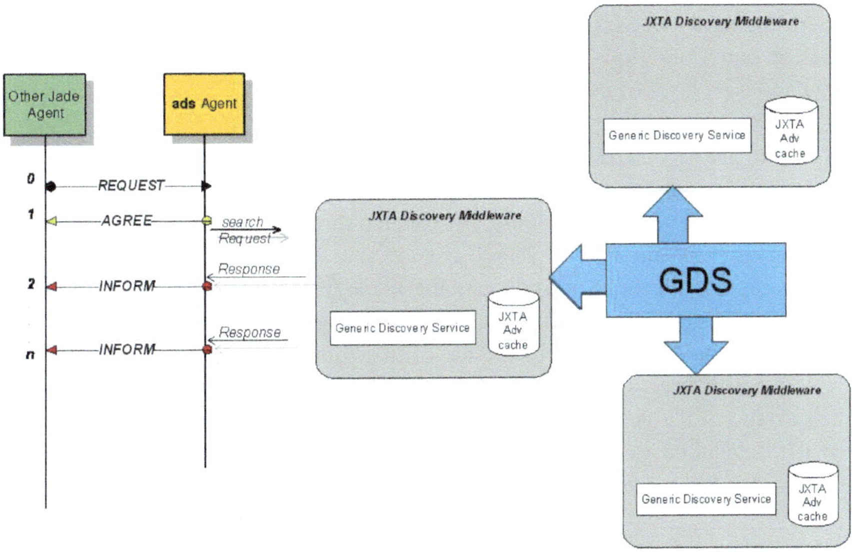

Fig. 2 Discovery of agents in a JXTA network of JADE platforms

Moreover, to facilitate the communication between agents of different platforms, we realized a JXTA implementation of the message transport protocol that FIPA defined for allowing the interconnection among agents of different platforms. This implementation allows the exchange of messages between the agents of two platforms through JXTA pipes. These pipes are dynamically bound to specific endpoints (typically an IP address and a TCP port). JXTA pipes are advertised on the network in the same way as other services offered by peers, and provide a global scope to peer connectivity.

5 RAIS

We think that the file-sharing approach of actual peer-to-peer networks is inadequate when applied to documents that cannot be completely described by their title or some associated metadata. On the other hand, sharing such documents would be an interesting application in various contexts, but current tools do not provide the required features for a true content sharing approach. When we refer to "content sharing", we mean the capability to share documents in a peer-to-peer network with the search power of a desktop search application (e.g., Google Desktop Search). Therefore, we refer to the meeting point between peer-to-peer and desktop search applications, taking into consideration the advantages of each approach: the distributed file sharing of the former and the indexing capabilities of the latter.

Moreover, the use of a software framework for the realization of peer-to-peer multi-agent systems allow us to realize "intelligent content sharing" applications where intelligent means, for example, the possibility to retrieve content on the basis of user's preferences, to allow the access to the content only on the basis of right and trust level of the user.

Starting from the previous premises and, in particular, taking advantage of the JXTA extension of JADE we realized RAIS (Remote Assistant for Information Sharing) that is a peer-to-peer multi-agent system supporting the sharing of information among a community of users connected through the internet. RAIS offers a similar search power of Web search engines, but avoids the burden of publishing the information on the Web and guaranties a controlled and dynamic access to the information. Moreover, the use of agent technologies simplifies the realization of three of the main features of the system: i) the filtering of the information coming from different users on the basis of the previous experience of the local user, ii) the pushing of the new information that can be of possible interest for a user, and iii) the delegation of access capabilities on the basis of a network of reputation built by the agents of the system on the community of its users.

5.1 System Architecture and Agents

RAIS is composed of a dynamic set of agent platforms connected through Intenet. Each agent platform acts as a peer of the system and it is based on three kinds of agents: a personal assistant, an information finder and a directory facilitator. Another agent, called personal proxy assistant, allows a user to access her/his agent platform from a remote system. Figure 3 shows the architecture of the RAIS system.

A personal assistant (PA) is an agent that allows the interaction between the RAIS system and the user that can send queries and view the related results through a graphical user interface (see figure 4). This agent receives the user's queries, forwards them to the available information finders and presents the results to the user. Moreover, a PA allows the user to subscribe her/him to be notified about new documents and information on some topics in which she/he is interested. Finally, a PA maintains a profile of its user preferences. In fact, the user can rate the quality of the information coming from another user for each search keyword (the utility of this profile will be clear after the presentation of the system behavior).

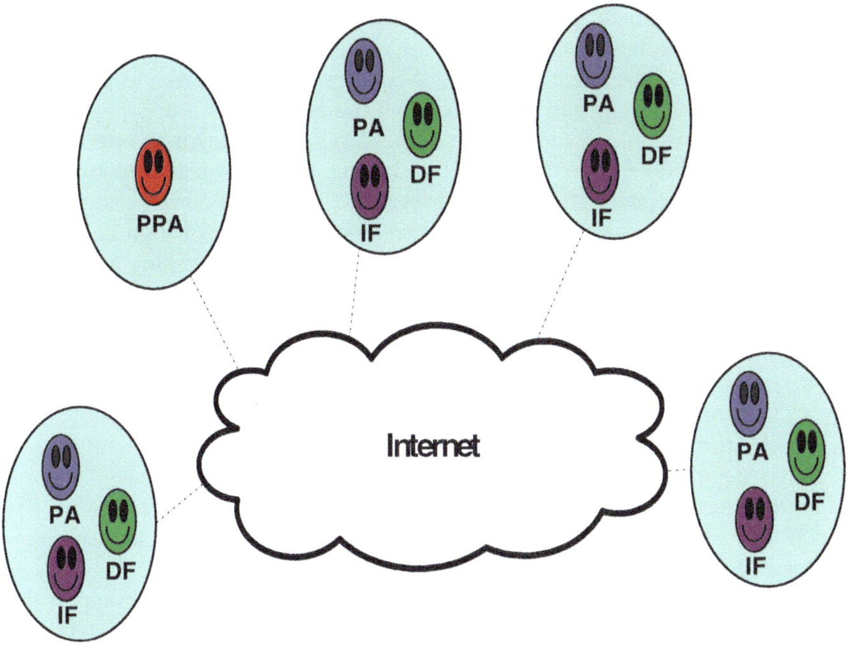

Fig. 3 Architecture of the RAIS system

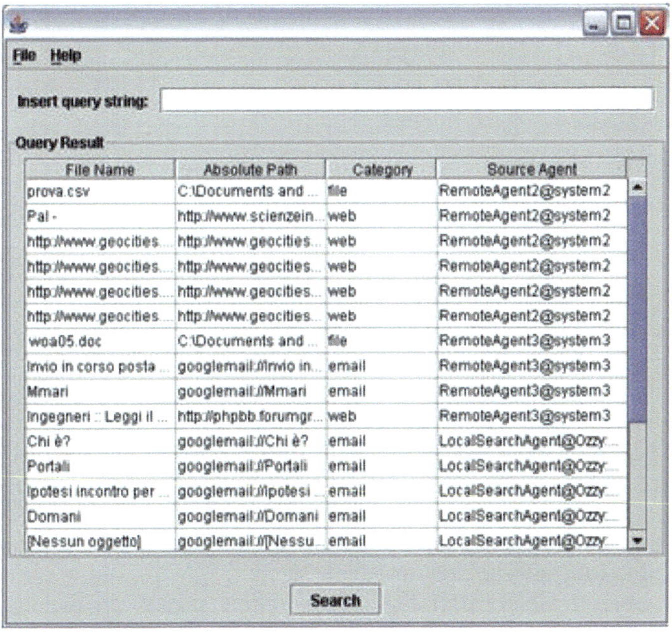

Fig. 4 View of the GUI allowing a user to communicate with her/his personal agent

An information finder (IF) is an agent that searches information on the repository contained into the computer where it lives and provides this information both to its user and to other users of the RAIS system. An IF receives users' queries, finds appropriate results and filters them on the basis of its user's policies (e.g. results from non-public folders are not sent to other users). An IF also monitors the changes in the local repository and pushes the new information to a PA when such information matches the subscriptions made by the user corresponding to this PA. Another task of the IF is to maintain an index of the information to be shared with the other users. The if performs it through a custom desktop searcher we developed taking advantage a of the well-known Lucene [27] indexing technology, together with Nutch [28], a Lucene subproject that provides add-ons for indexing the most diffused file formats.

A personal proxy assistant (PPA) is an agent that represents a point of access to the system for users that are not working on their own personal computer. A PPA is intended to run on a pluggable device (e.g. a USB key), on which the PPA agent is stored together with the RAIS binary and the configuration files. Therefore, when the user starts the RAIS system from the pluggable device, her/his RPA connects to the user's PA and provides the user with all the functionalities of her/his PA. For security reasons, only a PA can create the corresponding PPA and can generate the authentication key that is shared with the PPA to support their communication. For a successful connection, the PPA has to send the authentication key, then the user must provide her/his username and password.

Finally, the directory facilitator is responsible for registering the agent platform in the RAIS network. The DF is also responsible for informing the agents of its platform about the address of the agents that live in the other platforms available on the RAIS network. As introduced above, the RAIS system has been implemented taking advantage of the JXTA extension of JADE and so the directory facilitator uses the Agent Discovery Service for getting the information about the other RAIS platforms on the Internet.

5.2 Searching and Pushing of Information

In order to understand the system behavior, we present two practical scenarios. In the first, a user asks her/his PA to search for some information, while in the second the user asks to subscribe her/his interest about a topic. In both cases the system provides the user with a set of related information.

In the first scenario, the system activity can be divided in four steps: i) search, ii) result filtering, iii) results sending and presentation, and iv) retrieval.

Search: the user requests a search to her/his PA indicating a set of keywords and the maximum number of results. The PA asks the DF for the addresses of available IF agents and sends the keywords to such agents. The information finders apply the search to their repositories only if the querying user has the access to at least a part of the information stored into them.

Results filtering: each IF filters the searching results on the basis of the querying user access permissions.

Results sending and presentation: each IF sends the filtered list of results to the querying PA. The PA orders the various results as soon as it receives them, omitting duplicate results and presenting them to its user.

Retrieval: after the examination of the results list, the user can ask her/his PA for retrieving the information corresponding to an element of the list. Therefore, the PA forwards the request to the appropriate IF, waits for its answer and presents the information to the user.

In the second scenario, the system activity can be divided in five steps: i) subscription, ii) monitoring and results filtering, iii) results sending and user notification, iv) results presentation and v) retrieval.

Subscription: the user requests a subscription to her/his PA indicating a set of keywords describing the topic in which she/he is interested. The PA asks the DF for the addresses of available IF agents and sends the keywords to such agents. Each IF registers the subscription if the querying user has the access to at least a part of the information stored into its repository.

Monitoring and result filtering: each IF periodically checks if there is some new information satisfying its subscriptions. Then, the IF filters its searching results on the basis of the access permissions of the querying user.

Results sending and user notification: each IF sends the filtered list of results to the querying PA. The PA orders the various results as soon as it receives them, omitting duplicate results and storing them in its memory. Moreover, it notifies its user about the new available information sending her/him an email.

Results presentation: the first time the user logs into the RAIS system, the PA presents her/him the new results.

Retrieval: in the same way described in the previous search scenario, the user can retrieve some of the information indicated in the list of the results.

After receiving the results, the PA has the duty of selecting N results to send to its user (as said above the user can impose a constraint on the number of results to provide) and ordering them. Since the IFs create a digest for each result sent to the PA, the PA is able to omit duplicate results coming from different IF agents. Then, the PA orders the results and, if the total number of results exceeds the user's constraint, the PA only selects the first N. The ordering of results coming from different IFs is a complex task. Of course, each IF orders the results before sending them to the PA, but the PA has not the information on how to order results from different IF agents. Therefore, the PA uses two solutions on the basis of its user request: i) the results are fairly divided among the different sources of information, ii) the results are divided among the different sources of information on the basis of the user preferences. User preferences are represented by triples of the form <source, keyword, rate> where: source indicates an IF, keyword a term used for searching information and rate a number representing the quality of information (related to the keyword) coming from that IF. Each time a user gets a result, she/he can give a rate to the quality of the result and, as consequence, the PA can update her/his preferences in the user profile.

5.3 Mobile Users Support

People traveling for work may often be in need of access from a remote system their own computer. In this situation, a solution could be to install a VNC server on the desktop computer and to find a system with a VNC client while traveling. This solution has the advantage that the user gains the complete control on his remote PC, but it has also two main drawbacks: it's not easy to find computers with VNC clients available and the VNC connects only to one computer, not to the whole set of files and information of a workgroup. For users that don't require a complete control over a remote computer, but need to search and access a distributed set of documents, we have included in our system a remote search feature. The user can ask his/her PA to create a PPA on a pluggable device, e.g., an USB key or a removable hard disk. The PA copies on the device the RAIS run-time, the RPA and the authentication key shared by the PPA and the PA itself. When the user inserts the pluggable device on another computer, he can immediately launch his PPA and connect to its corresponding PA. Therefore, the way of using the RAIS system is analogous to the situation in which the user works on her/his own computer, except for the interactions between the RPA and the PA, that, however, are transparent to the user. In fact, at the initialization, the PPA sends an authentication key to the PA. If the key matches those of the PA, the user can provide his/her username and password and enter the system (this step must be done by the user when she/he uses the RAIS system from her/his own computer too). After these two steps, the PPA acts as a simple proxy of the remote PA.

5.4 Security

The information stored into the different repositories of a RAIS network is not accessible to all the users of the system in the same way. In fact, it is important to avoid the access to private documents and personal files, but also to files reserved to a restricted group of users (e.g. the participants of a project). The RAIS system takes care of users' privacy allowing the access to the information on the basis of the identity, the roles and the attributes of the querying user defined into a local knowledge base of trusted users. In this case, the user defines who and in which way can access to her/his information. Furthermore, the user can allow the access to unknown users enabling a certificate based delegation built on a network of the users registered into the RAIS community. In this sense, the system completely adheres to the principles of trust management. For instance, if the user U_i enables the delegation and grants to the user U_j the access to its repository with capabilities C_0 and U_j grants to the user U_k the access to its repository with the same capabilities C_0, then U_k can access U_i's repository with the same capabilities of U_j.

The security architecture, in particular, is founded on a more generic framework implementing a distributed RBAC model [29]. In particular, the theory of RAIS delegation certificates is founded on SPKI/SDSI specifications [30], though the certificate encoding is different. As in SPKI, principals are identified by their public keys, or by a cryptographic hash of their public keys. Instead of s-expressions, RAIS uses XML signed documents, in the form of SAML assertions [31], to convey identity, role and property assignments. As in SPKI, delegation is made possible if the delegating principal issues a certificate whose subject is a name defined by another, trusted, principal. The latter can successively issue other certificates to assign other principals (public keys) to its local name. In this sense, local names act as distributed roles [32].

6 Conclusions

In this paper we presented how JADE, one of the most known software frameworks for the development of multi-agent systems, has been extended with a JXTA based module that supports a global index of resources and services realized through the use of generic discovery advertisements and allow a more reliable communication between remote platforms through the use of JXTA pipes. Moreover, we presented a system, called RAIS, that offers a set of advanced services for content sharing and retrieval that can be considered a reference example of how the integration of multi-agent and peer-to-peer systems – in particular, the use of the integration of JADE with the JXTA technology – allows an easy realization of applications for communities of users connected through the Internet.

The JXTA module has been experimented in the realization of some other applications providing services to communities of users connected through the Internet and, in particular, to support the recommendation of experts on Java programming and to support collaborative mail filtering.

The current implementation of the RAIS system is an extension of the system presented in [33] thanks to the JXTA extensions of JADE. RAIS has been experimented by a set of staff members and students of our department that used the system both inside the department (i.e., the platforms are connected through a local network) and remotely through the Internet. Qualitative experimentations, carried on in a local area environment, have not shown a significant performance loss in the normal functioning of the system, using JXTA pipes instead of http connections. The search of remote platforms is performed only at startup, or on user request, and requires to wait for a configurable amount of time. A comparative and quantitative experimentation, in a wide area environment, instead, showed the same performances, but few problems at a startup and in any situation where there is a reconfiguration of the network of platforms.

Our current and future activities is oriented to: i) the extension the RAIS system with a service that allows to communities of users, that share a set of domain ontologies, to use them for semantic search over the content of their documents, and ii) the experimentation of the JXTA module for the realization of business process applications as, for example, e-market and customer care applications.

Acknowledgments. Many thanks to the colleagues and the students that were and are involved in the developed of the JXTA JADE extension and of the RAIS system. This work is partially supported by the Italian MIUR (Ministero dell'Istruzione, dell'Università e della Ricerca).

References

1. Koubarakis, M.: Multi-agent Systems and Peer-to-Peer Computing: Methods, Systems, and Challenges. In: Klusch, M., Omicini, A., Ossowski, S., Laamanen, S. (eds.) CIA 2003. LNCS (LNAI), vol. 2782, pp. 46–61. Springer, Heidelberg (2003)
2. Langley, B.K., Paolucci, M., Sycara, K.: Discovery of infrastructure in multi-agent systems. In: 2nd International Joint Conference on Autonomous Agents and Multi-agent Systems (AAMAS 2003), pp. 1046–1047 (2003)
3. Sycara, K., Paolucci, M., Van Velsen, M., Giampapa, J.: The RETSINA MAS Infrastructure. Autonomous Agents and Multi-Agent Systems 7(1-2), 29–48 (2003)
4. Koubarakis, M., Tryfonopoulos, C., Raftopoulou, P., Koutris, T.: Data Models and Languages for Agent-Based Textual Information Dissemination. In: Klusch, M., Ossowski, S., Shehory, O. (eds.) CIA 2002. LNCS (LNAI), vol. 2446, pp. 179–193. Springer, Heidelberg (2002)
5. Babaoglu, O., Meling, H., Montresor, A.A.: Anthill: A Framework for the Development of Agent-Based Peer-to-Peer Systems. In: 22nd international Conference on Distributed Computing Systems (ICDCS 2002), Vienna, Austria, pp. 15–22 (2002)
6. JXTA Web Site (2009), `https://jxta.dev.java.net`
7. Li, T., Zhao, Z., You, S.: A-peer: an agent platform integrating peer-to-peer network. In: 3rd IEEE/ACM International Symposium on Cluster Computing and the Grid, Tokyo, Japan, pp. 614–617 (2003)
8. Bertolini, D., Busetta, P., Nori, M., Perini, A.: Peer-to-peer multi-agent systems technology for knowledge management applications. An agent-oriented analysis. In: WOA 2002, Milano, Italy, pp. 1–6 (2002)
9. Gorodetsky, V., Karsaev, O., Samoylov, V., Serebryakov, S.: P2P agent platform: Implementation and testing. In: Joseph, S., Despotovic, Z., Moro, G., Bergamaschi, S. (eds.) AP2PC 2007. LNCS, vol. 5319, pp. 41–54. Springer, Heidelberg (2010)
10. FIPA P2P Nomadic Agents Functional Architecture Specification - Draft 0.12 (2005), `http://www.fipa.org/subgroups/P2PNA-WG-docs/P2PNA-Spec-Draft0.12.doc`
11. Androutsellis-Theotokis, S., Spinellis, D.: A survey of peer-to-peer content distribution technologies. ACM Computing Surveys 36(4), 335–371 (2004)
12. Lopes, A.L., Botelho, L.M.: Improving Multi-Agent Based Resource Coordination in Peer-to-Peer Networks. Journal of Networks 3(2), 38–47 (2008)
13. Zhang, H., Croft, W.B., Levine, B., Lesser, V.: A Multi-Agent Approach for Peer-to-Peer Based Information Retrieval System. In: 3rd International Joint Conference on Autonomous Agents and Multiagent Systems (AAMAS 2004), New York, NY, pp. 456–463 (2004)
14. Mine, T., Matsuno, D., Takaki, K., Amamiya, M.: Agent Community Based Peer-to-Peer Information Retrieval. In: 3rd International Joint Conference on Autonomous Agents and Multiagent Systems (AAMAS 2004), New York, NY, pp. 1484–1485 (2004)
15. Kungas, P., Matskin, M.: Semantic web service composition through a P2P-based multiagent environment. In: Despotovic, Z., Joseph, S., Sartori, C. (eds.) AP2PC 2005. LNCS (LNAI), vol. 4118, pp. 106–119. Springer, Heidelberg (2006)
16. Zhang, Z.: E-Commerce Based Agents over P2P Network. In: International Conference on Management of E-Commerce and E-Government, pp. 77–81 (2008)

17. Bellifemine, F., Poggi, A., Rimassa, G.: Developing multi agent systems with a FIPA-compliant agent framework. Software - Practice & Experience 31, 103–128 (2001)
18. Bellifemine, F., Caire, G., Poggi, A., Rimassa, G.: JADE: a Software Framework for Developing Multi-Agent Applications. Lessons Learned. Information and Software Technology 50, 10–21 (2008)
19. FIPA Specifications (2000), http://www.fipa.org
20. JADE Software Framework (2009), http://jade.tilab.com
21. BlueJade software (2003), http://sourceforge.net/projects/bluejade
22. Bergenti, F., Poggi, A., Burg, B., Caire, G.: Deploying FIPA-Compliant Systems on Handheld Devices. IEEE Internet Computing 5(4), 20–25 (2001)
23. Chmiel, K., Gawinecki, M., Kaczmarek, P., Szymczak, M., Paprzycki, M.: Efficiency of JADE agent platform. Scientific Programming 2(2005), 159–172 (2005)
24. Zimmermann, R., Winkler, S., Bodendorf, F.: Supply Chain Event Management with Software Agents. In: Kirn, S., Herzog, O., Lockemann, P., Spaniol, O. (eds.) Multi-agent Engineering - Theory and Applications in Enterprises, pp. 157–175. Springer, Berlin (2006)
25. FIPA Agent Discovery Service Specification (2003), http://www.fipa.org/specs/fipa00095/PC00095.pdf
26. FIPA JXTA Discovery Middleware Specification (2004), http://www.fipa.org/specs/fipa00096/PC00096A.pdf
27. Lucene project (2009), http://lucene.apache.org
28. Nutch project (2009), http://lucene.apache.org/nutch
29. Poggi, A., Tomaiuolo, M.: XML-based Trust Management in MAS. In: WOA 2007, Genova, Italy, pp. 126–131 (2007)
30. Ellison, C., Frantz, B., Lampson, B., Rivest, R., Thomas, B., Ylonen, T.: SPKI Certificate Theory. RFC 2693 (1999)
31. SAML - Security Assertion Markup Language (2009), http://xml.coverpages.org/saml.html
32. Li, N., Mitchell, J.M.: RT: A Role-based Trust-management Framework. In: 3rd DARPA Information Survivability Conference and Exposition (DISCEX III), Washington, DC, pp. 201–212 (2003)
33. Mari, M., Poggi, A., Tomaiuolo, M.: A Multi-Agent System for Information Sharing. In: Proc. of ICEIS 2006, Paphos, Cyprus, pp. 147–152 (2006)

Intelligent Advisor Agents in Distributed Environments

Agnese Augello, Giovanni Pilato, and Salvatore Gaglio

Abstract. The chapter presents a Distributed Expert System based on a multi-agent-architecture. The system is composed of a community of intelligent conversational agents playing the role of specialized advisors for the government of a virtual town, inspired to the SimCity game. The agents are capable to handle strategic decision under uncertainty conditions. They interact in natural language with their owners, obtain information on the current status of the town and give suggestions about the best strategies to apply in order to govern the town.

1 Introduction

Traditionally the notion of decision support has involved two aspects: the organization of decision making and the techniques of interactive computer systems. In last years, it has become an autonomous research area of research concentrating on computerized system supporting decision making activities [1].

A decision support system (DSS) can be defined as a software program that provides information in a given domain of application by means of analytical decision models [2]. The idea of offering decision support always arise when timely decisions must be made in complex domains, in complex and dynamic environments, and where multiple actors are present. Support in this context means "assistance for structuring the problem, and for analyzing and verifying the obtained structure"[3], [4].

Agnese Augello · Salvatore Gaglio
DINFO (Dipartimento di Ingegneria Informatica,
Viale delle Scienze, Ed.6
e-mail: augello@dinfo.unipa.it, gaglio@unipa.it

Giovanni Pilato
ICAR Italian National Research Council,
Viale delle Scienze, Ed.
e-mail: g.pilato@icar.cnr.it

Soro et al. (Eds.): Inform. Retrieval and Mining in Distrib. Environments, SCI 324, pp. 109–124.
springerlink.com

Intelligent systems are being applied to larger, open and more complex problem domains, and many applications are found to be more suitably addressed by multi-agent systems [5] [6] paradigm that has emerged to better formalize problems in Distributed Articial Intelligence (DAI) [7]. An agent is capable of autonomous action in a given environment in order to meet its design objectives [8]. According to this definition, an intelligent agent can be extended with three additional characteristics: reactivity, proactivity and social ability [6]. An agent usually performs complex actions on behalf of their users [9]; this approach is particularly effective in a distributed and dynamic environment (potentially on a web-wide scale). In general proactive agents must take decisions, and decision making is not a simple event but a process leading to the selection of a course of action among several alternatives. Agents need to select and to compose different actions in order to make a decision [10]. On the other hand, unpredictable environments, characterized by highly decentralized, up-to-date data sets coming from various sources, are natural examples of mission-critical decision making [11].

Consider a complex problem domain, affected with uncertainty: a set of agents can process it in order to infer the true state of the domain. Generally agents can be charged with many possible tasks according to the peculiarity of the domain. Usually each agent has only a limited point of view of the world: it has only knowledge on a local subdomain and can only get local observations [12].

Agent based paradigm and Computational Intelligence (CI) techniques have been successfully applied also in economic fields. The field of Agent-based Computation in Economics (ACE) describes and studies this complex domain in order to catch the effects deriving by the interaction between different agents. Merging experimental economics and ACE disciplines brings to the development of validation tests about economic theories substituting human agents with software agents [13]. In order to accomplish this task, it is important to take into consideration heterogeneous software agents, characterizing each one of them with its own cognitive capability, a personality and a different level of risk attitude [13]. Evolutionary algorithms, neural networks, fuzzy systems, and Bayesian networks are very common CI techniques in economics and finance [14], [15], [16], [17].

A useful approach to study the effects of economic policies is recreating typical scenarios of the real world exploiting simulation games. Many simulation games exist where the player task is to take strategic decisions with the aim of reaching specific goals. As an example Global Economics Game [18] is an educative game where the task is to define fiscal, economic and trading policies of a nation. The goal is to promote the economic growth without causing excessive pollution, mantaining high the values of employment and low inflation.

The chapter describes a decision support system composed of a community of intelligent conversational agents, playing the role of intelligent advisors, which are able to provide support to human beings in their activities.

The agents are able to retrieve information and provide useful directions through a reasoning process which makes use of the retrieved information.

The architecture of the system will be explained and a case of study will be presented. The case of study will regard the creation of a distributed expert system for the government of a virtual City inspired to SimCity game [19].

Users assume specific administrative roles and can be supported by personal advisors to accomplish their tasks and to take important decisions. The roles will be related to economic/management issues, such as the imposition and variation of taxes, building of structures, garbage management, energy, and so on...

The agents, interacting in natural language with the user, and exchanging messages each other obtain information about the current state of the city and give, as a consequence, suggestions about the best strategies to apply.

The key role in the City is played by the agent which represents the mayor of the town. It has the main role of coordinating the activities of the town. The other agents will try to accomplish their own goal respecting the main interest of the city population but following their specific utilities which are modeled according to the personalities of the players that they represent. The different political and personal orientation and consequently the choices of the agents will influence the evolution of the City.

2 Knowledge Representation Models and Agents Learning

Among different knowledge representation tools, bayesian networks play a prominent role in analyzing information characterized by uncertainty, which cover most part of the problems that usually occur in the real world. For this reason, we focus our attention on this kind of tools.

In [20] Kyoung-Min Kim proposed a conversational agent that that makes use of semantic Bayesian networks (SeBN). This kind of tool makes the agent capable to manage context and uncertainty, and to infers users intentions. Besides, it reduces the complexity of the conversational agent. In [10] the authors propose an ontology driven uncertain model: Onto-Bayes, consisting of knowledge and decision model parts. The former is the integration of ontologies and Bayesian Networks (BN) while the latter can describe different decision models.

The possibility to dynamically build knowledge representation models of agents improves the capability of the agents of adapting their behavior to different scenarios. In [21] a methodology and a tool for transferring DM-extracted knowledge into newly created agents has been presented. The approach exploits both the deductive and the inductive reasoning paradigm; the former for facilitating the well-known and established processes and functionalities of the multi agent systems, the latter for the discovery and the exploitation of previously unknown knowledge. Data mining is used to generate knowledge models that can be dynamically embedded into the agents; this enhances the adaptability, reusability and versatility of multiaent systems. In [22] the author offer a perspective on distributed data mining algorithms in the context of multi-agents systems. They provide a high-level survey of distributed data mining, then focuses on distributed clustering algorithms and some potential applications in multi-agent-based problem solving scenarios. In [23] authors summarize the main functionalities and features of an agent service and data mining symbiont, named F-Trade. That provides services of trading evidence back-testing, optimization and discovery, as well as plug and play of algorithms, data and system modules for financial trading and surveillance with online connectivity to huge quantities of global market data.

An agent-based system, therefore, can get many advantages in terms of adaptability and reusability if its knowledge representation models can be inferred by data automatically acquired from the domain of interest. In this work we exploit bayesian networks structures learning which can be embedded in agent knowledge models. In literature different algorithms are available to learn the structure of a bayesian network and the evaluation of the parameters starting from a set of observations. Structure learning can be divided into two main categories: constraint-based methods and score based methods [26]. The first category is composed of algorithms that build the network on the basis of conditional independency constraints inferred from observations. The second category is related to those methods that treat the learning of the structure of a bayesian network as an optimization problem of a particular objective function (distinguishing between local and global optimization metrics), searching in the space of all direct acyclic graphs and selecting the one with the highest score. The results heavily depend on the characteristics of the observations, and on the completeness of the observations dataset. For a detailed analysis refer to [26].

3 Intelligent Advisor Agents in Distributed Environments

Decision support systems in complex environments are characterized by different people who take decisions and require the definition of adequate tools capable to adapt their behavior in an efficient and flexible manner in order to reflect and manage the environment to model as much as possible.

These kind of environments are generally characterized by the presence of different agents that concur to the organization and the management at different levels responsibility. Often two or more agents take part in processes that can affect each other, requiring the capability of the system to solve possible conflicts that can arise.

We present an architecture in this chapter for the construction of a decision support system, made of a community of intelligent conversational agents, able to provide support to human beings in decision-making processes. The agents take the role of personal intelligent advisors, able to give information and provide useful directions through a reasoning process which makes use of information retrieved and data analysis on the domain.

The architecture counts two different kind of agents:

- *Advisor agents*: they interact with users in natural language and support the user in the decision-making process. They are characterized by a complex knowledge management architecture. Advisor agents can be divided into "task-oriented" advisors agents and "mediator" advisor agents, depending on the hierarchy they belong to (i.e. roles, levels, etc...). Task-oriented agents are autonomous agents,

specialized for a particular task. Mediator agents control and coordinate the task-oriented agents, with the main goal of solving possible conflicts within the strategic decisions carried on by each task-oriented agent.
- *Service agents*: this kind of agents accomplish the task of providing services to advisor agents. In particular, the architecture is composed of message broker agents and the data collector agents. The former ones are finalized to the management of the communications between the agents of the community; the latter ones are interfaces with the external world in order to retrieve and organize useful information for the community of agents.

3.1 Advisor Agents

3.1.1 Main Functionalities and Architecture

An advisor agent has the task to support the user who needs to get information and take decisions for a particular aspect of the domain. The domain of interest is subdivided into areas, and it is characterized by a set of strategic variables that can affect one or more areas. Each agent is responsible for a specific area of the domain, and is usually associated to a single user, who is the owner of the agent. Of course more agents can share the same owner.

Each agent has information about the strategic variables that affect the area for which the agent is responsible for. The information can be acquired:

- through communication with other agents of the community;
- through a dialogue with the user;
- through observations on the environment.

Information is organized within the knowledge representation module of each agent and is exploited to define the most appropriate strategies to suggest to the user. The agent tries to imitate the personality of its owner; for this reason the agent acquires the preferences of the user through an interactive process.

The distributed and multi-goal nature of the DSS could lead to possible conflicts among the different strategies proposed by different "task-oriented" agents. Conflicts arise when different agents propose contrasting strategies based on their own observations, beliefs, and reasoning; ie when the estimated values of the same strategic variable differ in a significant manner for at least two task-oriented agents.

Conflicts are solved by mediator agents. Mediators act as soon as a conflict is detected. In order to detect conflict situations, mediators register themselves on the message broker agent to request information about values of specific strategic variables. If a conflict is detected, the mediator requires information to the task-oriented agents involved in the conflict about the solution that each of them proposes. This phase activates the decisional process of the mediator that tries to solve the conflict.

The knowledge representation model of the agent is a hybrid model, constituted by the integration of a traditional ontology-based reasoning system with a probabilistic one. This choice is due to the requirement of managing deterministic information and also information that is characterized by uncertainty, which cannot be treated by ontology-based systems. The integration of ontologies with bayesian networks grants the advantages of both models.

Three areas, named deterministic area, probabilistic area, and linguistic area, compose the agent architecture, shown in figure 1 and in more detail in figure 2.

The core of the system is given by a module, called "corpus callosum" that allows the interaction between the agent and the user, and the communication between the deterministic and the probabilistic areas. These areas will be referred, for short, as "knowledge representation areas". The agent builds its own knowledge model according to a subset of strategic variables that characterize the domain and that are related to the decisional process that involves the agent. The decisions of the agent affect the values of these variables.

The system is designed in order to manage dynamic situations. For this reason the agent obtains information about the current state of the domain and updates the knowledge models of both its deterministic and probabilistic areas.

Below we describe more in detail the main components of the agent brain.

3.1.2 Linguistic Area

The agent interacts with its owner to inform him about the current state of the domain and to support him during the decisional process. Interaction is made possible thanks to the linguistic area, which includes a dialogue module implemented with a conversational agent based on the ALICE technology [25]. This module is aimed at understanding the dialogue with the user through the definition of rules, described using a mark-up language called AIML (Artificial Intelligence Mark-up Language).The rules are coded in question answer modules, named *categories*. User questions are processed by the chatbot engine by means of a pattern matching algorithm. The user request is caught and processed by the linguistic area and sent to the corpus callosum, which actuates the most proper action to execute on one of the two modules or even on both of them. The possible consequences of the action are analyzed and a sentence is consequently composed *ad hoc* by the linguistic area in order to answer and inform the user about the possible repercussions of the move. User

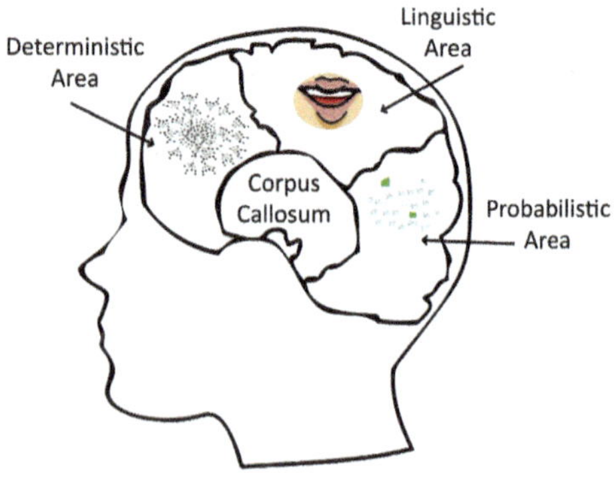

Fig. 1 Advisor Agent Brain

queries could include support request for a given decision policy, information about the domain, a refinement or revision of information represented in the deterministic and decisional areas.

3.1.3 Deterministic Area

The deterministic area is oriented to a detailed description of the domain in an ontological model, described in Web Ontology Language (OWL). The main concepts, the taxonomy, the properties and rules of the domain are described in the ontology. An inferential deterministic engine allows the agent to reason about the knowledge formalized in the ontology. It gives support to the user inferring facts about the domain.

The deterministic area supports the user giving him information related to the domain, answering to his requests, and modeling the knowledge of the agent according to the user preferences. The deterministic area describes the strategic variables and their properties. Rules that can help the agent to take decisions that reflect the user preferences are also defined for each variable.

3.1.4 Probabilistic Area

The capability of the system to manage situations characterized by uncertainty is given by the probabilistic area. The core of this area is constituted by a bayesian decision network (BDN) that is inferred by the agent from data about the domain through an interactive process with the user. Each agent analyzes data coming from observations of the domain variables that are strictly related to the task the agent is

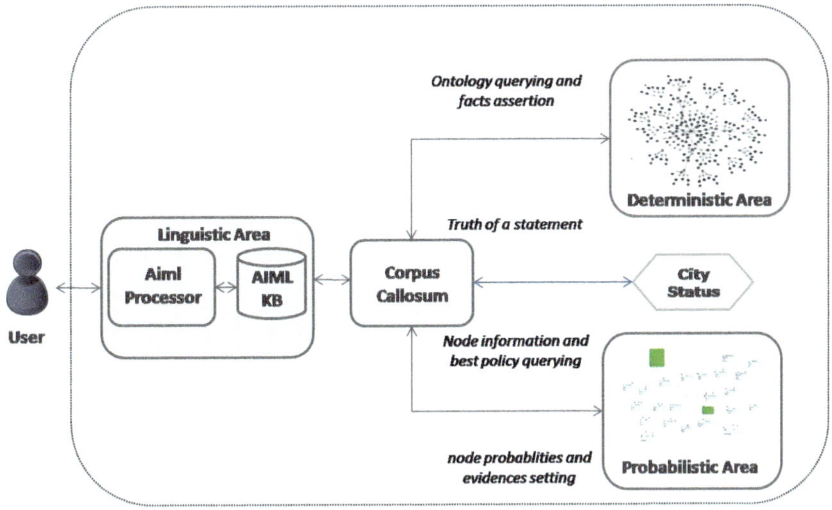

Fig. 2 Advisor Agent Architecture

oriented to. These data are then processed and analyzed in order to determine the causal relationships between the variables in order to learn one or more bayesian networks that are embedded into the probabilistic knowledge representation model of the agent.

Since a multitude of algorithms present in literature are oriented to learn the structure of a bayesian network starting from a set of observations, our approach is to give the possibility to the user to choose the network which is more adequate to describe the decisional task. In this phase the user has the possibility to determine the decision nodes and to add the desired utility nodes in order to transform the bayesian network into a decision network.

Thanks to the decision network, the agent can reason about uncertain information, and suggest to the user the decisions that it believes to be the best to take in order to reach a specific goal. In particular the agent can estimate the benefits and the potential risks resulting from the adoption of different strategies, and, as a consequence, to evaluate the effects on the variables of interest of the analyzed domain. The decisions are suggested taking into account also the preferences of the user.

The decisional analysis is conducted through a direct comparison among the utility values corresponding to all possible choices. Causal relations among concepts of the domain are described in the network, which has been developed with GeNIe [24], a free environment for building of decision-theoretic models, developed at the University of Pittsburgh.

3.1.5 Corpus Callosum

The corpus callosum links information coming from the different areas of the agent brain. It communicates with the linguistic area in order to understand the meaning of user sentences. It coordinates actitivies like asserting facts and activating queries, inferring complex relations starting from known facts and properties represented inside the ontology, setting and updating the beliefs and evidences in the Bayesian decision network and realizing a decisional analysis in order to show to the user the most suitable decision strategy to accomplish the desired targets.

Through the corpus callosum it is possible to manage the influence of an area over the other one. As an example, a formal reasoning that leads to asserting a new fact may have as a consequence setting evidence in the decisional network. Evidences in the network change the values of the expected utility associated to the decisions. This has a direct consequence on what can be considered the best decision to take according to the bayesian model. Vice versa, an inference on the decisional network can trigger a deterministic reasoning or a modification of the ontology. The corpus callosum analyzes the current status of the domain, and uses this information to update the knowledge stored in both areas and reasoning about the user requests, e.g. the agent can explore the decisional network to estimate the consequences of possible decisions starting from the current status of the game. As soon as the user takes his own decisions, the status of the game changes consequently, thus dynamically modifying the decisional path.

3.2 Service Agents

The distributed architecture of the Decision Support System requires that each agent alerts the community about the choices that the agent intends to carry on, since that a single decision could significantly change the environment.

This requires a persistent information exchange among agents. The presence of dedicated agents, named "Message broker" grants to avoid the proliferation of messages among agents and the simplification of the community management. Message broker agents manage the message queues according to the publish/subscribe model [27]. This paradigm is based on two kind of agents: *publishers*, who send messages and *subscribers*, who receive messages. The former ones publish information for a particular area of interest while the subscribers register themselves on the communication channel related to the area of interest in order to receive the associated notifications.

The message broker agent constitutes the medium between the publisher and the subscribers. According to this schema, agents can register themselves to the message brokers as publisher or subscriber. In our case of study we have chosen to identify each area of interest with the main variables that characterize the domain of interest. Each message broker is specialized to manage information regarding a specific strategic variable All task oriented advisor agents that want to work on this variable must register as publishers into the message broker agent, informing it about their intentions. The mediator, which is responsible for the resolution of conflicts on the particular strategic variable, registers itself as subscriber into the message broker agent, which has information regarding the strategic variable of interest. In this manner the mediator agent will be permanently informed by the message broker about the intentions of the different task oriented agents and it will detect possible conflict situations.

Messages sent by the task oriented agents will declare the decision variable on which they act and the estimated value that the variable will assume as a consequence of their decisions.

Figure 3 shows the interaction between the agents in the publish/subscribe model.

Data collector agents are interfaces with the environment in order to retrieve information of interest of the community of agents. Each data collector agent sends the retrieved information to a message broker agent that sends them to the interested agents. In theory the advisor agents could retrieve automatically information from the environment, but this would require their specialization for the data source to query. Besides, without data collector agents, if more advisor agents require the same information, each of them should implement the specific functionalities to access these data. The presence of the message broker grants the reduction of communications between agents.

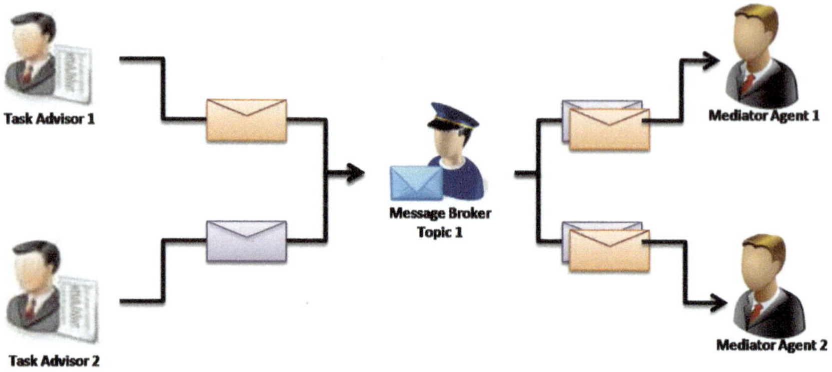

Fig. 3 Publish/Subscribe Schema

4 City Policies Support: A Case Study

The goal of the proposed system is to provide a support for taking political and economic decisions. To this aim, we have developed, as a test-bench, an agent-based simulation of the a town. The agent analyzes data arising from the simulation, which represent the current state of the virtual town. Data can be used to train the agent and its beliefs. The proposed architecture can be easily adapted on other problems.

The case of study regards the creation of a distributed decision support system for the governance of a hypothetical City. Task-oriented advisor agents are created to support City administrators, each one having a specific administrative role. Obviously the key role in the City is played by the mediator agent which represents the mayor of the town. It has the main role of coordinating the activities of the town. The other agents will try to accomplish their own goal respecting the main interest of the city population but following their specific utilities which are modeled according to the personalities of the user that they represent (their owner). The different political and personal orientation and consequently the choices of the agents will influence the evolution of the City. The agents, interacting in natural language with the user, and exchanging messages each other can obtain information about the current state of the city and give, as a consequence, suggestions about the best strategies to apply.

We have hypothesized the possible actions for the management of the city. Each user responsible for a particular administrative sector receives a support from his personal advisor regarding the decisions about the specific administrative sector of the town. Since each agent is suited to reflect the user's personality, it tries to advise respecting, as much as possible, his preferences. Proper utility functions are defined for giving decision support in this scenario. From the analysis of the domain, we have derived the knowledge models present in the knowledge representation areas and the necessary rules to understand natural language expressions, managed by the linguistic area. The implementation of the knowledge representation areas is described below. The Decision Support System is composed by five advisor agents: four task oriented agents whose tasks are:

- taxation policy;
- energy policy;
- pollution and garbage management;
- public transport management;

and one mediator agent representative of the mayor of the town. The task-oriented agents suggest strategies to invest the economic resources of the city trying to control parameters of interest (e.g. level of pollution, criminality index, well being of citizens and other strategic variables). The mayor coordinates other agents as soon as conflicts arise between different strategic decisions proposed by the task oriented agents. The mayor forces the best strategy according to its own beliefs and preferences.

4.1 Agents Ontology

Each agent has a model of the environment mapped in an ontology describing concepts and facts of the domain, in particular the set of information regarding the town, like variables bound to the city status (pollution level, energy availability), or variables bound to the inhabitants behavior (criminality level, happiness, healthiness, and so on).

The ontology describes also the different kinds of buildings that can be constructed in the town, according to particular administrative sectors. As an example, for the energy policy we have the classification of different Power Building, like Natural Gas Plant, Nuclear Power Plant, Solar Power Plant, Wind Power Plant.

The ontology describes also the possible administrative policies that can be carried on by the agent. As an example, for the garbage management the use of incinerators, recycle policy, waste material exporting, and building of garbage dumps are all possible policies.

The ontology describes the characteristics that govern the life of the town, like the fact that Citizens have to contribute to town budget paying taxes, and rules about specific administrative sectors.

The agent exploits this information in order to perform deterministic reasoning that take into account also the current status of the town. As an example, the agent can suggest the user to build power plants if the energy required for the city is not sufficient. Otherwise, the agent can propose a priority for the construction of buildings. Examples of reasoning could be: a town should have at least an hospital, a city hall, a fire station; if the level of population exceeds a given threshold, it is necessary to build another hospital; if a building is not connected with any street, it is useless; it is preferable connect strategic buildings, like hospitals, to main streets; the houses of the citizens should be connected to workplaces, and so on...

Rules regarding strategic variables grant the definition of the utility of the user with respect to the values assumed by the strategic variables. For example it is possible to assert that the user, and as a consequence the agent which represents him, prefers an higher value of budget or an higher value of well-being of citizens.

4.2 An Example of Probabilistic Area Induction

Decision networks of the agents are created in an interactive manner, starting from a set of observations, from the user preferences and from the suggestions of the users.

The first step is to learn the structure of the bayesian decision network and the related contingency tables. Both of them can be changed and refined afterwards, as soon as the number of observations increases.

For this task the user can use the main algorithms for Learning the Structure of Bayesian Networks and parameter estimation offered in the Weka Data Mining tool [28]. A set of induced networks are proposed to the user, who chooses the one which best models the environment and which is best suited for the required decision process.

Figure 4 shows an example of network induced by observations for the "public transport management". The bayesian network is induced through the observations on the strategic variables like the costs for the enhancement of the public transportation systems, the cost of the tickets of the public transport, the wellbeing of the citizens, and variables like pollution and the traffic congestion level, considered before and after the decision. The network has been obtained with the Weka Tool selecting a *Bayes net* classifier learned with the *Simulated Annealing* algorithm and specifying as options *markovBlanketClassifier*.

The user chooses the structure which best models the world. The network is transformed into a decision network through an interactive process, by defining the necessary decision and utility nodes. Figure 5 shows the decision network obtained starting from the structure previously induced with the Weka Tool.

The choice of particular decisions implies a different evolution of the network. The presence of a "well-being" utility node makes possible to set a utility function

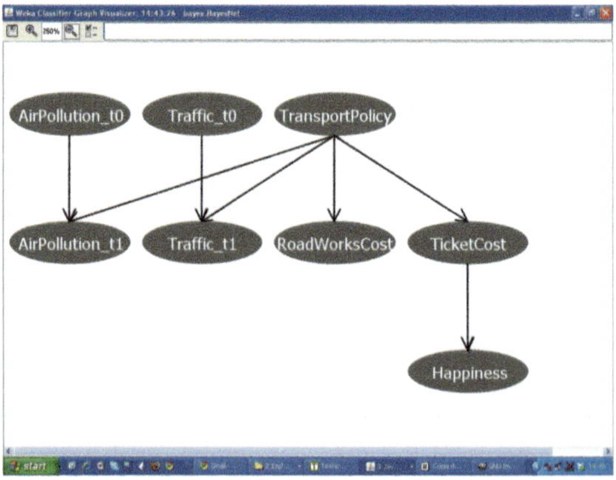

Fig. 4 An example of Bayesian Network induction in Weka

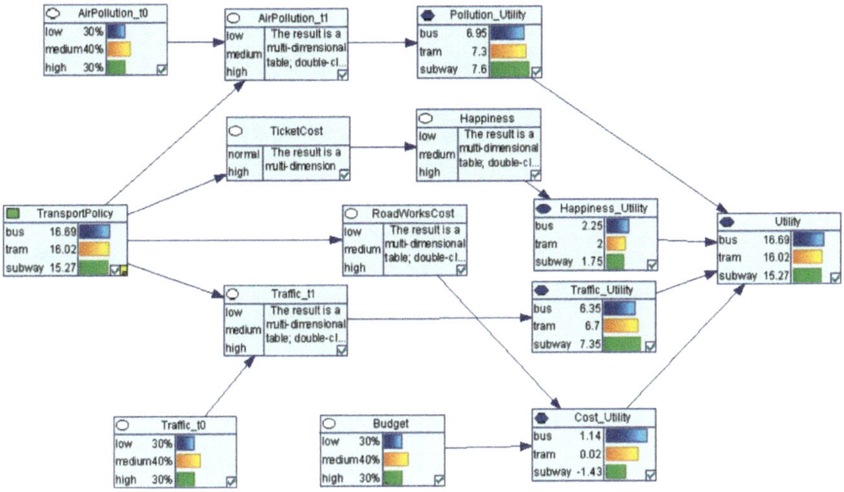

Fig. 5 The decision network for the public transport management

that takes into account the temporal evolution of these values. Starting from this information we can estimate the expected values for strategic variables at time t_1,after which a decision can be taken (forecasting phase).

4.3 A Scenario Example

The scenario describes the behavior of the community of agents that copes with a critical situation. This happens whenever one of the strategic variables reaches a critical level. As an example in our model the variable pollution can assume values like "high", "medium", "low" and its status is monitored by the pollution and garbage management (environmental policies) advisor agent. Let suppose that the pollution level is "high". This situation is detected by the environment agent that starts a decision process with the goal of reducing the pollution rate of the town.

The agent retrieves the values of CO_2 (high), the level of traffic (high) and the budget of the town (that we suppose "low"), and sets the evidences inside the "airpollutionnet" decision network. The network models the consequences of the policies regarding the management of the traffic of the town and the pollution that generates. In particular the policies that the agent can carry on are: growth of pedestrian areas, limitation of traffic according to even and odd numbers of license plates, enhancement of public transportation system, and creation of green areas. The agent estimates the most convenient decision in terms of utility. In this case the utility is 15.56 obtained increasing the pedestrian areas. This decision implies a lower probability to have a "high" pollution. This scenario is shown in figure 6.

Let us suppose that a politic crisis arises and that an energy crisis is foreseen because of a crisis of the main energy supplier of the town. The energy management

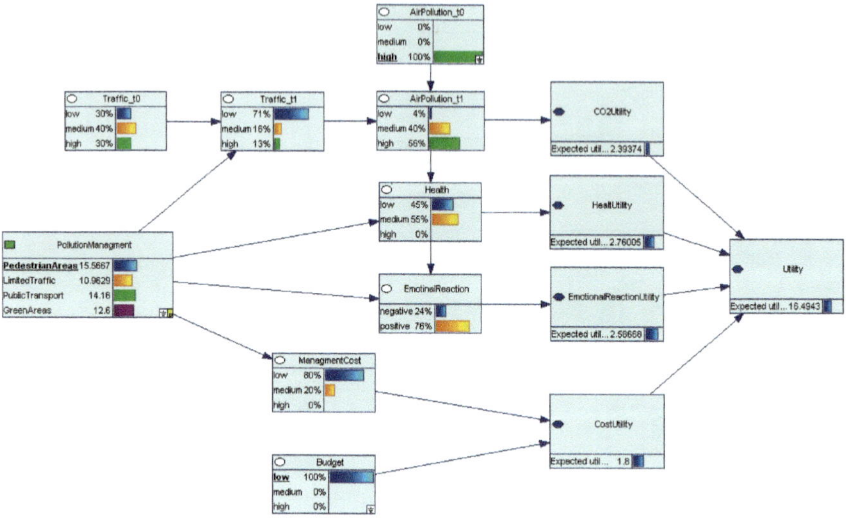

Fig. 6 Best Pollution policy for the described scenario

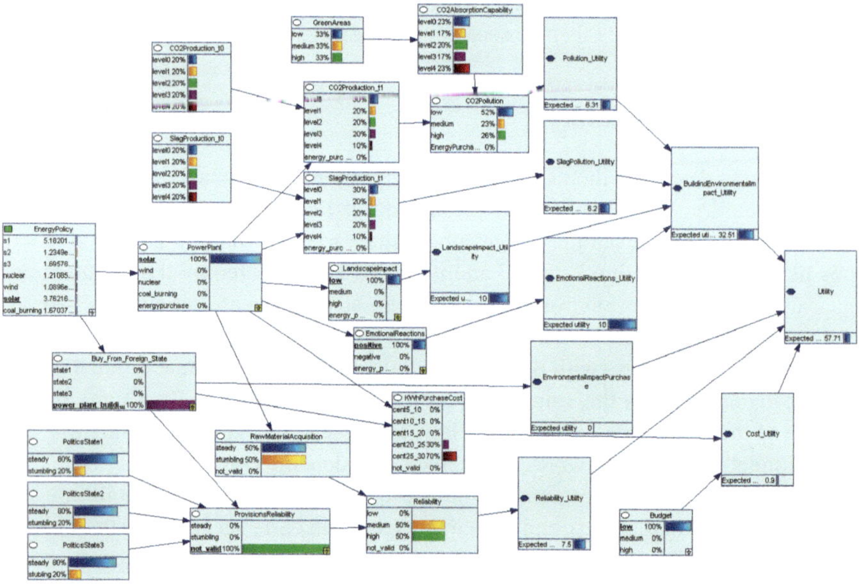

Fig. 7 Best Energy Policy for the described scenario

agent starts a decisional process that brings it to the decision of building a new power plant. The probabilistic reasoning leads to the consideration that this solution would generate a relevant increase of the pollution level, with a probability of 0.7. On the other side, the decision carried on by the environmental policies agent prospects a drop in pollution level.

This conflict situation is detected by the mediator agent, that decides to decrease the pollution level and constraints the energy policies agent to take another decision, according to the constraint. The agent therefore suggests to build a solar power plant which leads to an estimated value of pollution as "low" with a probability of 0.52. This scenario is shown in figure 7.

5 Conclusion

In this chapter we have presented a work-in-progress Multi-Agent System, with embedded knowledge representation and probabilistic reasoning capabilities, aimed at supporting decisions in city administration scenarios. The use of such an agent that suggests the best strategies to adopt, even in uncertainty situations, in order to manage a virtual city is a good test-bench for using knowledge representation and reasoning in real life economics and management. As a matter of fact the approach is general and it can be applied on more general fields. Future developments of the architecture presented in this chapter will regard the definition of a more detailed model, a better analysis of different strategies, and a deeper assessment of the effectiveness of the approach.

References

1. Power, D.J.: A brief history of decision support systems. DSSResources.COM (2003)
2. Klein, M.R., Methlie, L.B.: Knowledge-Based Decision Support Systems: with Applications in Business, 2nd edn. John Wiley and Sons, Inc., Chichester (1995)
3. Marakas, G.M.: Decision Support Systems in the Twenty-First Century. Prentice-Hall, Inc., Englewood Cliffs (1999)
4. Turban, E.: Decision Support and Expert Systems: Management Support Systems, 3rd edn. Prentice Hall PTR, Englewood Cliffs (1993)
5. Sycara, K.P.: Multiagent systems. AI Magazine 19(2), 79–92 (1998)
6. Wooldridge, M., Jennings, N.R.: Intelligent agents: theory and practice. Knowledge Eng. Rev. 10(2), 115–152 (1995)
7. Weiss, G.: Multiagent Systems: A Modern Approach to Distributed Artificial Intelligence. MIT Press, Cambridge (1999); Within the agent oriented abstraction. In: Proc. International Conference on Cyberworlds (CW 2004), pp. 224–231. IEEE Computer Society, Los Alamitos (2004)
8. Wooldridge, M.: Intelligent agents. In: Gerhard, W. (ed.) Multiagent Systems: A Modern Approach to Distributed Artificial Intelligence. ch. 1, pp. 27–78. The MIT Press, Cambridge (1999)
9. Hendler, J.: Agents and the semantic web. IEEE Intelligent Systems 21(2), 30–37 (2001)

10. Yang, Y., Calmet, J.: From the OntoBayes Model to a Service Oriented Decision Support System. In: International Conference on Computational Inteligence for Modelling Control and Automation and International Conference on Intelligent Agents Web Technologies and International Commerce (CIMCA 2006), p. 127 (2006)
11. Houari, N., Far, B.H.: Intelligent Agents based Architecture for Distributed Decision Support to Manage Projects Lifecycle Proceedings of ATS (2003)
12. Lesser, V.R., Erman, L.D.: Distributed Interpretation: A Model and Experiment. IEEE Transactions on Computers C-29(12), 1144–1163 (1980)
13. Chen, S.-H.: Software-Agent Designs in Economics: An Interdisciplinary [Research Frontier]. Computational Intelligence Magazine 3(4), 18–22 (2008)
14. Olsen, R.: Computational Finance as a Driver of Economics [Developmental tools]. Computational Intelligence Magazine 3(4), 35–38 (2008)
15. Rice, A.J., McDonnell, J.R., Spydell, A., Stremler, S.: A Player for Tactical Air Strike Games Using Evolutionary Computation. In: IEEE Symposium on Computational Intelligence and Games, pp. 83–89 (May 2006)
16. Chen, S.: Editorial: Computationally intelligent agents in economics and finance. Inf. Sci. 177(5), 1153–1168 (2007), `http://dx.doi.org/10.1016/j.ins.2006.08.001`
17. Tseng, C.-C.: Influence diagram for investment portfolio selection. In: Proceedings of 7th Joint Conference on Information Sciences, pp. 1171–1174 (2003)
18. Global Economics Game, `http://www.worldgameofeconomics.com/`
19. SimCity, `http://simcitysocieties.ea.com/index.php`
20. Kim, K.-M., Hong, J.-H., Cho, S.-B.: A semantic Bayesian network approach to retrieving information with intelligent conversational agents. Information Processing and Management 43, 225–236 (2007)
21. Symeonidis, A.L., Chatzidimitriou, K.C., Athanasiadis, I.N., Mitkas, P.A.: Data mining for agent reasoning: A synergy for training intelligent agents. Engineering Applications of Artificial Intelligence 20(8), 1097–1111 (2007) ISSN 0952-1976, doi:10.1016/j.engappai.2007.02.009
22. da Silva, J.C., Giannella, C., Bhargava, R., Kargupta, H., Klusch, M.: Distributed data mining and agents. Engineering Applications of Artificial Intelligence 18(7), 791–807 (2005) ISSN 0952-1976, doi:10.1016/j.engappai.2005.06.004
23. Cao, L., Zhang, C.: F-trade: an agent-mining symbiont for financial services. In: Proceedings of the 6th international Joint Conference on Autonomous Agents and Multiagent Systems AAMAS 2007, Honolulu, Hawaii, May 14-18, pp. 1–2. ACM, New York (2007), `http://doi.acm.org/10.1145/1329125.1329443`
24. GeNIe environment for decision-theoretic model, `http://genie.sis.pitt.edu/`
25. Alice Chatbot, `http://www.alicebot.org`
26. Jensen, F.B., Graven-Nielsen, T.: Bayesian Networks and Decision Graphs Series: Information Science and Statistics, 2nd edn., vol. XVI, p. 448 (2007) ISBN: 978-0-387-68281-5
27. Eugster, P.T., Felber, P.A., Guerraoui, R., Kermarrec, A.: The many faces of publish/subscribe. ACM Comput. Surv. 35(2), 114–131 (2003), `http://doi.acm.org/10.1145/857076.857078`
28. Weka, `http://www.cs.waikato.ac.nz/~ml/index.html`

Agent-Based Search and Retrieval in Virtual World Environments

Joshua Eno, Susan Gauch, and Craig W. Thompson

Abstract. Virtual world and other 3D Web content has been treated as a separate domain from the traditional 2D Web, but is increasingly being integrated with broader web content. However, search engines do not have the ability to crawl virtual world content directly, making it difficult for users to find relevant content. We present an intelligent agent crawler designed to collect user-generated content in the Second Life and related virtual worlds. The agents navigate autonomously through the world to discover regions, parcels of land within regions, user-created objects, and other users. The agents also interact with objects through movement or 'touch' to trigger scripts that present dynamic content in the form of note cards, chat text, landmark links, and web URLs. The collection service includes a focused HTML crawler for collecting linked web content. The experiments we performed are the first which focus on the content of a large virtual world. Our results show that virtual worlds can be effectively crawled using autonomous agent crawlers that emulate normal user behavior. Additionally, we find that the collection of interactive content enhances our ability to identify dynamic, immersive environments within the world.

1 Introduction

Virtual worlds like Second Life[1] have often been regarded as separate applications existing outside the realm of web content. However, as virtual worlds have gained in popularity, users have begun to link to them directly from the traditional web. From the other direction, virtual worlds contains many links back out to the web because it presents an accessible way to present text content. The closed nature of Virtual Worlds is also breaking down as technologies are developed that allow

Joshua Eno · Susan Gauch · Craig W. Thompson
University of Arkansas
e-mail: {jeno,sgauch,cwt}@uark.edu

[1] www.secondlife.com

Soro et al. (Eds.): Inform. Retrieval and Mining in Distrib. Environments, SCI 324, pp. 125–143.
springerlink.com

users to travel directly between Second Life and emulator worlds based on the open source OpenSim project[2].

Although most locations in virtual worlds are stylized versions of reality, some have worked to recreate virtual mirror worlds. The potential for realistic virtual mirror worlds becomes clear when we consider the possibility of linking to them from geospatial browsers like Google Earth or Microsoft Virtual Earth. It is easy to envision zooming in on a location, viewing the outside of a business in Street View, then entering the business, browsing, and purchasing items in a virtual world version. Considering that Google has already dabbled in the Virtual World space briefly with the Lively[3] service, it would not be surprising if a future version of Google Earth provides such functionality directly. However it happens, businesses and individuals who create content for this mirror world will want to create the kind of dynamic, immersive environments that now exist in virtual worlds.

Although we cannot predict whether Second Life, Google Earth, some open 3D web platform, or some combination of these will become the dominant platform, we can be confident that it will incorporate and expand upon current virtual world functionality. However, this functionality presents several challenges for content collection, even when a single authority controls access to the content within the world. If, as seems more likely, there is no central database with content to gather content from, then a system much like today's web crawlers will be needed to gather the content.

One of the challenges to crawling a 3D world is that the content is only presented to users as they move close to the content's location. However, movement within most worlds is constrained in terms of speed and direction. Although both of these constraints could change, they provide two functions that are desirable and that are unlikely to go away. On one hand, they provide a sense of realism to the worlds that providers strive for. Games such as World of Warcraft have intentionally limited movement speeds to maintain the feeling of immersion. Jumping around randomly within a 3D space can be disorienting. The second advantage is the natural limit on bandwidth consumption that these constraints enforce. For these reasons we find it unlikely that future virtual worlds will relax these constraints, so a crawler that moves and navigates naturally through the world will be necessary.

The second challenge is that content is more compelling when it reacts dynamically to the actions of users. Although Opensim and Second Life allow content creators to attach names and descriptions to virtually everything in the world, most of this text is never seen by users. Instead users are presented with dynamic content such as note cards, landmarks, text messages, and web pages when they enter an area or click on an object that looks interesting. Collecting this content has not been attempted by any existing search services or research. Gathering it is a difficult problem to solve, requiring a crawler to intelligently identify, navigate to, and interact with dynamic objects.

The final challenge we deal with is accomplishing movement and interaction without disrupting either the server or other users. Just as a web crawler must

---

[2] `www.opensimulator.org`

[3] `www.lively.com`

avoid crashing a server by limiting the number of requests to any given server, we must limit our demands on the virtual world servers we visit. If we fail to do this, our agents will be banned and our IP address blocked from connecting. An issue unique to virtual worlds is the reality that our crawlers inhabit the world with tens of thousands of other avatars. We would like our agents to be as unobtrusive as possible to avoid offending or distracting these other users, lest they be reported and banned.

Our approach to addressing these problems is to emulate the client protocol with intelligent crawling agents. These agents interact with the content just as a normal avatar would, move around the world, touch objects, and record the content presented. As much as possible, we emulate normal user behavior both to make it more likely that we will discover dynamic content aimed at normal users and so that we blend in from both the server and other user perspectives. Our goal is to determine how much of the name and description information currently indexed by the internal Second Life search engine we can retrieve through an external crawl. We also want to determine how much extra information is available beyond the entity names and descriptions and evaluate how useful this information is in determining the content and quality of a location.

2 Related Work

The virtual world crawler combines several aspects of traditional web search, multimedia search, and 3D Web technologies. Collecting and searching text-based content parallels traditional web crawlers in processing and retrieving results. The interactive component has parallels with the discovery and indexing of hidden web content such as databases only accessible from form interfaces. The problem of exploring a world, virtual or otherwise, has been studied by robotics and artificial intelligence researchers. Finally, inasmuch as this work is at root an exploration of the content of virtual worlds, there has been related work in studying usage and content of virtual worlds, though in aggregate terms rather than for the purpose of search and retrieval.

2.1 Traditional Web Crawlers

We have divided the previous work on web crawlers into those that crawl the traditional flat web by following hypertext links and those that attempt to collect information from the hidden web, which can only be accessed through web forms or other dynamic interfaces.

2.1.1 Flat Web Crawlers

The flat web consists of HTML content, either static or dynamically generated, that is accessed by requesting an HTTP URL obtained from a hyperlink (as opposed to sending a GET or POST request through a web form). The research in web crawlers has consisted of improving component parts such as agents, indexer,

URL manager, inverted index, and query server. The agents receive a set of URLs from the URL manager and send the retrieved content to the indexer. The indexer parses out any links and sends them to the URL manager. The query server receives user queries, submits them to the index, and formats the results to present them to the user.

Current agents gather content by sending HTTP requests to servers just like a normal web browser, retrieving the full content of a page, and sending it on to the indexer process. The Harvest system [1] took an approach of using Gatherers running on the servers to generate summary contents for documents. While the distributed approach taken by Harvest was useful for a wide range of information sources and for server performance reasons, it failed to gain traction compared to the approach taken by Pinkerton's WebCrawler [2] and later crawlers. The WebCrawler approach was to use standard HTTP protocols to retrieve the entire contents of a page, rather than generated summaries. Pinkerton found that full-text indexing produced much better results than the competing approach of summaries or a combination of titles and anchor text in links referring to the page as found in [3].

While much of the paper was devoted to the novel PageRank algorithm, there was also a discussion of the crawler, indexer, and URL manager aspects of the Google search engine [4]. The Google crawlers would each have approximately 300 concurrent connections open to retrieve pages. After the Google paper, there was renewed interest from the research community in addressing the problems of fast DNS lookups, politeness in ordering requests, setting URL priorities to collect important or fast-changing sites early or frequently in the crawl, duplicate detection, and other issues [5-7]. The stated intention of the IRLbot crawler [8] is to scale to a very large collection (hundreds of billions of pages) while sustaining a page processing rate of at least thousands of pages/s while using a fixed amount of resources on a single server.

2.1.2 Hidden and Dark Web Indexing

While traditional web crawlers build their collection through following links from a seed set of documents, this fails to discover content that is not directly linked through a hyperlink URL. There are three information classes that are missed in this process: interactive dynamic content, uninterpretable content, and form-accessible content. Interactive dynamic content refers to client or server content that is presented in response to browsing events, such as a mouse-over or button click. Uninterpretable content refers to content such as Flash or Second Life that may be linked or embedded in a page, but cannot be interpreted by the indexer. Finally, form-accessible content refers to content that is presented in response to form inputs on a web page.

The current generation of mainstream web crawlers do not attempt to gather interactive dynamic content, instead expecting web authors to put anything they want indexed into plaintext, often pages optimized for search engine crawlers. Likewise, current crawlers do not include tools to retrieve content from within Flash or virtual worlds once they discover such a resource. This may change soon for some uninterpretable content. Adobe has added functionality for Flash and

Shockwave that will enable content creators to expose content information to search engine crawlers [9].

Content behind web forms has received much more research attention than other forms of hidden web content. One of the first papers on hidden web crawlers by Raghavan and Garcia-Molina [10] addressed the problem of generating valid queries for web form database interfaces. The focus of their work was to use human assistance to create hidden web crawlers for task-specific rather than generic applications. Since then, researchers have attempted to minimize or eliminate the need for human input [11,12].

There have been two approaches to the problem of what to do with content behind a site's search form. One approach, taken by the ProFusion search engine [13], is to send queries to a set of search utilities, then process and merge the results before presenting them to the user. Another approach is to mine the searchable index and import the results into a larger index. Ntoulas, Zerfos, and Cho demonstrated a probabilistic approach to maximizing database coverage while minimizing the number of queries [14].

2.2 Multimedia Search

3D environments commonly include image and video content in addition to 3D models, often in combination such as when a face of a prism displays an image or video. In this sense, 3D environments present a version of multimedia search with the addition of extra proximity and orientation information. There are two broad areas of multimedia search we examine: image and video search, and 3D model search.

The problem of indexing and searching image/video databases has received a lot of interest over the last two decades, and numerous techniques have been proposed to extract meaningful information from this multimedia data. Early work in this area was summarized by Yang [15] and current research and future trends are presented by Lew [16] and Datta [17]. We will gather color and texture descriptors similar to those used by QBIC [18] and VisualSEEK [19] to search images/videos to classify texture maps and other images.

One difficulty in image and video search is that queries are often text, making direct comparisons to the image content impossible. As such, there has been research into using text associated with or extracted from multimedia content to enable text-based search. An early system involved extracting content hints from HTML 'alt' attributes, label text, or other context clues [20]. Although this is still common, newer systems have utilized voice recognition to extract text from videos and optical character recognition to obtain text content more directly [21].

Searching for 3D models has been an active field recently, with good summaries given by Iyer et al [22] and Tangelder and Veltkamp [23]. In addition to systems that search against models directly, other systems extract text content extracted from surrounding titles or other HTML hints to support text-based queries. As tagging systems have become more prominent in the web, some work has been done to enable tagging of 3D objects to aid retrieval [24]. Opensim and Second Life do not natively provide a means for third-party tagging of content, but we have experimented with creating a tagging application that users can use to tag objects created by others.

2.3 Virtual World Research

The integration of user-created content into virtual worlds is increasingly common. The first virtual worlds consisted of static content programmed into the code. In the late 1980s, programs like VMS Monster[4] began to allow users to create puzzles and areas for other users to interact with. These systems presented for the user tools for content creation, including scripting languages. Because user-designed graphics present a much greater challenge both in their creation and viewing, similar functionality did not come to 3D graphical worlds until 1997 in the form of Active Worlds[5], which is now accessible through a browser plug-in. The current market leader in terms of active users is Second Life, a virtual world created by Linden Lab[6]. The client/server protocol used by Second Life forms the basis for the OpenSim project.

There are three sub-areas of virtual world research that relate the proposed research. The first is the technology used to implement the virtual worlds themselves. This technology determines what functionality a crawler can implement, and indicates future features a crawler may need to take advantage of. The second area involves the current approaches to search and ranking in the monolithic server world. Finally, the area of exploring synthetic worlds has been a research area as a means to test real-world exploration techniques in a far less expensive simulation.

2.3.1 Virtual World Technology

Depending how broadly one defines a virtual world, there are few commonalities among all virtual world technology implementations. A virtual world such as The Sims consisted of a single program on a PC. In contrast, the current king of massively multiplayer online role playing games, World of Warcraft, uses a multi-tiered client server technology, where players can usually only see other players on their own server, but can compete against players on a subset of other servers for player-vs-player combat. This section will explore the technologies used along the full spectrum of size and dynamic interactivity among online, multiplayer virtual worlds.

In client/server scaling to the size of a large or crowded world requires some level of partitioning to limit the demands on a single server. The strategy for this generally depends on the acceptability of multiple instances of the world and the computational cost of each concurrent user. Second Life has a peak concurrency of just over 70,000 users [25]. In contrast, World of Warcraft has topped 1.5 million concurrent users on US and EU servers, but the peak concurrency for any

4 `www.skrenta.com/monster`

5 `www.activeworlds.com`

6 `lindenlab.com`

individual realm is on average around 2,300[7]. Beyond separating users into separate servers, servers are also partitioned along geographic and task boundaries. So for instance, a server may handle tracking positions and actions of all avatars within a grid area or world region. Other servers may handle the item database or authentication tasks for the entire world. It is difficult for a client or user to determine where these divisions exist, other than noting from experience that when some things fail, others continue to work.

2.3.2 Search in Monolithic Virtual Worlds

While search in worlds without player-created content, such as World of Warcraft, devolves to a simple database query, communities within the games have developed social information sharing sites to augment that information. The content itself is accessed by decoding the static binary game files stored on a client computer. These are then posted to sites such as Thottbot[8] that allow users to comment, providing screenshots, opinions, and extra information for completing quests, etc. Community opinion and knowledge is an important resource for search services, but may not scale well to dynamic worlds with vastly more content that changes frequently.

Searching through player-created content and ranking results effectively is a difficult problem even when the search program has full information. The current Second Life search within the game allows a user to perform keyword searches against classified ads, events, land sales, people, groups, and places. The search service is accessible both through the game and via a web interface, with similar or identical results.

Parcels returned by a Second Life place search are found based on the parcel name and description as well as the name and description of objects on the parcel that are flagged for search indexing. The results are ranked in order of descending dwell. Dwell is a popularity metric that resembles the PageRank of a web page. Parcels can have dwell added based on avatar traffic at the parcel, avatar picks pointed at the parcel, classified ads referring to the parcel (allowing users to buy popularity), and other factors[9]. Because avatar traffic is the largest component of dwell, landowners wishing to rank high on search results have resorted to several strategies to inflate these numbers, including paying people to sit in chairs at their parcel and creating avatar bots to masquerade as legitimate visitors.

One useful aspect of the Second Life place search is the ability to see a list of names of objects located at a given location. This gives searchers an indication of the kind of things they are likely to see if they visit the location. However, because users rarely see the names and descriptions of objects, these text fields have become a place to pack keywords where they will not detract from the content the user sees. In one instance, we found every section of the floor in a building had names and descriptions packed with the keywords the owner wanted to associate

[7] `www.warcraftrealms.com`

[8] `thottbot.com`

[9] `wiki.secondlife.com/wiki/SearchAPI`

with the parcel, rather than a more accurate name of 'floor'. This problem is endemic to search engines that rely on content that the average user does not see, as evidenced on the web with the now-nearly-useless meta-keyword.

The OpenSim search service is still a work in progress. The current system provides a means to take a snapshot of the content of a location and expose it through an external API [26]. The snapshot only captures name and description content information. This would allow for a centralized collection of contents from all OpenSim worlds created directly from API calls. In both Second Life and OpenSim it is possible to associate an image with a location that will be presented to the user with the search result. MetaverseInk is a commercial company that has implemented a search service based on both OpenSim snapshots and SecondLife content extraction. Their approach appears to be similar to ours for collecting internal content, without the benefit of dynamic content. MetaverseInk has focused on presenting their search with graphics and other context that allows users to gain more knowledge about the results than the more text-based and minimal Second Life search interface. Another feature they have built into their search engine is privacy controls that allow landowners to exclude all the crawlers by adding a single account to a banned list. They also do not collect information from any area marked as residential.

There has been some research into creating avatar crawlers for the Second Life world, but the goal has primarily been to explore user actions in the world, rather than to study or collect the content of the world. La and Pietro created a crawler to study user movement within the world and found that it was similar to patterns found in real-world studies [27]. Varvello et. al. created avatar agents to study user behavior with the goal of learning how to better optimize the client/server communication [28].

2.3.3 Exploration

Because our crawlers will need to navigate to within the search radius of every possible inch of the world, this research presents a challenging world exploration problems in obstacle avoidance, path planning, and crawler coordination. The closest parallel to our current work is research on navigating virtual worlds by Gayle and Manoche [29]. Their system was designed to guide Second Life avatars through the world using global high-level objectives and local obstacle avoidance techniques, with the goal being collision-free motion. They used an earlier version of the same OpenMetaverse[10] client library we use, but only dealt with navigation in a static 2D walking surface space.

A similar two-tiered approach to route planning and obstacle avoidance is used in [30]. A social forces model is used to provide collision avoidance at the local agent level, but the global path planning uses a modified Voronoi graph approach. In a traditional Voronoi graph, edges represent boundaries between objects that are equidistant from two objects. An avatar travelling through this space could follow these edges to maintain a maximal distance between obstacles.

[10] `openmv.org`

3 Our Approach

3.1 Crawler Architecture

To approach the problem of crawling a virtual world is to emulate the client/server protocol using the open source LibOpenMetaverse library. We tested the crawler agents by performing a set of experiments to test how successfully they traversed the world, how much static content they gathered, and how much dynamic content they triggered and collected.

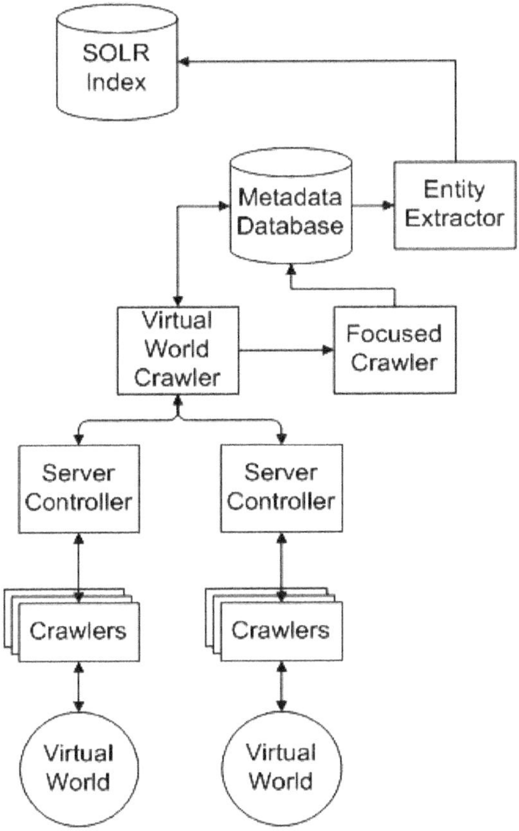

Fig. 1 Crawler Architecture

The server manager starts the crawls, tasks individual agents with crawling specific regions, keeps track of completed and queued regions, and signals the various components to shut down if a crawl needs to be terminated. The controller builds a list of regions to crawl either by retrieving a stored list from the database or by tasking a connected agent with requesting existing grid regions from the world's grid manager. This list is checked against the list of recently crawled regions to determine if any new or outdated regions need to be crawled. If so, the server manager dispatches a crawler to the region.

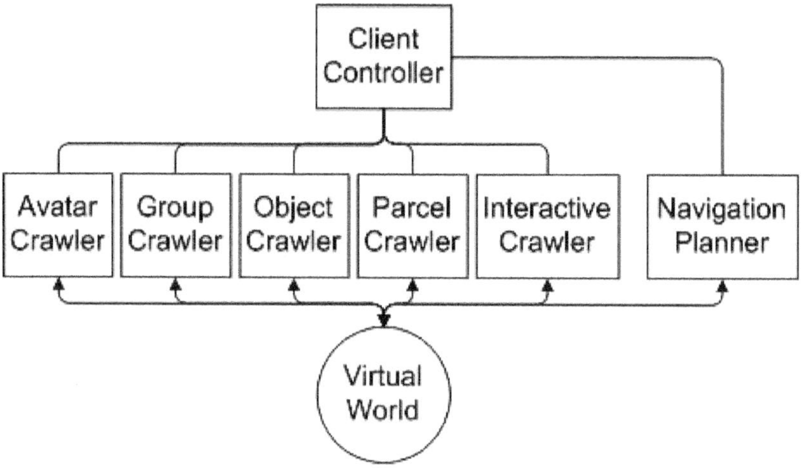

Fig. 2 Client Controller Architecture

As shown in Figure 2, the crawler agents have several subsystems designed to gather information of different types and to provide navigation support. Each component independently implements handlers for client events related to different types of content. They work independently and can be turned on or off depending on what type of crawl we are performing. For instance, because avatars have lists of groups they belong to and groups have lists of avatars who are members, it is possible to perform a crawl of the virtual world social network simply by starting with a seed avatar or group and requesting all of the groups and avatars associated with those seed values. The process continues until no more groups or avatars can be found. For the location-based crawls described in this paper, we turned off group crawling so that this extra request traffic would not add to the server load.

3.2 Agent Behavior

When a crawler agent receives a call to crawl a region, it attempts to teleport to the region and flags the region as inaccessible if the teleport fails. If it succeeds, it performs one of several crawling steps depending on the current crawler configuration. If the configuration calls for object positions to be stored, then as soon as the agent arrives in a location, it will begin storing the object positions as they are automatically sent from the server. If the configuration calls for the name and description properties to be stored, each root object is added to a queue for object properties requests which also begin immediately.

Once a client arrives in a region, there are two major subsets of agent behaviors that require planning and strategies to mitigate the impact on the server and other users. Navigation requires obstacle avoidance and path planning. It may also impact other users' experiences if the crawler avatar's movement is distracting or violates accepted behavioral norms. We would like our crawlers to be as unobtrusive as possible. The second behavior is collecting interactive content, which may also impact other users by causing the entire region to slow down as it executes interactive scripts or by causing the entire region to change as scripts move or change objects.

3.2.1 Navigation

As object position information is received by the client, it is added to an octree index with a minimum cell size configurable based on the memory available and precision required. A list is maintained in each cell to store the primitives within the cell. The index is used to determine if navigation waypoints are clear of obstructions and as a navigation mesh for path planning. Because object locations tend to cluster around ground level, the octree cells for higher altitudes tend to be large and empty, making path planning less computationally intensive.

The first active step after arriving in a region and preparing to navigate is to request all of the region parcels and their dwell values. As soon as all the parcels are received, the agent checks to see if a region survey is called for. If so, it uses its navigation and movement modules to get a set of clear waypoints that it can traverse without running into any objects or being stopped by parcel access restrictions. We always aim for a location in the air because our agents can move more quickly flying than walking, even if no obstructions exist. The agent flies directly up to the given altitude and searches in an expanding spiral over the waypoint grid. The size of the grid is a configurable parameter that we can vary to trade off speed and comprehensive coverage.

Finding a way to reach unobstructed airspace to do the region survey, while a small subset of the general movement problem, can still present difficulties. Many regions specify where an incoming agent will arrive regardless of where the agent specifies. In some cases, the arrival point is inside buildings or under structures that prevent the agent from directly reaching the first waypoint. Navigating out of such structures can be expensive at best, but in some cases simple movement will still fail to get the agent out of the room. For these cases we have several strategies. One approach is to search for a location that will allow direct teleportation, but these are relatively rare. Another option is to attempt to discover an object that will transport the avatar to a different location, which is often the case when an avatar is placed in a sealed room on arrival. A final option is to use a scripted object attachment to apply upward force to the avatar and 'break through' the roof. This only works if the parcel owner has not disabled visitor scripts. If none of these options is available, the region is flagged as inaccessible.

3.2.2 Interactive Crawl

After the region survey is completed, the interactive crawl begins. During the survey, any root primitives with a scripted touch action are added to a list of interactive primitives. We do not begin to touch these until after all movement is stopped so that we can determine which scripted objects give us content based on touching rather than based on movement. Objects are touched one at a time, with a short interval in between so that we do not overtax the server that must run the scripts to react to each touch event.

If the object reacts by offering a note card or landmark and the agent accepts, the content is added to the agent's inventory. Although the agent is notified when the inventory is added, there is no indication from the notification where the inventory came from. This restricts our flexibility in responding to multiple objects simultaneously. Rather than wait for the entire process of touch-offer-accept-receive to play out for each object, we keep a queue of offers and respond to each one sequentially as content is received. Since most objects do not make a content offer but instead respond to touching with some other action, we can touch multiple objects at once, initiating the offer response only to those objects that respond with a content offer.

Throughout this process we take several measures to lessen the burden on the region server. Requesting object properties is somewhat problematic it terms of being polite to the servers. Regions can have as many as 15,000 primitive objects, each of which could require a separate call to request the properties which include the name and description. Because most users do not request this information, the servers do not appear to be optimized to handle a large number of requests. One way we limit the number of requests is by requesting several objects in a single request packet.

The other way we minimize requests is by only requesting properties for the root primitive of a linked set. Primitives objects can be linked together into larger objects such as chairs, houses, or vehicles. In such link sets, the name and description that are displayed when a user looks at the details are the name and description of the root primitive. Although the child primitives can also have this information, it is harder for a normal user to get at it so it is often left unmodified by content creators. This optimization significantly decreases the number of object requests we make. In addition to the limits we place on object property requests, we only touch root objects when searching for interactive content. It is relatively rare for a child primitive to have a different scripted reaction than the parent object.

Finally, we prevent our agents from entering and leaving regions quickly by setting a minimum visit time to each region. This is done to minimize the amount of information about agent movement that must be exchanged between region servers and the central agent management and authentication system. After all of the configured crawling steps have been completed and the minimum time has passed, the agent notifies the server manager that it has finished crawling the region. The server manager then adds it to the list of crawlers that are ready to crawl another region, and the dispatch thread instructs it to crawl a new region.

3.3 Experimental Design

We ran two experiments that would both determine the effectiveness of our crawler and help us characterize the content that currently exists in Second Life, since it provided the largest single collection of user-created content. The first test is a broad survey of as many regions in Second Life as we could identify and visit. The broad survey was conducted over the course of several days in April, 2009 by a set of ten crawler agents. The only content collected was the region parcel information and any object information found near the agent teleport landing point. The minimum visit was set to one minute. Our goal with the survey was to get a broad sense of how many parcels are accessible to the public, how many object descriptions are available for collection, how many scripted objects are typical, and other broad measures of the Second Life world.

The second experiment involved performing detailed surveys of a small sample of regions that were representative of the kinds of regions we encountered in the survey. This was closely supervised to ensure that the crawler agent did not interfere with other users and to identify any problems with obstructions or other unanticipated issues. The agent performed every step of a full crawl, from the initial parcel information request to the movement survey and the interactive survey. Our goal was to get a sense of how much extra information is gained by moving around and interacting within a region. We would also like to see if our anecdotal sense of interactive content being useful is backed up by a detailed examination of the actual content received.

4 Survey Crawl

We obtained a list of 27,055 distinct regions using the OpenMetaverse utility that retrieves every simulator that shows up on the mainland grid map. This includes both Linden Lab administered servers and privately leased islands hosted by Linden Lab but controlled by private users, educational institutions, or corporations willing to pay the setup charge and yearly fee. Of the identified regions, we successfully connected to 23,621 and found 3,362 that denied access for a variety of reasons summarized in Table 1.

In 1,917 cases, the region owner had restricted access by anonymous users. Another 1,311 failed because no valid parcels could be found to teleport to. We think this indicates an outdated region, but it may also indicate a new region that has not been set up by the owner or that all of the parcels restrict anonymous or non-affiliated users. We also found 40 regions that were Linden Lab orientation regions that do not allow users to go back after they have entered the wider world. We only found 4 regions that were full and would not allow more avatars, and the remainders were otherwise flagged as invalid destinations, possibly because the server was down or had been removed before we could attempt to visit. Even with the high numbers of restricted or inaccessible parcels, we still successfully visited 87% of identified regions.

Table 1 Teleportation Failure Reasons

Failure Reason	Number	Percent of Total
No Access	1917	7.1%
No Parcel	1311	4.8%
Help Island	40	0.1%
Full	4	0.0%
Invalid	91	0.3%

While visiting the regions, we gathered some statistics on the number of agents, objects, active scripts, and permissions for the visited simulators. Table 2 summarizes our findings on avatars and objects.

Table 2 Region Population Statistics

Attribute	Max	Mean
Avatars	97	3.47
Objects	15000	6597.9
Scripted Objects	7696	336
Active Scripts	60701	2747

Although objects can only be detected if they are in close proximity to the crawler, stationary crawlers were more successful at detecting a higher proportion of all avatars within a region. We found an average of 2.8 server-reported avatars other than our crawler per region, slightly more than the actual population based on avatars detected by the crawler. The discrepancy is probably caused by avatars moving in and out while our avatar is present, resulting in more sightings than show up at the single time we record the simulator statistics. There also seems to be a lag in the server updating its agent count, since we found a small number of regions with zero listed agents, even though our crawler was connected. By way of comparison, we found only 26% of the objects in our survey crawl, suggesting that simulators distribute avatar information in a wider radius than object information. Another possible explanation is that avatars tend to congregate near the teleport landing point, while objects are more uniformly distributed. The number of scripted objects is fairly low for the typical region. Only 541 regions had more than 1000 scripted objects. This is not surprising, since it is far easier to build static content than to create dynamically scripted content.

The other focus of our survey crawl was the parcels on each region. We found a total of 311,443 parcels, or an average of 13.2 parcels per region, with the average area of 5030 square meters. Because both the recorded average area and the expected area based on total area divided by number of parcels agree, we conclude that we have obtained a full survey of the regions we surveyed. Of the parcels surveyed, 87% do not allow direct teleportation, suggesting that our approach of moving rather than teleporting within regions will be necessary to reliably collect content from most locations.

MetaverseInk has found that owners of residential parcels would prefer not to have their content collected and searched, so we analyzed how frequently parcels are categorized as residential, and more generally what the distribution of category types is between parcels.

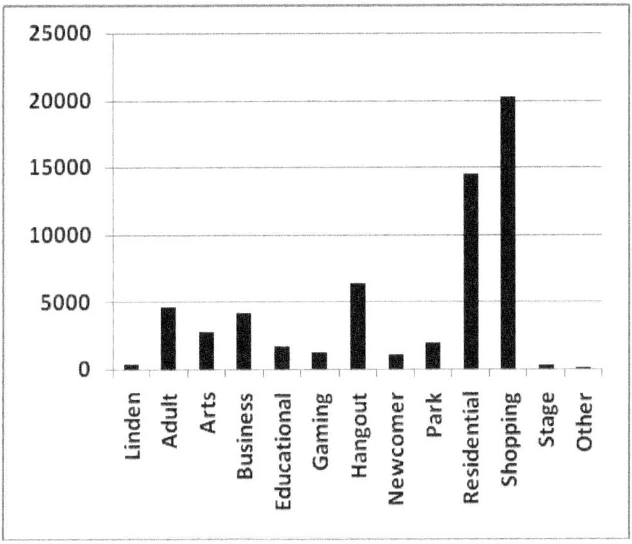

Fig. 3 Parcel Categories

Figure 3 does not include the majority of parcels, which were categorized as either none or any. Only 19% of parcels had a specific category assignment, with a total of only 4.6% flagged as residential. Even with privacy-sensitive collection, that leaves most of the world open to crawling.

For the most part we found the parcel names and descriptions to be useful, though there were a fair number of blank or repetitive values. About two thirds of parcel names were unique, and only 22,963 (7%) had names left blank. There appear to be several individuals or groups that purchase multiple parcels of land throughout Second Life and name them identically. Descriptions are more frequently left blank, with over two thirds left blank or the default value. Other than those, the descriptions had about as many duplicates as the names did. Many of the duplicates were commercially oriented, with "High Visibility & Traffic! Drive Customers to YOUR Biz!" being a typical example.

Although we did not move throughout the regions we visited during the survey, we did still record the information for objects we could see from the teleport landing point. We found nearly 48 million objects, of which 80% were children in a link set. Of the root objects for which we retrieved object properties, we found that creators give names to a majority (73%) of the objects they create. Descriptions are less commonly provided, with only 31% of object properties having a useful description.

A final aspect we examined during our survey was the number of messages, note cards, and landmarks we were offered during the crawl. Although we did not touch objects or otherwise solicit note cards and landmarks, we received welcome note cards at a rate of one for every 64 regions visited. Landmarks were more common, coming in at a rate of one for every 26 regions. The note cards provided quite a lot of useful information, including region FAQs, rules, and descriptions. They averaged 2320 characters long, with the longest stretching to nearly 15,000 characters. The landmarks pointed to a broad range of locations, with more destination regions represented than source regions.

5 Detailed Crawl Results

Our detailed crawl focused on two regions that were typical of the average region and three regions that looked interesting because of high popularity or a high level of scripting. For each region, we had both a human-controlled supervisory avatar and an autonomous crawler agent present. We saw no signs of either the server or other users being disrupted by the crawler.

The agent performed a full crawl in each region, including moving around the regions and triggering any objects with touch-sensitive scripts. While we found that only 26% of a region's objects are discovered through observations at the teleport landing point, we were able to find at least 97% of the objects using a 10 meter distance between subsequent passes. Using a 20 meter grid required three independent passes to achieve the same level of coverage. In some instances we found more objects while moving through the region than were reported by the simulator statistics for the region, yet all of the objects found had unique identifiers and were legitimate objects. We suspect these are transient objects created then destroyed by scripts.

Doing detailed crawls of regions was a time-consuming process. Simply traversing the more than 6,000 meters required by the crawl takes 12 minutes. Regions with thousands of touchable objects add several minutes to that. We estimate that 20 agents would require 16 days to crawl all known regions at the current rate.

Table 3 Interactive Crawl Resuls

Region	URLs	Note Cards	Landmarks	Text Messages
Quark	1	1	2	9
Horsa	1	12	10	1
Iml	15	13	0	0
Sliterary	27	7	45	15
REEF	0	1	0	2

We also found that our agents will need to learn to deal with more complex interactions than simple note card, landmark, and URL acceptance. Some scripts require a user to select between multiple options when presenting content. The ability to deal with these dialogs is not yet built into the OpenMetaverse library, so we will need to add it. Our other detailed crawl results are summarized in table 3.

Quark was chosen because it had the parcel with the highest dwell in the overall survey. When we did the detailed crawl, though, we discovered that the dwell was being artificially inflated by 87 zombie avatars floating in the sky above the building. The parcel attempting to inflate its dwell ranking was a commercial shop selling rather plain furniture. The simulator itself was stale and did not offer much interesting content.

Horsa was chosen because it had the highest dwell of any parcel with the educational category. It was a location devoted to teaching Second Life users how to create scripts, but despite that it had relatively few scripted interactions. Most of the content was presented embedded in images on the walls. There were however several useful note cards given to the crawler. It also had several other avatars present chatting and moving around, indicating that its high dwell was probably justifiably earned.

Iml was one of the most highly scripted regions and was designed by the Institute for Multimedia Literacy at the University of Southern California. It was the most highly complex and obstructed region we crawled, and required us to modify our altitude selection algorithm slightly to avoid the numerous airborne obstructions. Although we did find several useful URLs and note cards in the region, we also found that there was a lot of video content that our current crawler ignored. Although we could not gather the content of the videos, we have added a feature to note the presence of videos as an indication of a high level of interactivity at the location.

Sliterary and REEF were chosen because they were within 10 scripted objects and active scripts of the average values for all regions. As the table shows, similar characteristics can lead to very divergent contents. Sliterary was devoted to users interested in literature. It had note cards about literature, landmarks to such places as a simulated Globe Theatre of Shakespearean fame, and notifications of upcoming artistic events. REEF was a beach resort with just a few static buildings and very little interesting content. The only dynamic content the agent found was a note card explaining an available fishing game. Comparing these two regions reinforces our view that interactive content is highly valuable when determining both the content of a region and its intangible qualities. As a proxy for showing how interesting a location is, it is much more indicative than the easily gamed dwell metric. Although it too could be gamed, at least a search that relied on interactive content would be more transparent to users.

6 Conclusions

One thing the detailed crawl sampling indicates is that it is difficult to tell which regions have the most content before the agent performs the interactive crawl. After the crawl it was clear that the regions devoted to scripting, multimedia and literature provided much more content to visiting avatars. This dynamic content is all information that is hidden from current collection systems. Although making crawler agents intelligent enough to autonomously explore the world is difficult and it takes a long time to interact with each region, the benefits justify the effort.

We have demonstrated that a crawler that emulates normal user behavior can successfully collect both static and interactive user-created content. This extra interactive content is very useful in differentiating interesting, immersive environments from stale and static locations. We have shown that our agents can accomplish their tasks without adversely affecting either other users or the servers. As virtual worlds continue to grow and real-world services such as Google Earth and Microsoft Virtual Earth begin to incorporate even more user-generated content, we expect to find a need to add functionality to the agents to cope with varied content and environments. Going forward, we will be studying how best to integrate dynamic content with the static names and descriptions to provide enhanced search services. We will also be expanding the scope of dynamic content we collect to include scripted interactions and multimedia content.

References

1. Bowman, C.: The Harvest Information Discovery and Access System. Computer Networks and ISDN Systems 28, 119–125 (1995)
2. Pinkerton, B.: Finding What People Want: Experiences with the WebCrawler. In: Proc. 2nd Intl. WWW Conf. (1994)
3. McBryan, O.: GENVL and WWWW: Tools for Taming the Web. In: Proc. 1st Intl. WWW Conf. (1994)
4. Brin, S., Page, L.: The Anatomy of a Large-Scale Hypertextual Web Search Engine. Computer Networks and ISDN Systems 30, 107–117 (1998)
5. Heydon, A., Najork, M.: Mercator: A Scalable, Extensible Web Crawler. In: World Wide Web, vol. 2, pp. 219–229 (2004)
6. Boldi, P., Codenotti, B., Santini, M., Vigna, S.: UbiCrawler: A Scalable Fully Distributed Web Crawler. Software: Practice and Experience 34, 711–726 (2004)
7. Cho, J., Garcia-Molina, H.: Proc. 26th Intl. Conf. on Very Large Databases, pp. 200–209 (2000)
8. Hsin-Tsang, L., Leonard, D., Wang, X., Loguinov, D.: IRLbot: Scaling to 6 Billion Pages and Beyond. ACM Transactions on the Web 3(3) (2009)
9. Adobe Systems Inc.: Adobe Advances Rich Media Search on the Web (2008), `http://www.adobe.com/aboutadobe/pressroom/pressreleases/pdfs/200806/070108AdobeRichMediaSearch.pdf` (accessed January 14, 2010)
10. Raghavan, S., Garcia-Molina, H.: Crawling the Hidden Web. In: Proc. 27th Intl. Conf. on Very Large Databases, pp. 129–138 (2001)
11. Wu, W., Yu, C., Doan, A., Meng, W.: An Interactive Clustering-Based Approach to Integrating Source Query Interfaces on the Deep Web. In: Proc. 2004 ACM SIGMOD Intl. Conf. on Management of Data, pp. 95–106 (2004)
12. He, H., Meng, W., Yu, C., Wu, Z.: Wise-Integrator: An Automatic Integrator of Web Search Interfaces for E-Commerce. In: Proc. 29th Intl. Conf. on Very Large Databases, pp. 357–368 (2003)

13. Gauch, S., Wang, G., Gomez, M.: Profusion: Intelligent Fusion from Multiple, Distributed Search Engines. Journal of Universal Computer Science 2(9), 637–649 (1997)
14. Ntoulas, A., Zerfos, P., Cho, J.: Downloading Textual Hidden Web Content through Keyword Queries. In: Proc. 5th ACM/IEEE-CS Join Conference on Digital Libraries, pp. 100–109 (2005)
15. Yang, Z., Kuo, C.: Survey on Image Content Analysis, Indexing, and Retrieval Techniques and Status Report on MPEG-7. Tamkang Journal of Sci. and Eng. 2(3), 101–118 (1999)
16. Lew, M., Sebe, N., Djeraba, C., Jain, R.: Content-based Multimedia Information Retrieval: State of the Art and Challenges. ACM Trans. on Multimedia Computing, Communications and Applications 2(1), 1–19 (2006)
17. Datta, R., Joshi, D., Li, J., Wang, J.: Image Retrieval: Ideas, Influences, and Trends of the New Age. ACM Computing Surveys 40(2), 1–60 (2008)
18. Niblack, W., et al.: QBIC Project: Querying Images by Content, Using Color, Texture, and Shape. In: Proc. SPIE (2004), doi:10.1117/12.143648
19. Smith, J., Chang, S.: VisualSEEK: A Fully Automated Content-Based Image Query System. In: Proc. 4th ACM Intl. Conf. on Multimedia, pp. 87–98 (1997)
20. Ortega-Binderberger, M., Mehrotra, S., Chakrabarti, K., Porkaew, K.: WebMARS: A Multimedia Search Engine. In: Proc Intl. Society for Optics and Photonics, vol. 3964, pp. 314–321 (1999)
21. Yan, R., Hauptmann, A., Jin, R.: Multimedia Search with Pseudo-Relevance Feedback. In: Proc. 2nd Intl. Conf. on Image and Video Retrieval, pp. 238–247 (2003)
22. Iyer, M., Jayanti, S., Lou, K., Kalyanaraman, Y., et al.: Three Dimensional Shape Searching: State-of-the-Art Reviews and Future Trends. Computer Aided Design 5(15), 509–530 (2005)
23. Tangelder, J., Veltkamp, R.: A Survey of Content-Based 3D Shape Retrieval Methods. Multimedia Tools and Applications 39(3), 441–471 (2007)
24. Rodriguez-Echavarria, K., Morris, D., Arnold, D.: Web Based Presentation of Semantically Tagged 3D Content for Public Sculptures and Monuments in the UK. In: Proc. 14th Intl. Conf. on 3D Web Technology, pp. 119–126 (2009)
25. Au, W.: New World Notes: Second Life Concurrency Exceeds 70K – Is SL's User Growth Plateau at an End, Too? New World Notes (2008), `http://nwn.blogs.com/nwn/2008/09/second-life-con.html` (accessed January 15, 2010)
26. OpenSimulator.org, OpenSim.Region.DataSnapshot. OpenSimulator (2009), `http://opensimulator.org/wiki/OpenSim.Region.DataSnapshot` (accessed January 15, 2010)
27. La, C., Pietro, M.: Characterizing User Mobility in Second Life. Institute Eurocom Technical Report RR-08-212 (2008)
28. Varvello, M., Picconi, F., Diot, C., Biersack, E.: Is There Life in Second Life? Thomson Technical Report CR-PRL-2008-07-0002 (2008)
29. Gayle, R., Manocha, D.: Navigating Virtual Agents in Online Virtual Worlds. In: Proc. 13th Intl. Symposium on 3D Web Technology, pp. 53–56 (2008)
30. Sud, A., Anderson, E., Curtis, S., Lin, M., Manocha, D.: Real-Time PAth Planning in Dynamic Virtual Environments Using Multiagent Navigation Graphs. IEEE Transactions on Visualization and Computer Graphics 14, 526–538 (2008)

Contextual Data Management and Retrieval: A Self-organized Approach

Gabriella Castelli and Franco Zambonelli

Abstract. Pervasive computing devices are able to generate enormous amounts of distributed data, from which knowledge about situations and facts occurring in the world should be inferred for the use of pervasive services. However accessing and managing effectively such a huge amount of distributed information is challenging for services. In this paper after having outlined these challenges, we propose a self-organized agent-based approach to autonomously organize distributed contextual data items into sorts of knowledge networks. Knowledge networks are conceived as an alive self-organized layer in charge of managing data, that can facilitate services in extracting useful information out of a large amount of distributed items. In particular, we present the W4 Data Model we used to represent data and the self-organized approach to build Knowledge Networks. Some experimental results are reported to support our arguments and proposal, and related research work are extensively discussed.

1 Introduction

Pervasive and mobile computing scenarios consider the possibility of providing users with ubiquitous and on the move access to digital services, and of supporting users interactions with their surrounding environments. For this possibility to become practical, services should be able to understand situations occurring in the surrounding physical context and autonomously adapt their behavior to the context from which they are requested [11].

Pervasive devices are already able to generate an overwhelming amount of data, from which knowledge about situations and facts occurring in the world should

Gabriella Castelli · Franco Zambonelli
DISMI - Dipartimento di Scienze e Metodi dell'Ingegneria
University of Modena and Reggio Emilia
Via Amendola 2 , 42100 Reggio Emilia, Italy
e-mail: {gabriella.castelli,franco.zambonelli}@unimore.it

Soro et al. (Eds.): Inform. Retrieval and Mining in Distrib. Environments, SCI 324, pp. 145–162.
springerlink.com

be inferred for the use of pervasive services. Accordingly, the real challenge for future pervasive services is the investigation of principles, algorithms, and tools, via which this growing amount of distributed information can be properly organized so as to facilitate the successful retrieval by pervasive services [6]. We envision that the access by services to contextual information does no longer occur directly, but rather via a knowledge network layer. Such layer should encapsulate mechanisms to analyze and self-organize contextual information into sorts of structured collections of related data items, i.e. knowledge networks. Thus, services are left with the only duty of exploiting such data to reach specific functionalities.

Traditional (e.g., centralized and/or deterministic) approaches to data organization and aggregation are not practical in this scenario because of the enormous amount of generated data and the decentralized, dynamic, and unpredictable nature of pervasive systems. Accordingly, in this paper, we propose a distributed self-organized approach to organize, link, and aggregated, related items of contextual information. In the proposed approach a multitude of simple agents continuously analyze the data space in order to link together isolated pieces of data. The overall result is the self-organization of the data space in Knowledge networks. We also show with evaluation results that accessing such an organized and distributed data space is beneficial for services because knowledge management and knowledge access costs are reduced.

The remainder of this paper is organized as follows. Section 2 motivates the need for middleware layer in charge of analyzing and managing the contextual knowledge. Section 3 briefly summarizes the W4 data model that is used to represent contextual data provided by pervasive devices and the middleware architecture. In Section 4 we introduce the W4 Knowledge Networks idea, describe the algorithmic approach to organize isolated and distributed pieces of paper into networks of correlated data items. Section 5 presents some preliminary performance evaluation. Section 6 discusses related work in the area, and finally Section 7 concludes.

2 Motivations and Challenges

Ubiquitous computing considers the possibility for users to access general digital services from everywhere and on the move. Pervasive computing additionally considers exploiting pervasive networks, made up of both sensing and data-consumer devices, and possibly actuating infrastructures for the provisioning of innovative services for on-line monitoring of surrounding world and interacting with it, as well as services for enhancing our social experiences in an environment by enabling novel models of localized social interactions. In both cases, it is clear that pervasive services have to collect information about situations around and acting accordingly, i.e. they should be context-aware. Moreover, given the intrinsic dynamics and decentralization of pervasive scenarios, autonomic behavior is necessary to ensure services continuity without forcing costly and hard to be managed human intervention.

A number of technologies that contribute producing large amounts of contextual information already exists. The produced items of contextual information (i.e., "data

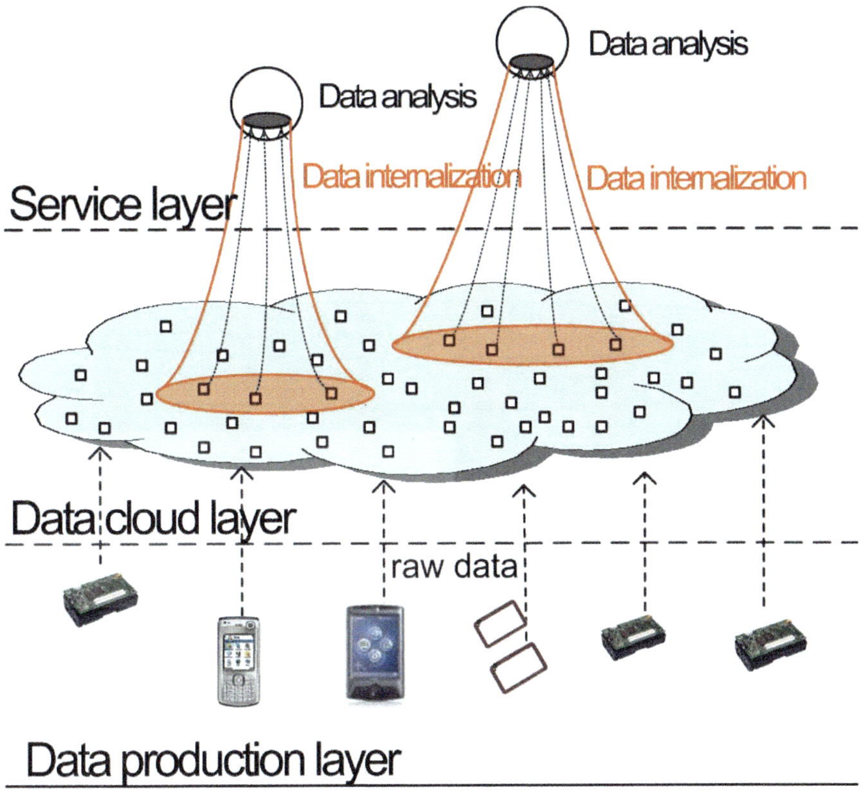

Fig. 1 Pervasive devices and sensors make available to services a sort of "data cloud layer", fed with large amounts of heterogeneous data atoms. To serve their purposes, a service needs to retrieve contextual information (i.e., internalize data atoms from the cloud), analyze it to properly understand situations, and finally exploit such knowledge as needed for their own goals.

atom"), contribute populating a large cloud of data atoms and at making it available to services (see Fig. 1). A service in need of understanding what is happening around can access (i.e., internalize) the needed data atoms and analyze them to understand what is the current situation of its context. Unfortunately, such description is far too simplistic and does not emphasize a number of complexities inherent in it: the process of data internalization can lead to high communication and computational costs for a service, in that it may require accessing large amounts of data atoms, and the process of analyzing retrieved contextual data atoms and turning them into useful knowledge may be non-trivial.

In this paper we claim that there must be an evolution from a model of simple context-awareness, in which services access isolated pieces of contextual data and

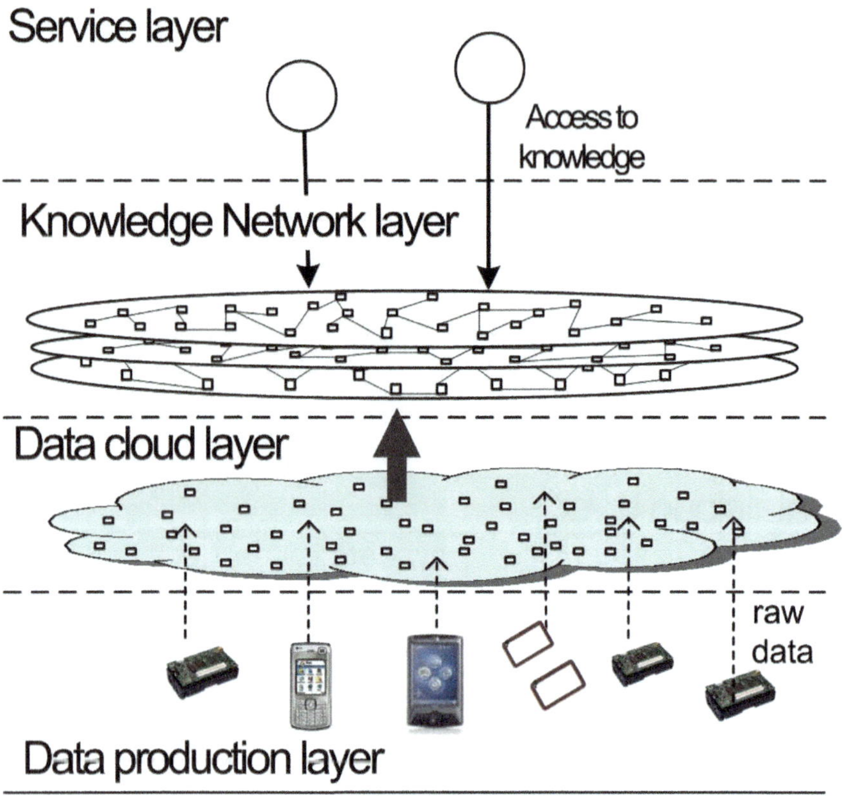

Fig. 2 By exploiting a knowledge network layer, services are no longer forced to access the raw data cloud layer. Knowledge organization and analysis is externalized in the middleware, and services are given access to pre-digested information, with a notable complexity reduction.

are directly in charge of digesting them, towards a model of situation-awareness, in which services access properly structured and organized information reflecting comprehensive knowledge that is related to a "situation" of interest. With reference to Figure 2, we envision that the access by services to contextual information does no longer occur directly, but rather via a knowledge network layer. Such layer should encapsulate mechanisms and tools to analyze and self-organize contextual information into sorts of structured collections of related data items, i.e. knowledge networks.

From the software engineering viewpoint, an approach based on knowledge networks has the advantage of providing a clear separation of concerns between data analysis and data exploitation. While data analysis and organization is delegated to the knowledge network layer, services are left with the only duty of exploiting such data to reach specific functionalities.

For the knowledge networks to be attainable and become a useful tool, both in the case study and in general pervasive scenarios, a number of challenges have to be faced. In particular:

- *Data Model.* There is the need for a simple, general-purpose and uniform model to represent contextual information as individual data atoms as well as their aggregates.
- *Access to data.* It is necessary to identify a suitable API by which services can be given access to the knowledge network layer and the information within.
- *General approaches for data aggregation.* The knowledge networks should be a live layer continuously and autonomously analyzing information to aggregate data atoms, relate existing knowledge atoms with each other, and extract meaningful knowledge from the available data.
- *Application specific views.* Specific services may require the dynamic instantiation within the knowledge networks of application-specific algorithms for knowledge analysis.

In the following Section the W4 data model that is used to represent contextual and the W4 API will be presented. Then, in Section 4 the W4 Knowledge Networks and the algorithms used to organize data will be introduced. We also discuss how dynamically instantiating specific knowledge networks makes it possible to realize application specific views.

3 The W4 Approach

Our proposal for a novel, simple yet effective, data model for expressing contextual knowledge about the world starts from the consideration that any elementary data atoms as well as any higher-level piece of contextual knowledge, in the end, represents a fact which has occurred. Such facts can be expressed by means of a simple yet expressive four-fields tuples (Who, What, Where, When): *"someone or something (Who) does/did some activity (What) in a certain place (Where) at a specific time (When)"*.

3.1 Knowledge Representation

More in particular each of the four-fields (*Who*, *What*, *Where*, *When*) of the W4 data model describes a different aspect of a contextual fact.

- The *Who* field associates a subject to a fact. The *Who* field is represented by a type-value pair, in the form of a string, with an associated namespace that defines the *type* of the entity that is represented. E.g., *student: Gabriella*.
- The *What* field describes the activity performed by the subject. It is represented as a string containing a predicate:complement statement. E.g., *attending:Computer Foundation class*, *read:temperature = 23*.

- The *Where* field associates a location to the fact. In our model the location may be a physical point represented by its coordinates, a geographic region, or it can also be a place label. In addition, context-dependent spatial expressions can be used, e.g., *here*.
- The *When* field associates a time or a time range to a fact. This may be an exact time/time range or a context-dependent expression, e.g., *now*.

The way it structures and organizes information makes the W4 data model able to represent data coming from very heterogeneous sources and simple enough to promote ease of management and processing (although we are perfectly aware that it cannot capture each and every aspect of context, as freshness of data, reliability, access control, etc).

3.2 Data Generation and Data Access

In the W4 model, we rely on the reasonable assumption that software drivers are associated with data sources and are in charge of creating W4 tuples and inserting them in some sorts of shared data spaces.

Knowledge atoms are stored in the form of W4 tuples in a shared data space (or in multiple data spaces), we took inspiration from tuple-space approaches [1] to define the following API:

```
void inject(KnowledgeAtom a);
KnowledgeAtom[] read(KnowledgeAtom a);
```

The *inject* operation is equivalent to a tuple space out operation: an agent accesses the shared data space to store a W4 tuple there.

The *read* operation is used to retrieve tuples from the data space via querying. A query is represented in its turn as a W4 template tuple. Upon invocation, the read operation triggers a pattern-matching procedure between the template and the W4 tuples that already populate the data space. Pattern-matching operations work rather differently from the traditional tuple space model and may exploit differentiated mechanisms for the various W4 fields.

In [6] we provide several examples of knowledge representation and knowledge generation using the W4 Data Model.

3.3 Architecture and Implementation

Figure 3 depicts the overall architecture of a W4 system.

At the bottom there are diverse data sources that produce data formatted according the W4 data model and feed a number of W4 tuple spaces. We assume that software drivers gather information from all the available devices (e.g., RFID tags, GPS devices, Web services), and combine them with the goal of producing a W4 tuple as accurate and complete as possible. In any case the system can effectively deal with incomplete tuples too.

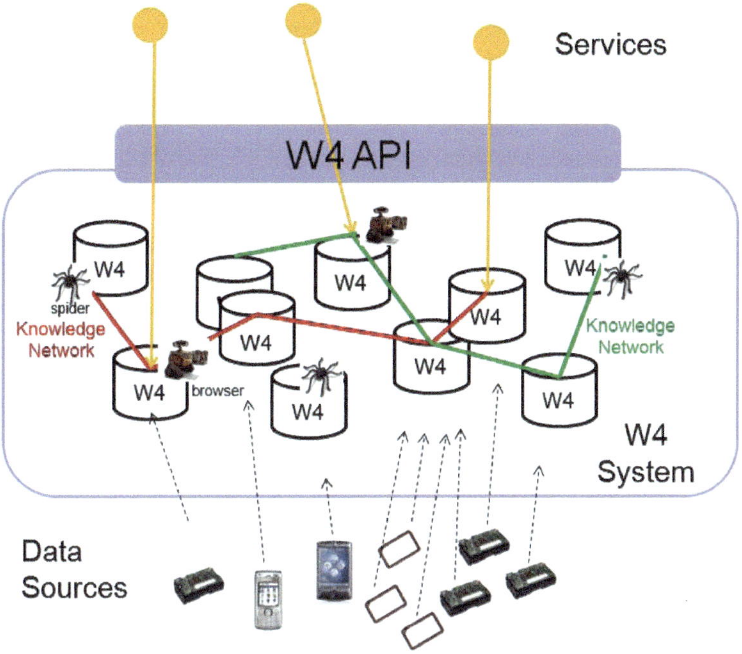

Fig. 3 The W4 System Architecture

The W4 system is made up by a number of distributed W4 tuple spaces. Those tuple spaces can be both local tuple spaces hosted by personal devices (e.g., PDAs, smartphones, laptops) and shared tuple spaces acting as public accessible servers. In the systems there are a variety of agents that access the tuple spaces via the W4 API in order to organize the data layer. In particular:

- *Spiders*: can access the tuples in the W4 tuple spaces and are able to jump from a tuple space to another and link tuples that are related into knowledge networks
- *Browsers*: can browse a knowledge network to solve a query and to infer new tuples

Many W4 Knowledge Networks can be realized and coexist in the W4 system, each realizing a specific view over the data. Those agents (i.e., spiders and browsers) and the algorithms to create and manage knowledge networks will be presented in Section 4.

Finally, at the top there are the various services that access the W4 system to retrieve data, to whom the internal W4 system and data location are completely transparent. Indeed they can act over the system submitting queries to the closest W4 tuple space via the W4 API.

We developed a prototype implementation of the described architecture in a small pervasive computing testbed by extending the LighTS Tuple Space [2], a light weight tuple space implementation particularly suitable for context-aware application, and by realizing spiders and browsers as simple Java agents.

The implemented W4 middleware runs on laptops and on PDAs equipped with wireless interface and J2ME-CDC (Personal Profile) Java virtual machine.

4 W4 Knowledge Networks

Although pattern-matching techniques proved rather flexible to retrieve context information, our idea is to exploit the W4 structure to access the context repository in a more sophisticated and flexible way. More specifically, we propose general-purpose mechanisms and policies to link together knowledge atoms, and thus form W4 knowledge networks in which it will be possible to navigate from a W4 tuple to the others. Moreover, new information could be produced and combined aggregating existing tuples while navigating the space of W4 tuples.

4.1 The Knowledge Networks Approach

The W4 knowledge networks approach is based on the consideration that a relationship between knowledge atoms can be detected by a relationship (a pattern-matching) between the information contained in the atoms fields. In particular, for the W4 model, we can identify two types of pattern matching correlations between knowledge atoms:

- *Same value – same field*: We can link together W4 tuples belonging to the same user, about the same place, activity or time (or, more in general, those W4 tuples in which the values in the same field match according to some pattern-matching function). Matching two or more *same value – same field* relationships, we can render complex concepts related to groups of W4 tuples, e.g. *All students (same subject) who are attending a class (same activity) at the same room (same location)*.
- *Same value – different field*: We can link atoms in which the same information appears in different fields. This kind of pattern matching can be used for augmenting the expressive level of the information contained in the W4 tuples. For example, a knowledge atom having *When: 18/09/2008* can be linked with another atom like *Who: Fall Class Begin* , *When: 18/09/2008* to add semantic information to that date.

Table 1 summarizes the basic relationships between knowledge atoms. On the principal diagonal, it is represented the "same value – same field" pattern matching. By reading the table by columns, it is possible to find all relationships between one particular atom with all other atoms in a knowledge network. For example, looking at the first column on the left, we are comparing all atoms with the same subject. The first cell is on the diagonal, so it is a "same value – same field" pattern

Table 1 Relations between the fields of the W4 Knowledge atoms

	who	what	where	when
who	Same subject	All subject who performed a particular activity	Atom describing an indoor location	Atom describing a logical time
what	Different activity performed by the same subject	Same activity	All activities performed in the same location	All activity performed at the same time
where	All locations in which a subject has been	All location in which an activity has been performed	Same location	All location occupied at the same time
when	Same subject-different time: a living diary	All times in which an activity has been performed	All times in which an activity has been performed	Same time

matching. The 2nd row, 1st column cell identifies all atoms containing the different activities performed by the same subject. Then we have all atoms containing the different locations where the same subject has been, the last cell is a particular case: all atoms generated for the same user.

Exploiting those correlations makes it possible to find all relationships between one particular W4 tuple with all other tuples in the data space which may then be used as the basis for more elaborated inference and reasoning, even for eventually creating new W4 tuples.

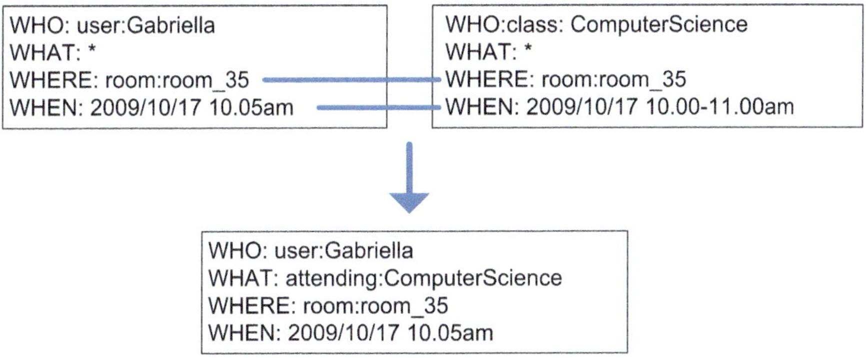

Fig. 4 W4 knowledge network data inference

For instance (see also Fig. 4), suppose that Gabriella's PDA, at a certain time, creates the following tuple: (student:Gabriella, - , room:room_35, 01/09/2008 10:05 am), where - means an empty field. Simple agents in the system continuously analyze the data spaces and find a correlation with the following tuple: (class: Computer Foundation, - ,room: room_35, 01/09/2008 10:00-12:00 am). A new tuple carrying higher level logical information may be created: (student:Gabriella, attending: Computer Foundation,room:room_35, 01/09/2008 10:05 am).

4.2 The Self-organizing Algorithms

A self-organizing approach to generate and maintain the knowledge networks' layer is clearly required by the decentralized nature of pervasive computing systems and the overwhelming amount of generated data, which prevent the use of a centralized process for data management. To this end, we adopt an agent-based approach which relies on a two-phase process.

The first phase is the identification of all possible correlations between knowledge atoms, and the creation of links between W4 atoms. This can be done by a number of agents (we call *spiders* as they weave their webs between correlated tuples), in charge of identifying relationships between tuples. Spiders continuously surf W4 Tuple Spaces in order to retrieve tuples that fulfill the specific relationship, those tuples are virtually linked together thus creating a W4 knowledge network for the given relationship. The spiders' algorithm follows:

```
define:
    spider_rel;
    knet;

Main:
    Do forever:
        TS = choose_Random_TS();
        last_Knet_Ref = Analize_TS(TS, last_Knet_Ref);
    Done;

Analize_TS(current_TS, last_Knet_Ref){
    tuples_inRel[] = current_TS.read[spider_rel];
    /*non destructive read
    if (tuples_inRel NOT null){
        Add_to_Knet (knet, tuples_inRel);
        return current_TS;
        }
    else
        return last_Knet_Ref;
}
```

The second step is the generation of new knowledge atoms, by analyzing which of the identified link can lead to a new W4 atom as a process of merging related atoms. This is performed by another type of agents, called *browsers* because they are capable of browsing the knowledge networks trying to generate new W4 atoms. Each browser is capable of inferencing a specific type of relationship. The browsers' algorithm follows:

```
define:
    browser_rel;
    /* the relation that the browser is cable to infer

Main:
    Do forever:
        TS = choose_a_random_TS;
        t = choose_a_ramdom_tuple (TS);
        Generate_New_Knowledge(t);
    Done;

Generate_New_Knowledge(t){
    For each (Knet(t)):
        ti = get_Next_Tuple (Kneti, t);
        if (isInfereable (t, ti) IS true){
        new_tuple = inference (t, ti);
        add (new_tuple);
        }
}
```

The idea at the base of the W4 Knowledge Networks approach is that spiders and browsers continuously surf, analyze, correlate and infer new knowledge. In this way new tuples are linked to the knowledge networks of interest and new knowledge networks can be realized as soon as they become of interest for services that access the data middleware. At the same time, browsers can exploit the knowledge networks to browse among tuples that are somehow related and possibly infer new knowledge to be injected in the system in form of a W4 tuple.

Although the knowledge networks can be used as the basis for knowledge reasoning, even when new data are not generated, the web of links between atoms can be fruitfully used during querying to access and retrieve contextual information more effectually. When a query is submitted to the W4 tuple space system, a query-solving agent capable of browsing knowledge networks, i.e. a query solving browser created in order to solve a query, analyze the query template and determine one or more knowledge networks to which the matching tuples should belong. Then the query solving browser choose a random W4 tuple space in the system and scans it until he finds an entry point for one of the identified knowledge networks, i.e. a tuple belonging to one of those knowledge networks. When the entry point is found, the agent starts to jump from the entry point tuple to the other tuples in the identified knowledge network, checking if they matches the template and finally returns the retrieved tuples. This is beneficial for services because fewer read operations have to be performed when exploiting knowledge networks instead of a set of data spaces in which information is not related to each other.

5 Preliminary Evaluation Results

In order to evaluate the proposed approach, we conducted some experiments to determine how the services improve their data access costs exploiting the W4 knowledge networks infrastructure instead of accessing isolated pieces of information.

We developed a W4 tuple space implementation on top of LighTS Tuple Space [2]. Data stored in the W4 tuple spaces system come from the simulated environment based on the Repast framework [http://repast.sourceforge.net/]. We represented a location with a number of users each moving in the environment. The virtual environment is split in 100 zones, each of them holds a private W4 tuple space that stores all the tuples generated in it. Periodically a W4 tuple for each user is generated based on the current position and the current time.

In this scenario many tuples are stored in the W4 tuple spaces, and services may find difficult to access those data. We performed some experiments to measure how services may have benefits by accessing W4 knowledge networks.

We submitted to the system the following complex query: "Retrieve all the users that were near agent A5 was, on time 500". For a W4 System this means the following two queries should be subsequently resolved: "Find the position of user A5 on time 500" (let's call it locationX) and then "Retrieve all the users in locationX at time 500".

On this simulated environment, we compared the W4 Tuple Space System with the Exhaustive Search in Tuple Spaces and with Hash based Tuple Spaces.

The Exhaustive Search is performed on a Tuple Space that embeds the W4 Data Model but not the W4 knowledge networks mechanisms. When a query is submitted to this system, a query agent have to scan the whole system to solve the query.

Hash based Tuple Space is a well known and popular technique for data indexing in distributed environment. Here we follow an approach similar to [14] in which a single field of the tuple structure is used for the hashing operation and indexing purpose. In this simulations we considered two cases: hash performed on the who fields and hash performed on the where fields.

We run the simulations 10 times and depicted the average values. Figure 5 (a) shows the number of tuple spaces visited by the query-solving agent in the considered systems. The W4 Tuple Space Systems performs better than traditional approaches. Indeed, in the medium case, the Exhaustive Search has to query half the number of Tuple Spaces in the system to solve the first sub-query and the whole number of the Tuple Spaces to solve the second ones. The Hash approaches work better than the Exhaustive query because one of the sub-query is solved thanks to the hashing operation, nevertheless the other sub-query have to be solved traditional as in the case of the Exhaustive Search. However exploiting the W4 Knowledge Networks is even better because the number of accessed tuple spaces is determined by the number of tuple spaces involved in the knowledge networks of interest.

Figure 5 (b) shows the number of read operation performed. Also in this case the W4 Tuple Space System performs better then the other systems. As in the previous case, the Exhaustive Search have to access half the number of tuples in the systems to solve the first sub-query, and all the tuples in the system to solve the second one.

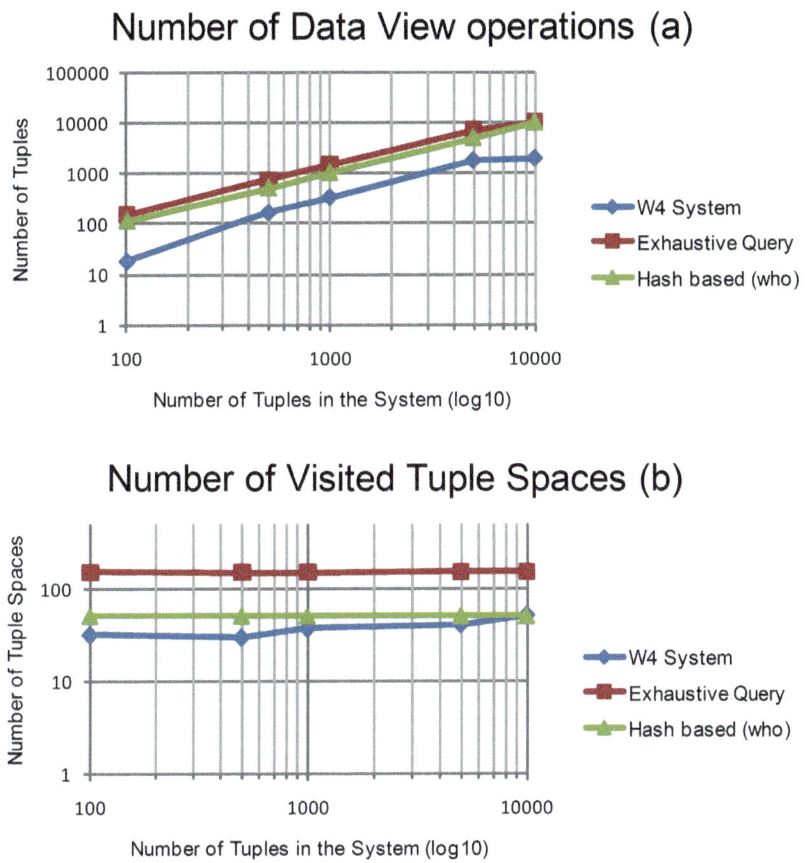

Fig. 5 (a) Number of view operations done by query-solving agents. (b) Number of view operations done by query agents

Here the performance of the Hash based systems are significantly different depending if the hashing is performed on the who field or on the where field. However the W4 System performs better because all the fields are considered important the same when building knowledge networks.

6 Related Work

In recent years, several models and middleware addressing contextual information and context-aware services have been investigated, and several knowledge networks infrastructure to deal with both contextual data and general knowledge have been proposed.

6.1 Related Context Models and Middleware

Context is a very fluid notion and although several researchers claim that it is very hard to abstract it in terms of variables and data models [9], it is also a widespread opinion that a more pragmatic perspective should be adopted. Early works in this area, as from Schmidt et al. [25] and Dey et al. [8], concentrates on the issue of acquiring context data from sensors and of processing such data but they generally miss in identifying a uniform model to describe the data and analyzing the issues at the middleware level. Some recent proposals, such as [27, 4] focus on providing models for contextual data that adopt a uniform well-defined structure. Indeed, our W4 proposal accounts for a very similar structuring for contextual information, and enriches it further with a well-defined API, and with the possibility of linking data atoms and of providing application-specific views to services.

An increasing number of research works get inspiration from tuple space middleware models [1] and propose representing and storing contextual information in the form of tuples to be stored in distributed tuple spaces. Egospaces [15] adopts this perspective, without committing to a specific pre-defined structure for context tuples, which can make it difficult for services to uniformly deal with tuples represented in different formats. Other proposal, such as The Context Fabric model [12] rely on well-structured context tuples. Recent proposals focusing on sensor networks, suggest exploiting a tuple-based approach to provide application-specific views on sensorial data [19]. In general we consider tuple-based approaches very suitable for organizing and accessing contextual information, but we also think that there is need of more structuring and flexibility than those exhibited by the existing approaches.

In the above described work, the issue of relating contextual data atoms with each other and of providing different views to different applications is not generally addressed. More recently, other proposals have adopted a similar endeavor but have considered the issue of adopting specific ontologies to model context information and enable other than efficient querying also efficient context-reasoning [24, 16]. Although such approaches tend to be too application-specific, they attribute the importance of linking independent atoms of contextual information (with ontological relations) and of reasoning not only on individual data items but also on their relations, an idea which is fully shared by our knowledge network vision. Other proposals experience different techniques for context reasoning. Many works, such as and [23], are focused on situation learning and situation relationships in smart environment. Other works, such as [22] propose predicate logic as an effective language for context-aware reasoning. The W4 Knowledge Networks approach we propose aims to be more general and proposes an approach different from traditional ones, considering self-organizing agents.

Campbell et al. [5] consider the possibility of extracting higher-level knowledge from raw sensed data merging feature vectors in an opportunistic fashion for people-centric application. The idea of merging and considering data coming from diverse sources is shared with the W4 Knowledge Networks approach. However in the W4 approach we go further considering multiple knowledge views that can be accessed by multiple services.

Obviously also other areas of research contributed towards the realization of our knowledge networks vision, in particular data mining and pattern discovery and granular computing. See [6] for a critical survey.

6.2 Related Knowledge Networks Approaches

Since the management of information has become a compelling need for complex systems, a variety of approaches for managing data and achieving high degree of knowledge awareness and self-adaptability has been proposed.

In the Knowledge Plane proposal [7] the idea is to couple the service layer with a heavyweight control plane where tools for the analysis of the knowledge are embedded together with actuating agents that exploit this knowledge. On the contrary, the W4 approach goes on the direction of a lightweight infrastructure which claims for self-organized algorithms and mechanisms to manage data.

In the last few years a number of approaches [13, 21, 10, 20] that goes in the direction of the Knowledge Networks arose, most of them are are limited to collection of data generated by mobile phones and sensors connected to them. However, even if they don't claim to realize knowledge networks, they are quite close to the idea. Among the others, BeTelGeuse [20] is particular interesting, it is a data collection system for multiplatform mobile phones that can infer high-level abstraction of user's location. Some ideas are shared with the W4 approach: gathering data from multiple sensors, the focus on context data, inferring higher-level knowledge and the idea that data can be used by diverse applications. However the W4 knowledge Networks considers data coming also from other data source and inferring information on more semantic dimensions.

The Knowledge Networks approach [3] developed by EU Project CASCADAS have a number of commonalities with the W4 proposal, such as the idea of a lightweight overlay networks and the need for self-organizing algorithms for the Knowledge Networks to autonomously manage themselves. However while the Knowledge Networks are at some extent based on a hierarchical organization of the knowledge, in the W4 proposal the W4 data atom organization is completely flat, indeed it is based on tuple spaces, and the self-organizing algorithms can exploit the W4 field structure giving the same relevance to the four fields.

Other approaches, such as [26], focus on providing inter-operability among different devices and technologies via the definition of a lingua franca that enables representing, using and managing knowledge for ubiquitous computing. This is achieved combining information extracted from ontologies and from models. In our opinion, there is the need for more light-weight and simple solutions since ontologies are computationally heavy for pervasive devices with limited resources.

Some approaches related to other application fields have been proposed too. In particular, [17] is oriented towards network management. The vision paper discusses how to manage future decentralized, dynamic and heterogeneous networks and proposes to use proper filtering and abstraction of knowledge to limit the amount of information that must be exchanged and optimization algorithms to adapt the network of networks to application and environment changes. The idea of using algorithms that involve only small part of the network and autonomously adapt to the changes in the networks is shared with the W4 approach.

7 Conclusion and Future Work

Despite the promising results achieved so far in the study of the W4 model and on the self-organized knowledge networks algorithms, several research issues still have to be faced.

Experimental results show that accessing pre-organized W4 Knowledge Networks instead of a flat or hash-based Tuple Space System greatly improves the system performance in terms of access costs. However more experiments should be done to evaluate properly the effectiveness, the scalability and the overlay costs. In particular it is worth noticing that the generation of Overlay Networks never comes without overhead costs. The idea behind W4 Knowledge Networks is to pre-organize data in a fashion that will be useful for a number of services and will decrease delay experiences by services. In order to keep Knowledge Networks feasible and manageable one should find a good trade-off between the number of knowledge networks to be build in the system and overhead costs associated, in general only knowledge networks that might be useful (i.e., accessed by agents) should be built and maintained.

Moreover, in the current implementation of the W4 System, the number of tuples stored in the system is constantly increasing as new data are injected in the system. There is the need for a "garbage collection" solution, we plan to experiment with a concept of knowledge tuple fading as introduced in [18].

References

1. Ahuja, S., Carriero, N., Gelernter, D.: Linda and friends. Computer 19(8-9), 26–34 (1986)
2. Balzarotti, D., Costa, P., Picco, G.P.: The LighTS Tuple Space Framework and its Customization for Context-Aware Applications. International Journal on Web Intelligence and Agent Systems 50(1-2), 36–50 (2007)
3. Bicocchi, N., Castelli, G., Mamei, M., Rosi, A., Zambonelli, F., Baumgarten, M., Mulvenna, M.: Knowledge networks for pervasive services. In: ICPS 2009: Proceedings of the 2009 International Conference on Pervasive Services, pp. 103–112. ACM, New York (2009)
4. Bravo, J., Hervs, R., Snchez, I., Chavira, G., Nava, S.: Visualization services in a conference context: An approach by rfid technology. Journal of Universal Computer Science 12(3), 270–283 (2006)
5. Campbell, A., Eisenman, S., Lane, N., Miluzzo, E., Peterson, R., Lu, H., Zheng, X., Musolesi, M., Fodor, K., Ahn, G.-S.: The rise of people-centric sensing. IEEE Internet Computing 12(4), 12–21 (2008)

6. Castelli, G., Mamei, M., Zambonelli, F.: Engineering contextual knowledge for autonomic pervasive services. International Journal of Information and Software Technology 52(8-9), 443–460 (2008)
7. Clark, D.D., Partridge, C., Ramming, J.C., Wroclawski, J.T.: A knowledge plane for the internet. In: SIGCOMM 2003: Proceedings of the 2003 Conference on Applications, Technologies, Architectures, and Protocols for Computer Communications, pp. 3–10. ACM, New York (2003)
8. Dey, A.K., Abowd, G.D., Salber, D.: A conceptual framework and a toolkit for supporting the rapid prototyping of context-aware applications. Human Computer Interaction 16(2), 97–166 (2001)
9. Dourish, P.: What we talk about when we talk about context. Personal Ubiquitous Computing 8(1), 19–30 (2004)
10. Froehlich, J., Chen, M.Y., Consolvo, S., Harrison, B., Landay, J.A.: Myexperience: a system for in situ tracing and capturing of user feedback on mobile phones. In: MobiSys 2007: Proceedings of the 5th International Conference on Mobile Systems, Applications and Services, pp. 57–70. ACM, New York (2007)
11. Henricksen, K., Indulska, J.: Developing context-aware pervasive computing applications: models and approach. Pervasive and Mobile Computing 2 (2005)
12. Hong, J.I.: The context fabric: an infrastructure for context-aware computing. In: CHI 2002 extended abstracts on Human Factors in Computing Systems, pp. 554–555 (2002)
13. Intille, S.S., Rondoni, J., Kukla, C., Ancona, I., Bao, L.: A context-aware experience sampling tool. In: CHI 2003: CHI 2003 extended abstracts on Human Factors in Computing Systems, pp. 972–973. ACM, New York (2003)
14. Jiang, Y., Xue, G., Jia, Z., You, J.: Dtuples: A distributed hash table based tuple space service for distributed coordination. In: Grid and Cooperative Computing, pp. 101–106 (October 2006)
15. Julien, C., Roman, G.-C.: Egospaces: facilitating rapid development of context-aware mobile applications. IEEE Transactions on Software Engineering 32(5), 281–298 (2006)
16. Lee, D., Meier, R.: Primary-context model and ontology: A combined approach for pervasive transportation services. In: Fifth Annual IEEE International Conference on Pervasive Computing and Communications Workshops PerCom Workshops 2007, pp. 419–424 (2007)
17. Manzalini, A., Deussen, P., Nechifor, S., Mamei, M., Minerva, R., Moiso, C., Salden, A., Wauters, T., Zambonelli, F.: Self-optimized cognitive network of networks. The Computer Journal (to appear)
18. Menezes, R., Wood, A.: The fading concept in tuple-space systems. In: Proceedings of the 2006 ACM Symposium on Applied Computing, Dijon, France, pp. 440–444. ACM Press, New York (2006)
19. Mottola, L., Picco, G.P.: Logical neighborhoods: A programming abstraction for wireless sensor networks. In: Gibbons, P.B., Abdelzaher, T., Aspnes, J., Rao, R. (eds.) DCOSS 2006. LNCS, vol. 4026, pp. 150–168. Springer, Heidelberg (2006)
20. Nurmi, P., Kukkonen, J., Lagerspetz, E., Suomela, J., Floréen, P.: Betelgeuse: A platform for gathering and processing situational data. IEEE Pervasive Computing 8(2), 49–56 (2009)
21. Raento, M., Oulasvirta, A., Petit, R., Toivonen, H.: Contextphone: A prototyping platform for context-aware mobile applications. IEEE Pervasive Computing 4(2), 51–59 (2005)
22. Ranganathan, A., Campbell, R.H.: An infrastructure for context-awareness based on first order logic. Personal Ubiquitous Comput. 7(6), 353–364 (2003)

23. Reignier, P., Brdiczka, O., Vaufreydaz, D., Crowley, J.L., Maisonnasse, J.: Context-aware environments: from specification to implementation. Expert Systems: The Journal of Knowledge Engineering 24(5), 305–320 (2007)
24. Roussaki, I., Strimpakou, M., Kalatzis, N., Anagnostou, M., Pils, C.: Hybrid context modeling: A location-based scheme using ontologies. In: IEEE International Conference on Pervasive Computing and Communications Workshops, vol. (1), pp. 2–7 (2006)
25. Schmidt, A., Aidoo, K.A., Takaluoma, A., Tuomela, U., Laerhoven, K.V., Velde, W.V.d.: Advanced interaction in context. In: Proceedings of the 1st International Symposium on Handheld and Ubiquitous Computing, vol. 8 (1999)
26. Strassner, J., O'Sullivan, D.: Knowledge management for context-aware, policy-based ubiquitous computing systems. In: MUCS 2009: Proceedings of the 6th International Workshop on Managing Ubiquitous Communications and Services, pp. 67–76. ACM, New York (2009)
27. Xu, C., Cheung, S.C.: Inconsistency detection and resolution for context-aware middleware support. In: Proceedings of the 10th European Software Sngineering Conference Held jointly with 13th ACM SIGSOFT International Symposium on Foundations of Software Engineering, pp. 336–345 (2005)

A Relational Approach to Sensor Network Data Mining

Floriana Esposito, Teresa M.A. Basile,
Nicola Di Mauro, and Stefano Ferilli

Abstract. In this chapter a relational framework able to model and analyse the data observed by nodes involved in a sensor network is presented. In particular, we propose a powerful and expressive description language able to represent the spatio-temporal relations appearing in sensor network data along with the environmental information. Furthermore, a general purpose system able to elicit hidden frequent temporal correlations between sensor nodes is presented. The framework has been extended in order to take into account interval-based temporal data by introducing some operators based on a temporal interval logic. A preliminary abstraction step with the aim of segmenting and labelling the real-valued time series into similar subsequences is performed exploiting a kernel density estimation approach. The prposed framework has been evaluated on real world data collected from a wireless sensor network.

1 Introduction

Sensor networks represent a powerful technology able to monitor many situations in the physical world including health, agriculture, emergency management, micro-climate and habitat, or earthquake and building health [3, 4, 11, 23]. The main objective of sensor networks is knowledge-gathering: each component (sensor node) acts to maximize its contribution by adding details and precision in order to completely outline the monitored situation and, by cooperating with the others, to understand phenomena *in situ* and in real time. Sensor nodes are small electronic components made up of a processing element, some measurement devices and a (wireless/wired)

Floriana Esposito · Teresa M.A. Basile · Nicola Di Mauro · Stefano Ferilli
Department Of Computer Science,
University of Bari, Italy
e-mail: {esposito,basile,ndm,ferilli}@di.uniba.it

Soro et al. (Eds.): Inform. Retrieval and Mining in Distrib. Environments, SCI 324, pp. 163–181.
springerlink.com

communication device. They are able to gather different types of information from the environment, such as temperature, light, humidity, radiation, the presence or nature of biological organisms, geological features and more. As a consequence, a great amount of data is available that if analyzed in an appropriate way might help to automatically and intelligently solve a variety of tasks thus making the human life more safe and comfortable.

A sensor network is usually made up of a set of nodes spatially distributed in the environment, i.e. each node i is located in an environment at the position p_i, and senses a set of properties $\mathcal{P}_i$ at every time instance t. In other words, each sensor produces a continuous time series describing its reading over time, hence we have an observation at every instant of time. Furthermore, the data generated by sensor nodes involved in a sensor network are type-related (the humidity depends on the temperature), time-related (the temperature may change over time) and spatio-related (topological arrangements of the sensors in the network). All these relations could be easily represented by using an interval-based relational language, such the one proposed in this work, as opposed to the point-based approach, trying to shift the basic time-series description language to a higher one.

Finally, a set of events on different dimensions (time, space, etc.) can take place in the physical environment that could influence the sensor behaviour and hence the observation, thus the contextual information could be taken into account as well. Hence, temporal and context-based relations must be combined into a heterogeneous language and mined with appropriate techniques in order to obtain useful knowledge.

In the last decade, some approaches were proposed to face the problem of extracting knowledge from sensor data. They focused either on the data representation (e.g., sensors clustering, discretization) or knowledge extraction (association rules, sequential patterns). Nevertheless, they usually do not consider contextual information or they generally consider events occurring in one dimension only or in a time instant, while, in some applications, like in sensor networks, data are environment related, high dimensional and may occur in time intervals.

In this work the exploitation of a relational language to describe the temporal evolution of a sensor network along with contextual information is proposed. Furthermore, it is presented the use of relational learning techniques to discover interesting and more human readable patterns relating spatio-temporal correlations with the contextual ones.

As regards the relational language, it is based on the work described in [9] where the authors proposed a framework for mining complex patterns, expressed in first-order language, in which events may occur along different dimensions.

Here, that framework is extended in order to take into account interval-based temporal data along with contextual information about events occurring in the environment. The extension concerns the introduction of interval-based operators based on the Allen's temporal interval logic [5] in the sequences. Specifically, firstly an abstraction step with the aim of segmenting and labelling the real-valued time series into similar subsequences is performed exploiting a kernel density estimation approach. Then the integration of such a new knowledge along with the relative interval-based operators, able to deal with it, in the relation pattern mining framework is carried out. Finally, in order to evaluate the validity of both the abstraction step and the general framework, an experimental session on real world data collected from a wireless sensor network deployed in the Intel Berkeley Research Lab [15] is presented.

2 Relational Pattern Mining

The framework we present in this chapter is based on the work described in [6, 9] where the authors proposed an algorithm for mining complex patterns, expressed in first-order language, in which events may occur along different dimensions. Specifically, multi-dimensional patterns were defined as a set of atomic first-order formulae in which events are explicitly represented by a variable and the relations between events were represented by a set of dimensional predicates. The algorithm has been extended in order to take into account interval-based temporal data. Finally, an automatic discretization algorithm based on the concept of kernel density estimation has been introduced and evaluated.

Datalog [28] is the language used as representation language for the domain knowledge and patterns. Sequences and patterns are represented by a set of logical atoms[1]. Thus, a *relational sequence* may be defined as an ordered list of atoms separated by the operator $<$: $l_1 < l_2 < \cdots < l_n$, while a relational pattern is defined as follows:

Definition 1 (Subsequence [16]). Given a sequence $\sigma = (e_1 e_2 \cdots e_m)$ of m elements, a sequence $\sigma' = (e'_1 e'_2 \cdots e'_k)$ of length k is a *subsequence* (or a pattern) of the sequence σ if:

1. $1 \leq k \leq m$
2. $\forall i, 1 \leq i \leq k, \exists j, 1 \leq j \leq m : e'_i = e_j$
3. $\forall i, j, 1 \leq i < j \leq k, \exists h, l, 1 \leq h < l \leq m : e'_i = e_h$ *and* $e'_j = e_l$.

Example 1. Let us now introduce an example to better explain the representation language and the operators introduced in the following. Suppose to have a wireless sensor network deployed inside a building, for example a conference building, to monitor people's motion over space and time and, accordingly, the temperature, light and voltage in the rooms.

[1] An atom $p(t_1, \ldots, t_n)$ is a predicate symbol p of arity n applied to n terms t_i (constants or variables).

In this setting, a 1-dimensional sequence, specifically a time dimension sequence, could be:

```
move(user1,room5) < in(user1,room5) < talk(user1,room5,machine_learning)
  < leave(user1,room5) < move(user1,room4)
```

and a possible pattern could be `move(X,Y) < talk(X,Y,Z)`.

As one can note, the exploitation of just one dimension is not sufficient to describe the environment, indeed information about sensors or events occurring in the environment cannot be easily represented. Hence, some modifications have to be introduced as reported in the following. □

In order to make the framework more general, the concept of *fluents* has been considered. Let a sequence be an ordered succession of events, a fluent is used to indicate that an atom holds for a given event. In this way we are able to distinguish in the sequence, and hence in the pattern, *dimensional* and *non-dimensional* atoms. Specifically, the first ones refer to the dimensional relations between events involved in the sequence while the non-dimensional atoms introduce an event and the objects involved in it (fluent atoms) or the properties and the relations of the objects already introduced by an event (non-fluent atoms).

Example 2. The introduction of such kind of atoms allows one to introduce the events occurring in a situation, as reported in the following:

```
move(entering1,user1,room5) (entering1 < entering2)
move(entering2,user1,room4) near(room5,room4)
```

It denotes a 1-dimensional relational sequence with three non-dimensional atoms (i.e., `move(entering1,user1,room5)`, `move(entering2,user1,room4)` and `near(room5,room4)`) and one dimensional atom. Specifically,

- `move(entering1,user1,room5)` denotes the fluent `move(user1,room5)` at the event `entering1`,
- `move(entering2,user1,room4)` denotes the fluent `move(user1,room4)` at the event `entering2`,
- `(entering1 < entering2)` indicates that the event `entering2` is the direct successor of `entering1`, and
- `near(room5,room4)` represents a generic relation between the objects `room5` and `room4`.

The choice to add the event as an argument of the predicates is necessary for the general case of n-dimensional sequences with $n > 1$. In this case, indeed, the operator $<$ is not sufficient to express multi-dimensional relations and we must use its general version $<_i$. Specifically, $(e_1 <_i e_2)$ denotes that the event e_2 is the next successor of the event e_1 in the dimension i, where i could be, for example, time or space. Hence, in our framework a multi-dimensional sequence is supposed to be a set of events, and a sequence of events corresponds to each dimension. □

However, the $<_i$ operator is not sufficient to represent the knowledge in the patterns. Hence, a further generalization of the framework consisted in the possibility to represent multi-dimensional relational patterns by introducing some dimensional operators able to describe general event relationships: a) $<_i$, *next step on dimension* i; b) $\lhd_i$, *after some steps on dimension* i; and c) $\bigcirc_i^m$, *exactly after* m *steps on dimension* i.

Now, the general definition of subsequence (Def. 1) can be cast in this new framework by modelling the *gaps* represented by the third condition of Definition 1 with the $\lhd_i$ and $\bigcirc_i^m$ operators as reported in the following definition.

Definition 2 (Multi-dimensional relational pattern). A multi-dimensional relational pattern is a set of atoms, involving k events and regarding n dimensions, in which there are non-dimensional atoms and each event may be related to another event by means of the operators $<_i$, $\lhd_i$ and $\bigcirc_i^m$, $1 \leq i \leq n$.

Example 3. With such extensions concerning both the representation language and the operators, it is possible to better represent the environment as reported by the following example.

`location(sensor1,room5)` $<_{topology\_r5}$ `location(sensor2,room5)` $<_{topology\_r5}$ `location(sensor3,room5)` ... `location(sensor1,room4)` $<_{topology\_r4}$ `location(sensor2,room4)` $<_{topology\_r4}$ `location(sensor3,room4)` ... `move(entering1,user1,room5)` `activity(talking,user1,room5)` `move(leaving,user1,room5)` `(entering1` $<_{time}$ `leaving)` `(entering1` $<_{time}$ `talking)` `(entering1` $<_{spatial}$ `entering2)` `(talking` $<_{time}$ `leaving)` `move(entering2,user1,room4)` `(talking` $<_{time}$ `entering2)` `activity(coffee_break,user1,room4)` `(leaving` $<_{spatial}$ `entering2)` `(entering2` $<_{spatial}$ `coffee_break)`

and the corresponding temporal patterns that may be true when applied to it could be:

- `move(entering1,user1,room5)` `(entering1` $<_{time}$ `talking)` `activity(talking,user1,room5)`
- `move(entering1,user1,room5)` `(entering1` $\lhd_{time}$ `leaving)` `move(leaving,user1,room5)`
- `move(entering1,user1,room5)` `(entering1` $\bigcirc_{time}^2$ `entering2)` `move(entering2,user1,room4)` if we consider 2 as hours

Note that the $<_i$ will describe the dimensional characteristics in the sequences, while all the three dimensional operators $<_i$, $\lhd_i$, $\bigcirc_i^m$, will be used to represent the discovered patterns. □

In particular, we are interested in mining maximal frequent patterns. Thus, let σ a sequence and p a pattern of σ. The frequency of p in σ is the number of different mappings from elements of p into the elements of σ such that the conditions reported in Definition 1 hold, and, p is *maximal* if there is no pattern p' of σ more frequent than p and such that p is a subsequence of p'.

Since the sequences and patterns are represented as a set of logical atoms, the frequency of a pattern over a sequence can be calculated as the number of subtitutions θ_i such that p subsumes σ, i.e. $p\theta_i \subseteq \sigma$ where subsumption and substitution are defined as follows.

Definition 3 (Subsumption). A a set of logical atoms c_1 θ-subsumes a set of logical atoms c_2 if and only if there exists a substitution θ such that $c_1\theta \subseteq c_2$.

A substitution θ is defined as a set of bindings $\{X_1 \leftarrow a_1, \ldots, X_n \leftarrow a_n\}$ where $X_i, 1 \leq i \leq n$ is a variable and $a_i, 1 \leq i \leq n$ is a term. A substitution θ is applicable to an expression e, obtaining the expression $e\theta$, by replacing all variables X_i with their corresponding terms a_i.

Example 4. Given the following sequence:

$S =$ `p(e1,a) q(a,t) q(a,s) (e1 < e2) p(e2,b) q(b,a)`

and the pattern

$P =$ `p(E,X) q(X,Y)`

there are 3 way to instantiate P from S in such a way that to different terms correspond different objects, i.e. :

1. $\theta_1 =$ {`E/e1, X/a, Y/t`},
2. $\theta_2 =$ {`E/e1, X/a, Y/s`},
3. $\theta_3 =$ {`E/e2, X/b, Y/a`}.

However, since θ_1 and θ_2 map the same constants to the variables of `p(E,X)` (the first literal of the pattern), the frequency of P on S is equal to 2. □

2.1 The Algorithm

The algorithm for frequent multi-dimensional relational pattern mining is based on the same idea of the generic level-wise search method, known in data mining from the APRIORI algorithm [2]. The generation of the frequent patterns is based on a top-down approach. Specifically, it starts with the most general patterns of length 1 generated by adding to the empty pattern a non-dimensional atom. Then, at each step it *specializes* all the frequent patterns, discarding the non-frequent patterns and storing the ones whose length is lesser than a user specified parameter *maxsize*. Furthermore, for each new refined pattern, semantically equivalent patterns are detected, by using the θ_{OI}-subsumption relation, and discarded.

In the specialization phase, the refinement of patterns is obtained by using a refinement operator ρ that maps each pattern to a set of specializations of the pattern, i.e. $\rho(p) \subset \{p' | p \preceq p'\}$ where $p \preceq p'$ means that p is more general of p' or that p subsumes p'.

The algorithm uses a background knowledge $\mathcal{B}$ (a set of Datalog clauses) containing the sequence and a set of constraints that must be satisfied by the generated patterns. In particular $\mathcal{B}$ contains the following predicates:

- `maxsize(M)`: maximal pattern length (i.e., the maximum number of non-dimensional predicates that may appear in the pattern);
- `minfreq(m)`: this constraint indicates that the frequency of the patterns must be larger than m;
- `dimension(next_i)`: this kind of atom indicates that the sequence contains events on the dimension i. One can have more that one of such atoms, each of which denoting a different dimension. In particular, the number of these atoms represents the number of the dimensions.

2.1.1 Constraints

Furthermore the background knowledge contains some constraints that are useful to avoid the generation of unwanted patterns. Specifically they are:

- `negconstraint(`[$p_1, p_2, \ldots, p_n$]`)`: specifies a constraint that the patterns must not fulfill, i.e. if the clause $\{p_1, p_2, \ldots, p_n\}$ subsumes the pattern then it must be discarded. For instance, `negconstraint([p(X,Y),q(Y)])` discards all the patterns subsumed by the clause {`p(X,Y),q(Y)`};
- `posconstraint(`[$p_1, p_2, \ldots, p_n$]`)`: specifies a constraint that the patterns must fulfill. It discards all the patterns that are not subsumed by the clause $\{p_1, p_2, \ldots, p_n\}$;
- `atmostone(`[$p_1, p_2, \ldots, p_n$]`)`: this constraint discards all the patterns that make true more than one predicate among $p_1, p_2, \ldots, p_n$. For instance, `atmostone([red(X),blue(X),green(X)])` indicates that each constant in the pattern can assume at most one of `red`, `blue` or `green` value.

Hence, the solution space is pruned by using some positive and negative constraints specified by the *negconstraint* and *posconstraint* literals. The last pruning choice is defined by the *atmostone* literals. This last constraint is able to describe that some predicates are of the same type.

2.1.2 Improving Efficiency

In order to avoid the generation of patterns containing not linked variables we used the classical types and modes declaration:

- `type(p)`: denotes the type of the predicate's arguments p;
- `mode(p)`: denotes the input output mode of the predicate's arguments p.

In this way we improve the efficiency of the algorithm, since it does not generate patterns containing unrelated atoms. These classical mode and type declarations specify a language bias indicating which predicates can be used in the patterns and to formulate constraints on the binding of variables.

Finally, the background knowledge contains the predicate `key(`[$l_1, l_2, \ldots, l_n$]`)` specifying that each pattern must have one of the predicates $l_1, l_2, \ldots l_n$ as a starting literal. Since each pattern a) must start with a non-dimensional predicate, or with a predefined key, and b) its frequency must be less than the sequence length, the frequency of a pattern can be defined as follows.

Definition 4 (Pattern Frequency and Support). Given a relational pattern $P = (p_1, p_2, \ldots, p_n)$ and S a relational sequence, the *frequency* of the pattern P is equal to the number of different ground literals used in all the possible SLD_{OI}-deductions of P from S that make true the literal p_1. The *support* of P on S is equal to the frequency of the pattern $\{p_1\}$ over the frequency of the pattern P.

Mining from more than one sequence the support is calculated as the number of covered sequences over the total number of sequences.

In order to improve the efficiency of the algorithm, for each pattern $P = (p_1, p_2, \ldots, p_n)$ the set Θ of the substitutions defined over the variables in p_1 that make true the pattern P is recorded. In this way, the support of a specialization P' of P is computed by firstly applying a substitution $\theta \in \Theta$ to P'. It is like to remember all the keys of a table that make true a query.

3 Interval-Based Relational Sequences

In this section we present the extension of the framework to the case of relational sequences including interval-based dependencies. Furthermore, since we are working on real-valued time series, an approach to subdivide/discretize the series into similar subsequences is presented. In particular, the aim is to segment a signal (assigning data to discrete categories) by looking for a sequence of measurements over which a property holds and to label this segment. After this discretization process, some interval relationships may be introduced in order to better describe the evolution of the data along the time.

3.1 Abstracting Using Kernel Density Estimation

A method to segment a sequence is to iteratively merge two similar segments based on the squared error minimization criteria. Another approach is using clustering, by firstly finding the set of subsequences with length w, by sliding a window of width w, and then clustering the set of all subsequences. A different symbol is associated with each cluster. Other approaches are based on using self-organizing maps.

The segmentation process has been obtained by adopting an unsupervised discretization method that uses non-parametric density estimators, as proposed in [7]. The algorithm searches for the next two sub-intervals to produce, evaluating the best cut-point on the basis of the density induced in the sub-intervals by the current cut and the density given by a kernel density estimator for each sub-interval. It uses cross-validated log-likelihood to select the maximal number of intervals.

Two classical techniques for unsupervised discretization are equal-width and equal-frequency binning, where continuous intervals are split into sub-intervals providing them the width or the frequency parameter. However, they require that the values follow a uniform distribution, with low accuracy in case of skew data. The method used here exploits density estimation methods to select the cut-points and their number is computed by cross-validating the log-likelihood.

3.1.1 Simple Binning and the Naive Estimator

The histogram, or simple binning, is the oldest and most popular density estimator. Given a set of training instances $x_1, \ldots, x_N$, let x_0 be an origin and w be the bin (class) width. The intervals (or bins) may be defined as follows

$$I_j = [x_0 + jw, x_0 + (j+1)w), j = 0, 1, \ldots$$

for which the histogram counts the number of instances x_i falling into each I_j, as reported in Figure 1. This procedure replaces the training data $x_1, \ldots, x_N$ with the smaller set $c_1, \ldots, c_g$, where c_j is the corresponding class (label) of the interval I_j.

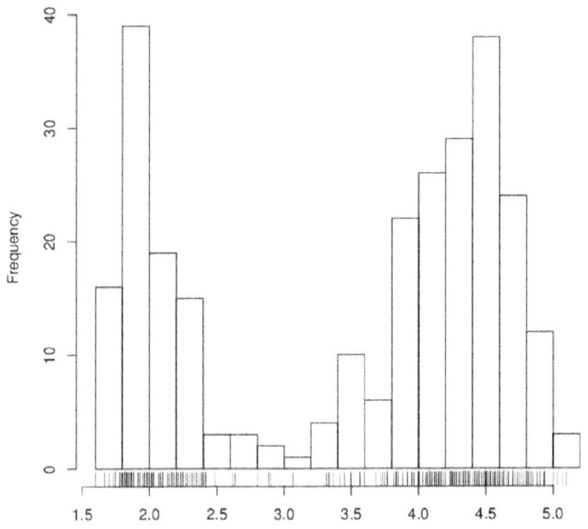

Fig. 1 Histogram for eruptions of the Old Faithful geyser in Yellowstone National Park, Wyoming, USA

In particular, let k be the number of intervals (bins) and $\mathbf{1}_{I_j}$ be the indicator function[2], the density function $\hat{f}(x)$ is computed by

$$\hat{f}(x) = \left(\sum_{i=1}^{M} \mathbf{1}_{I_i}(x) \sum_{j=1}^{N} \mathbf{1}_{I_i}(x_j) \right) (Nw)^{-1}.$$

[2] $\mathbf{1}_{I_j}(x)$ is equal to 1 when $x \in I_j$, 0 otherwise.

From the definition of a probability density, for any given h, it is possible to estimate the probability $P(x-h < X \leq x+h)$ by the proportion of the observations falling in the interval $(x-h, x+h]$. The naive estimator is given by choosing a small number h and setting

$$\hat{f}(x) = \frac{1}{2nh}[\text{no. of } X_i \text{ falling in } (x-h, x+h)].$$

3.1.2 The Kernel Density Estimator

The naive estimator is not a continuous function and hence it is interesting to consider its generalization. In particular, it is useful to consider the kernel estimator, using a smooth kernel function $K(\cdot)$, defined as

$$\hat{f}(x) = \frac{1}{nh}\sum_{i=1}^{n} K\left(\frac{x-X_i}{h}\right),$$

$$K(x) \geq 0, \int_{-\infty}^{+\infty} K(x)dx = 1, K(x) = K(-x).$$

The kernel function used in this paper is the Epanechnikov kernel function [22], see Figure 2, defined as

$$K(u) = \frac{3}{4}(1-u^2)\mathbf{1}_{|u|\leq 1}.$$

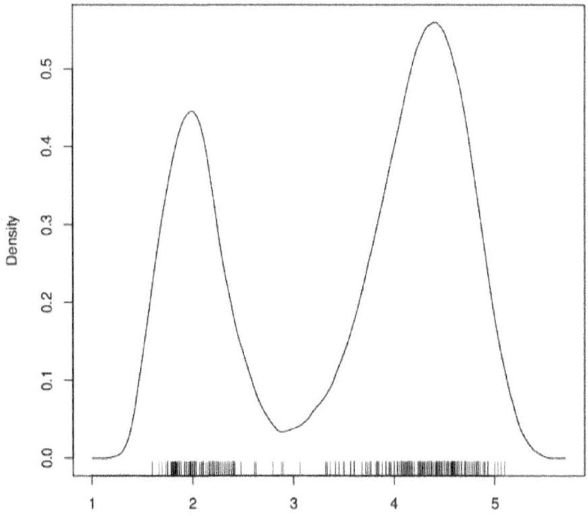

Fig. 2 Epanechnikov kernel (bandwidth = 0.2) for eruptions of the Old Faithful geyser in Yellowstone National Park, Wyoming, USA

3.1.3 The Scoring Function

The goal of discretization is to produce sub-intervals whose induced density over the instances best fits the available data. The cut points are the middle points between the instance values. On the other hand, the choice of the interval that should be split next, among those produced at a given step, is driven by an objective function capturing the significant changes of density in different separated bins.

All the possible cut-points are considered, and a score to each sub-interval is assigned. Given a single interval to split, any of its cut-points produces two bins and thus induces, upon the initial interval, two densities, computed using the simple binning density estimation formula. Every sub-interval produced has an averaged binned density that is different from the density estimated with the kernel function. The less this difference is, the more the sub-interval fits the data well, i.e. the better this binning is, and hence there is no reason to split it.

Hence, at each step of the discretization process, we must choose from different sub-intervals to split. In every sub-interval we identify as candidate cut-points all the middle points between the instances. For each of the candidate cut-points c_i we compute a score as follows:

$$score(T) = \sum_{x_i < c_i} (p(x_i) - f(x_i)) + \sum_{x_i > c_i} (p(x_i) - f(x_i)).$$

The density functions p and f are respectively the kernel density function and the simple binning density function, computed as

$$f(x_i) = \frac{m_i}{wN},$$

where m_i is the number of instances that fall in the bin (left or right) containing x_i; and

$$p(x) = \frac{1}{nw} \sum_{i=1}^{n} K\left(\frac{x - X_i}{h}\right),$$

where we set the bandwidth h to the value of the binwidth w.

3.1.4 The Stopping Criterion

In order to avoid overfitting and to define a stopping criterion, the log-likelihood has been used to evaluate the density estimators. Given a density estimator g and a set of test instances $y_1, \ldots, y_n$ the log-likelihood is computed as

$$LL(g|y_1, \ldots, y_n) = \sum_{i=1}^{n} \log g(y_i).$$

In order to have an unbiased estimation of how the model fits the real distribution, a cross-validation has been used. In particular, for a histogram having $n_{j_{train}}$ training instances in the I_j interval, let $n_{j_{test}}$ be the number of testing instances falling into the same interval, w the bin width, and N be the total number of training instances. Then, the log-likelihood on the test data is computed by

$$LL = \sum_j n_{j_{test}} \log \frac{n_{j_{train}}}{wN}.$$

3.1.5 Abstraction Rules

After having presented a method to discretize the data, that translates the initial sequence (with real-valued elements) to a segmented sequence made up of symbols taken from a given alphabet, now we can formalize the abstraction rules.

Given a real-valued time series $(t_i, x_i)_{1 \leq i \leq n}$, $x_i \in \mathbb{R}$, the goal is to transform it into a discrete series $(t_i, c_i)_{1 \leq i \leq n}$, $c_i \in \{1, \ldots, C\}$. In the case of a sensor network, made up of n nodes, each node i, located in the environment at the position p_i, senses a set of properties $\mathcal{P}$ at every time instance t. Our approach is to define some abstraction rules useful to shift the basic sensor description language into a more general one. In particular each sensor produces a time series, describing its reading over time, that is then divided into intervals.

Let $\mathcal{C}$ denotes the set of possible properties or descriptive labels, such as "temperature is high". Having a time series $(t_i, x_i)_{1 \leq i \leq n}$, denoted by $(t, x)_{1\_n}$, an abstraction rule is a function $\phi_a((t, x)_{1\_n})$ returning a set of m consecutive intervals of the time series. In particular,

$$\phi_a((t, x)_{1\_n}) = \{\delta_a(l, t_i, t_{i+h}, c_k) | t_j \in I_k^a, i \leq j \leq i + h \wedge c_k \in \mathcal{C}\}_{1 \leq l \leq m}$$

where $\delta(k, t_i, t_{i+h}, c_k)$ denotes an interval starting from t_i and ending to t_{i+h}, and I_k^a represents the domain of values for the function ϕ_a associated to the label $c_k \in \mathcal{C}$ extracted using the discretization process presented below. For instance, for the temperature time series in the wireless sensor network domain we firstly compute its discretization obtaining the following intervals

$$\begin{array}{ll} I_t^{vl} = \{x | x < 13\}, & I_t^l = \{x | 13 \leq x < 22\},\ I_t^m = \{x | 22 \leq x < 31\} \\ I_t^h = \{x | 31 \leq x < 40\},\ I_t^{vh} = \{x | x \geq 40\}. & \end{array}$$

Then we define the abstraction function as

$$\phi_t((t, x)_{1\_n}) = \{\delta_t(l, t_i, t_{i+h}, c_k) | t_j \in \mathcal{D}_k^t, c_k \in \mathcal{C}_t\},$$

with labels set to $\mathcal{C}_t = \{$ *very_low*, *low*, *medium*, *high*, *very_high* $\}$.

3.2 Relational Interval Sequences

Now that we have discretized the time series into intervals, we can extend the definitions of both sequences and patterns to the case of interval-based relational sequences.

Definition 5 (Relational Interval Sequence). Given a set $\mathcal{T}$ of time series and the sets $\mathcal{C}_1, \ldots, \mathcal{C}_{|\mathcal{T}|}$ of descriptive labels, a *relational interval sequence* is a sequence of relational atoms

$$\delta_{a_1}(id_1, b_1, e_1, v_1), \delta_{a_2}(id_2, b_2, e_2, v_2), \ldots, \delta_{a_n}(id_n, b_n, e_n, v_n)$$

where $v_j \in \mathcal{C}_i$ is a descriptive label, b_j and e_j represent, respectively, the starting and ending time, $id_j \in \mathbb{N}$ represents the interval identifier, and δ_{a_j} is the corresponding name of the time series $a_j \in \mathcal{T}$. (The interval $\delta(id, b, e, v)$ can be written also by means of three literals as $\delta(id, v)$, *begin*(*id*,*b*), *end*(*id*,e)).

In particular, a relational interval sequence can describe several labeled interval sequences into a single one, enabling one to take into account the multivariate analysis in case of different time series. Relations between time intervals are described adopting the Allen's temporal interval logic [5], as reported in Figure 3.

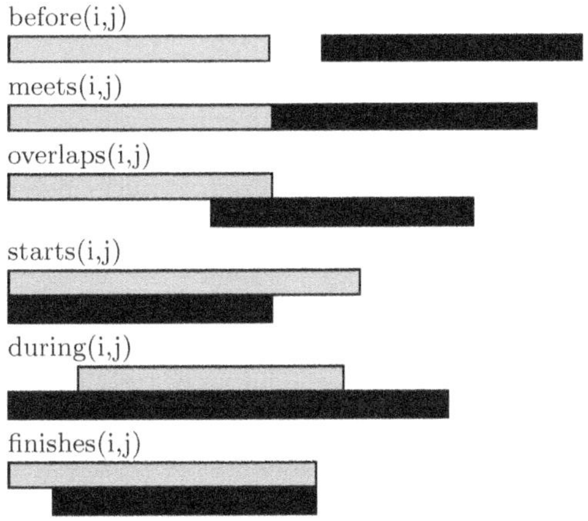

Fig. 3 Allen's temporal intervals [5]

Definition 6 (Relational Interval Pattern). Given $\mathcal{S}$, the set of interval relation symbols, a *relational temporal pattern* is a set of relational atoms

$$P = I \cup R = \{\delta_i(id_i, b_i, e_i, v_i)\}_{i=1\ldots n} \cup \{rel_j(id_j^1, id_j^2)\}_{j=1\ldots m}$$

where $rel_j \in \mathcal{S}$, and $\forall rel_j(id_j^1, id_j^2) \in R\ \exists \delta_h(id_h, b_h, e_h, v_h), \delta_k(id_k, b_k, e_k, v_k) \in I$ such that $id_j^1 = id_h$ and $id_j^2 = id_k$.

4 Related Work

The extraction of useful knowledge from raw sensor data is a difficult task and conventional tools might not be able to handle the massive quantity, the high dimensionality and the distributed nature of the data. For such reasons, in recent years a great interest emerged in the research community in applying data mining techniques to the large volumes of sensor data [1]. Specifically, the exploitation of data mining approaches could be a key issue for summarizing the data into events and for elaborating further adaptive tactical decisions or strategic policy.

Many works are presented in literature concerning the topic of distributed data mining [26], spatial data mining [10], and temporal data mining [27]. In a more general context, there is a growing interest in applying data mining techniques to sensor data [13] that operate mainly on a centralized data set, as we proposed, rather than providing mechanisms for in-network mining.

However, most of the existing techniques operate on an attribute-value descriptions of the (spatio, temporal and spatio-temporal) data and sensors involved in the network adopting a point-based event approach in order to discover useful patterns. On the other hand, very few works face the challenge of discovery temporal patterns using an interval-based approach but in some cases they do not use a relational language [17, 14, 21] and hence they cannot represent interval-based relations, and in other cases, even considering some kind of relations among temporal intervals [25] they are able to represent both intervals and their relations but they cannot completely describe the network, the sensors involved in it and the data (spatio, temporal and spatio-temporal) gathered from the sensors. Furthermore, spatial data mining and similarity-based approaches were designed to tackle into account complex representations with a relational language, however without considering temporal-based relations [8, 18, 12].

The work presented in this chapter can be related to that proposed in [19], optimized in [20]. In these works [19, 20], the authors represent a single sequence as a set of predicates and temporal relations. Each predicate is assumed to be hold in a given temporal interval, while the temporal relations are predicates expressing the Allen's temporal correlation between two predicates. Furthermore, each predicate is associated to a unique symbolic identifier indicating a specific temporal interval, and temporal relations are expressed between those identifiers. Hence, every time a predicate is used in a sequence, it is implicitly assumed that it corresponds to a fluent predicate. In this way, it is not possible to introduce predicates that only express a structural relation between objects, i.e. between sensors or (temporal, spatial) events involved in the network.

Furthermore, as reported in [19], the algorithm they presented is not applicable to real world problems due to its high complexity. Indeed, they specialize a pattern by adding a literal, or by variable unification, or by introducing k^n (where k is the number of different Allen's relation and n corresponds to the number of possible predicate pairs) temporal restrictions between predicate pairs leading to an exponential time complexity.

On the contrary, the framework we presented in this chapter can be used to solve complex temporal data mining tasks by using a relational interval-based description as shown by the outcome obtained by the application of the proposed framework to a real world wireless sensor network data (see Section 5). Furthermore, it is based on a powerful and general purpose multi-dimensional relational pattern mining system [9] and extends it with new dimensional operators thus allowing one to be able to represent and handle spatial (or other dimensional) information gathered form the network. Finally, the framework was extended to automatically provide an interval-based description of the temporal data.

As regards the language used to describe both sequences and patterns, it has some similarities with the Planning Domain Definition Language (PDDL) proposed in [24]. Adopting PDDL as a representation language could make our approach directly applicable to specific planning real world domains.

5 Experiments

In order to evaluate our approach, we used the data, freely available from [15], collected from a wireless sensor network made up of 54 Mica2Dot sensors deployed in the Intel Berkeley Research Lab and arranged in the laboratory as shown in Figure 4.

A sensor network node is a small autonomous unit, often running on batteries, with hardware to sense environmental characteristics, such as temperature, humidity and light. Such nodes usually communicate using a wireless network. A sensor network is composed of a large number of sensors deployed in a natural environment. The sensors gather environmental data and transfer the information to the central base station with external power supply.

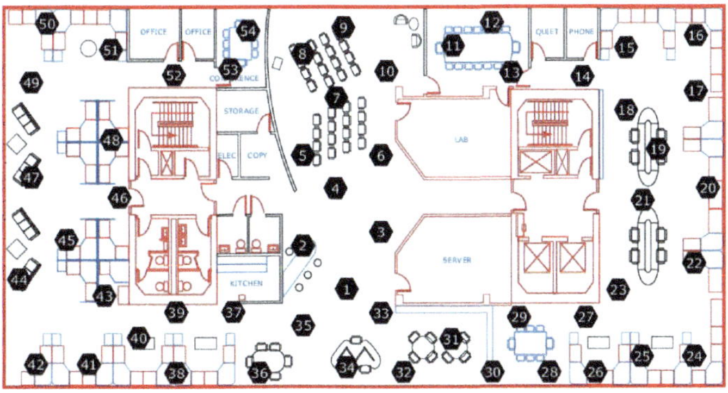

Fig. 4 Sensors in the Intel Berkeley Research lab

The 54 sensors have been monitored from February 28th to April 5th 2004, and the data, about 2.3 million readings, were collected using the TinyDB in-network query processing system, built on the TinyOS platform. Each sensor collected topology information, along with humidity, temperature, light and voltage values once every 31 seconds.

We selected the measurements (temperature, humidity, light and voltage) from the sensors 31, 32, and 34, for the time period from 2004-03-10 to 2004-03-13 corresponding to 16253 log rows. The aim is to discover some correlations between sensors and/or measurements useful for anomaly detection. For instance, there is a strong correlation between the temperature and humidity, as we can see from the Figure 5 that reports the corresponding graphs for the sensor 41. The first task is to discretize the time series corresponding to each information in order to obtain an interval-based temporal sequence, where each interval is labeled with a specific name.

The discretization step has been executed exploiting the functions ϕ_t, ϕ_h, ϕ_l, and ϕ_v with the corresponding domains $\mathcal{I}_i^j$, obtained with the algorithm presented in Section 3, where i is the time series name (temperature, humidity, light and voltage) and j is the descriptive label associated to the interval:

$$\begin{array}{lll}
\mathcal{I}_t^{t1} = \{x|x < 18.9\}, & \mathcal{I}_t^{t2} = \{x|18.9 \leq x < 21.3\}, & \mathcal{I}_t^{t3} = \{x|21.3 \leq x < 25.8\}, \\
\mathcal{I}_t^{t4} = \{x|25.8 \leq x < 30.6\}, & \mathcal{I}_t^{t5} = \{x|30.6 \leq x < 31.1\}, & \mathcal{I}_t^{t6} = \{x|x \geq 31.1\} \\
\mathcal{I}_h^{h1} = \{x|x < 32.6\}, & \mathcal{I}_h^{h2} = \{x|32.6 \leq x < 38.2\}, & \mathcal{I}_h^{h3} = \{x|38.2 \leq x < 42.9\}, \\
\mathcal{I}_h^{h4} = \{x|42.9 \leq x < 43.8\}, & \mathcal{I}_h^{h5} = \{x|43.8 \leq x < 46.3\}, & \mathcal{I}_h^{h6} = \{x|x \geq 46.3\}, \\
\mathcal{I}_l^{l1} = \{x|x < 2.7\}, & \mathcal{I}_l^{l2} = \{x|2.7 \leq x < 16\}, & \mathcal{I}_l^{l3} = \{x|16 \leq x < 84.6\}, \\
\mathcal{I}_l^{l4} = \{x|84.6 \leq x < 176.6\}, & \mathcal{I}_l^{l5} = \{x|x \geq 176.6\}, & \\
\mathcal{I}_v^{v1} = \{x|x < 2.5\}, & \mathcal{I}_v^{v2} = \{x|2.5 \leq x < 2.6\}, & \mathcal{I}_v^{v3} = \{x|x \geq 2.6\}
\end{array}$$

Adopting these functions we obtained a temporal sequence made up of 1249 intervals (138 for temperature, 427 for humidity, 117 for light and 612 for voltage). Then we added all the Allen's temporal relations between the intervals (836729 before, 1558 meets, 13714 overlaps, 122 starts, 11945 during, 134 finishes and 60 matches atoms) obtaining a relational sequence of about 868000 literals. The following literals represent a fragment of a sequence describing the relational representation of some time series, where each interval is described by three predicates

α(`sensor, interval, label`), `begin(interval, s)`, `end(interval, e)`
where $\alpha \in${`temperature, humidity, light, voltage`}.

```
temperature(31,i1,it3). begin(i1,0). end(i1,22).
humidity(31,i2,ih5). begin(i2,0). end(i2,30).
...
starts(i1,i2). before(i2,i3). ...
```

Table 1 reports the results of the algorithm when applied on the sequence previously described and using two different values for the minimum support. The fourth column reports the number of patters belonging to all the possible

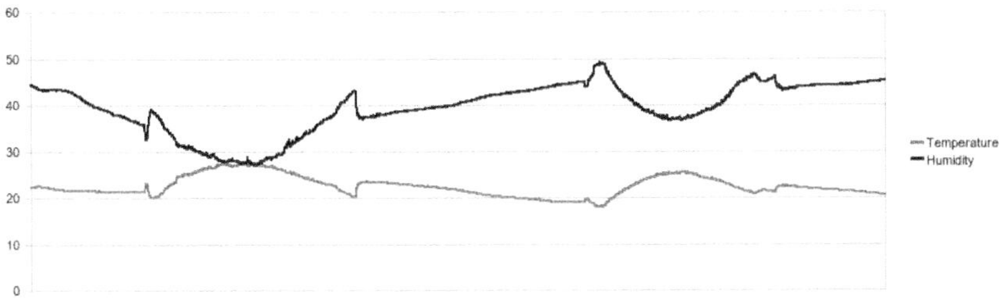

Fig. 5 Correlation between temperature (bottom) and humidity (top) time series

Table 1 Detailed results of the experiments

MinSupport	Level	Specializations	Candidates	Maximals	Time (secs)
10%	1	17	10	307	319.3
	2	166	40		
	3	666	170		
	4	2340	344		
	5	3864	200		
	6	1806	0		
15%	1	17	9	246	277.8
	2	150	34		
	3	568	141		
	4	1882	254		
	5	2784	141		
	6	1272	0		
20%	1	17	9	194	234.7
	2	150	33		
	3	555	122		
	4	1618	206		
	5	2293	55		
	6	494	0		

specializations whose support is greater than MinSupport. The fifth column reports the number of maximal patterns fulfilling all the constraints obtained by the algorithm.

Some interval-based patterns discovered by the algorithm and expressing the time correlation and the information correlation are:

temperature(_,A,B), before(A,C), temperature(D,C,E),
  18.95 $\leq$ B < 21.35, 21.35 $\leq$ B < 25.85, mote31(D) $[s = 28.4\%]$,
temperature(A,B,C), meets(B,D), temperature(A,D,E),
  18.95 $\leq$ C < 21.35, 21.35 $\leq$ E < 25.85 $[s = 24.8\%]$,
temperature(A,B,C), meets(B,D), temperature(A,D,E),
  21.35 $\leq$ C < 25.85, 18.95 $\leq$ E < 21.35 $[s = 26.2\%]$.

6 Conclusion

In this chapter a relational language useful to describe the temporal nature of a sensor network has been proposed, and a relational learning technique able to discover interesting and more human readable patterns relating spatio-temporal correlations has been implemented.

The framework, already presented in [9] has been extended in order to take into account interval-based temporal data along with contextual information about events occurring in the environment. The extension concerns the introduction of interval-based operators, based on the Allen's temporal interval logic [5], in the sequences. Firstly, an abstraction step with the aim of segmenting and labelling the real-valued time series into similar subsequences is performed exploiting a kernel density estimator approach. The knowledge is enriched by adding interval-based operators between the subsequences obtained in the discretization step, and the relation pattern mining algorithm has been extended in order to deal with these new operators.

In order to evaluate the validity of both the abstraction step and the general extended framework, an experimental session on real world data collected from a wireless sensor network has been presented.

References

1. International Workshop on Knowledge Discovery from Sensor Data (SensorKDD) (2007-2008-2009)
2. Agrawal, R., Manilla, H., Srikant, R., Toivonen, H., Verkamo, A.: Fast discovery of association rules. In: Fayyad, U., Piatetsky-Shapiro, G., Smyth, P., Uthurusamy, R. (eds.) Advances in Knowledge Discovery and Data Mining, pp. 307–328. AAAI Press, Menlo Park (1996)
3. Akyildiz, I., Su, W., Sankarasubramanian, Y., Cayirci, E.: A survey on sensor networks. IEEE Communication Magazine 40(8), 102–114 (2002)
4. Akyildiz, I.F., Su, W., Sankarasubramaniam, Y., Cayirci, E.: Wireless sensor networks: a survey. Computer Networks 38, 393–422 (2002)
5. Allen, J.: Maintaining knowledge about temporal intervals. Commun. ACM 26(11), 832–843 (1983)
6. Basile, T.M.A., Mauro, N.D., Ferilli, S., Esposito, F.: Relational temporal data mining for wireless sensor networks. In: Serra, R. (ed.) AI*IA 2009. LNCS, vol. 5883, pp. 416–425. Springer, Heidelberg (2009)
7. Biba, M., Esposito, F., Ferilli, S., Di Mauro, N., Basile, T.: Unsupervised discretization using kernel density estimation. In: Proceedings of the Twentieth International Joint Conference on Artificial Intelligence (IJCAI 2007), pp. 696–701 (2007)
8. Malerba, D., Lisi, F.: An ILP method for spatial association rule mining. In: Working notes of the First Workshop on Multi-Relational Data Mining, pp. 18–29 (2001)
9. Esposito, F., Di Mauro, N., Basile, T., Ferilli, S.: Multi-dimensional relational sequence mining. Fundamenta Informaticae 89(1), 23–43 (2008)

10. Ester, M., Kriegel, H.P., Sander, J.: Algorithms and applications for spatial data mining, vol. 1(Part 4), ch. 7, pp. 160–187. Taylor and Francis Group, Abington (2001)
11. Estrin, D., Culler, D., Pister, K., Sukhatme, G.: Connecting the physical world with pervasive networks. IEEE Pervasive Computing 1(1), 59–69 (2002)
12. Ferilli, S., Basile, T., Biba, M., Di Mauro, N., Esposito, F.: A general similarity framework for horn clause logic. Fundamenta Informaticae 90(1-2), 43–66 (2009)
13. Ganguly, A.R., Gama, J., Omitaomu, O.A., Gaber, M.M., Vatsavai, R.R.: Knowledge Discovery from Sensor Data. CRC Press, Inc., Boca Raton (2008)
14. Hoppner, F.: Learning dependencies in multivariate time series. In: Proc. of the ECAI Workshop on Knowledge Discovery in (Spatio-)Temporal Data, pp. 25–31 (2002)
15. Intel Berkeley Research Lab, http://db.csail.mit.edu/labdata/labdata.html
16. Jacobs, N., Blockeel, H.: From shell logs to shell scripts. In: Rouveirol, C., Sebag, M. (eds.) ILP 2001. LNCS (LNAI), vol. 2157, pp. 80–90. Springer, Heidelberg (2001)
17. Kam, P., Fu, A.W.: Discovering temporal patterns for interval-based events. In: Kambayashi, Y., Mohania, M., Tjoa, A.M. (eds.) DaWaK 2000. LNCS, vol. 1874, pp. 317–326. Springer, Heidelberg (2000)
18. Koperski, K., Han, J.: Discovery of spatial association rules in geographic information databases. In: Egenhofer, M.J., Herring, J.R. (eds.) SSD 1995. LNCS, vol. 951, pp. 47–66. Springer, Heidelberg (1995)
19. Lattner, A., Herzog, O.: Unsupervised learning of sequential patterns. In: ICDM Workshop on Temporal Data Mining: Algorithms, Theory and Applications (2004)
20. Lattner, A., Herzog, O.: Mining temporal patterns from relational data. In: Lernen Wissensentdeckung Adaptivität (LWA), GI Workshops, pp. 184–189 (2005)
21. Laxman, S., Unnikrishnan, K., Sastry, P.: Generalized frequent episodes in event sequences. In: 8th ACM SIGKDD Int. Conf. on Knowledge Discovery and Data Mining, Workshop on Temporal Data Mining (2002)
22. Li, Q., Racine, J.: Nonparametric Econometrics: Theory and Practice. Princeton University Press, Princeton (2007)
23. Mainwaring, A., Culler, D., Polastre, J., Szewczyk, R., Anderson, J.: Wireless sensor networks for habitat monitoring. In: Proceedings of the 1st International Workshop on Wireless sensor networks and applications, pp. 88–97. ACM, New York (2002)
24. McDermott, D., Hove, A., Knoblock, C., Ram, A., Veloso, M., Weld, D., Wilkins, D.: PDDL - The Planning Domain Definition Language. Yale Center for Computational Vision and Control (1998)
25. Papapetrou, P., Kollios, G., Sclaroff, S., Gunopulos, D.: Discovering frequent arrangements of temporal intervals. In: IEEE ICDM, pp. 354–361 (2005)
26. Park, B.H., Kargupta, H.: Distributed Data Mining: Algorithms, Systems, and Applications, pp. 341–358 (2002)
27. Roddick, J.F., Spiliopoulou, M.: A survey of temporal knowledge discovery paradigms and methods. IEEE Transactions on Knowledge and Data Engineering 14(4), 750–767 (2002)
28. Ullman, J.: Principles of Database and Knowledge-Base Systems, vol. I. Computer Science Press, Rockville (1988)

Content-Based Retrieval of Distributed Multimedia Conversational Data

Vincenzo Pallotta

Abstract. In this chapter we define the notion of multimedia conversational system and provide a classification schema for characterizing the type of content which can be produced, stored and retrieved with these systems. This classification schema will be used in reviewing the capabilities of a number of existing multimedia conversational systems and to assess requirements for advanced indexing and retrieval of conversational content. To meet these requirements, we present new types of indexing techniques and provide an evaluation of their effectiveness on real case studies.

1 Introduction

A large number of systems nowadays allow people to communicate and interact from different locations through the Internet. These systems provide both synchronous services such as Instant Messaging, VoIP, video-conferencing and asynchronous services such as social networks, blogs and discussion forums. What all these systems have in common is that they allow multiple users to participate in conversations. The availability of these systems allows not only the actual communication, which to a certain extent could replace face-to-face meetings, but also the possibility of recording the multimedia streams into the computer storage. With the proliferation of these communication means, designing new ways of searching large amount of multimedia content has become a compelling problem. Due to its non-narrative nature, multimedia conversational data can no longer be indexed with the same methods that have been proven effective for indexing textual narrative content (e.g. news articles, encyclopedias, products informations). Even if accurate transcriptions of dialogues were available, indexing only terms would not be sufficient for retrieving relevant information. Narrative language is

Vincenzo Pallotta
Department of Computer Science
Webster University, Geneva
e-mail: pallotta@webster.ch

Soro et al. (Eds.): Inform. Retrieval and Mining in Distrib. Environments, SCI 324, pp. 183–212.
springerlink.com

on the one hand, more informative and it has lower illocutionary force[1] than language used in conversations. In other words, it tells a story about a topic and it is not supposed to perform actions with language. Conversations, on the other hand, are aimed at pursuing (not necessarily informative) goals e.g., reaching an agreement in negotiations, reconciling different points of view, or elaborating an idea. As an example illustrating the different requirements for indexing conversational content, consider a conversation between two people who are talking about their political views. As a matter of fact, it is very difficult to assess if they share the same orientation or they have conflicting opinions, unless one fully reads and deeply understand the whole exchange.

Due to the pervasiveness of conversational content on the Internet, understanding conversations can be considered as a new outstanding challenge in modern information retrieval. The notion of relevance needs to be modified accordingly, not just because of the type of content (i.e. the conversational text) but also because of the type of queries that may be asked about this type of content. Queries are no longer sets of keywords that the user expects to be included in the text, but questions about what actually happened during the conversation and about the outcome of the conversational process. It is also apparent that classical information retrieval (IR) techniques for relevance ranking are no longer adequate (e.g. frequency-based, link-based). Moreover, there are communication phenomena that are hardly captured by simple term-based indexing such as speech overlaps, disfluencies, retractions, delays in answering and even silence (or long delays in replying), which can have, under certain circumstances, a perfectly defined meaning.

In this chapter we analyze multimedia conversational systems from two main perspectives:

1. The *type of interaction* that is enabled by a system, the type of media used, and the manner in which the interaction is captured.
2. The *type of access* suitable for the captured conversational content, and the technologies that might enable this type of access.
1.

The chapter is organized as follows. In section 2, we present a classification schema for characterizing the type of conversational content which is produced and recorded. This schema will then be used in section 3 in reviewing a number of multimedia conversation systems and to assess what types of processing and content analysis techniques are required to understand and answer the typical questions that the user might ask about the captured conversational content. In section 4, we propose a new definition of relevance that we prove to be more adequate in new types of retrieval tasks. We also present some new techniques of analysis especially tailored to meet the requirements of specific users when facing the retrieval of content from conversations. Section 5 concludes the chapter by providing a roadmap for future developments of this research domain.

---

[1] The illocutionary force of an utterance is defined as the effect that the speaker intends to achieve in producing that utterance.

2 Multimedia Communication Systems

A multimedia conversation system[2], henceforth a MCS, can be defined as a computer-mediated multi-party communication system that optionally provides ways to capture, store, index, retrieve and visualize the content of conversations.

If we keep this definition so general it might easily include a wide variety of computer systems including the whole Web. But there is a fundamental difference that we need to highlight here. A conversation is a temporal event where several parties interact for a limited amount of time, on a specific topic, and for a specific purpose. A Web page can support this type of activity, but it might also have other well-defined purposes such as, for instance, distributing information, selling products or playing games. Mutual (or bi-directional) information exchange is an essential constituent of conversations, whereas Web documents generally support unidirectional flow of information.

Another fundamental feature of conversations is that parties are performing actions by means of their participation. Even if they do not contribute actively to the exchange, their presence is already a form of interaction (e.g. they may be listening or watching). Temporal and physical co-location is not a pre-condition, but it might be necessary in some specific conversational systems. This means that communication can be synchronous or asynchronous, remote or co-located. Moreover, these categories are not mutually exclusive and can be combined in hybrid systems.

The term multimedia in MCS refers to the fact that the means of contributing to the conversation as well as those for capturing, storing, indexing and retrieving the content of the conversations can be supported by different media at the same time (e.g., speech, text, pictures, video, gestures, and hand-writing). However, when we talk about capturing the input we might want to use the term "modality". In essence, multimodal input is just one particular feature of a MCS where the parties can choose and use different media to express their contributions. The types of media supported by a given MCS essentially determine:

1. the richness of the conversational content, depending on the types of inputs that can be captured;
2. the ways in which the captured content can be stored, indexed, retrieved and visualized.

In many situations, the MCS need to convert one media into another in order to make the above processes possible. Media conversion is typically an information-loosing process and sometimes all the media (captured and derived from transformation) need to be preserved in the system for specific purposes. However, one captured media may be too rich for some specific retrieval purposes. Consider for instance the situation of an application that needs to count the number of male and

[2] The term "conversation system" has been recently used in literature with a different meaning than the one discussed in this chapter, namely to refer to what is more commonly called a "man-machine dialogue system" [Teng et al. 2008; Teng et al. 2009].

female callers to a call center for statistical purposes. An automatic system could just sample a few seconds of the calls and to easily determine the sex of the caller from the pitch of the voice without having to record the whole conversation. This transformation is clearly information-loosing but still effective in purpose.

In contrast, sometimes keeping all the media together, original and transformed, might be convenient and even necessary. Consider the scenario a conference call where one of the participants is hearing impaired. In order to provide easy access to all the participants, we might think of automatically converting speech to text to help the hearing impaired. We still want to keep the speech signal together with the generated text in order to let the non-impaired participants to use their more convenient modality, speech.

The above scenarios exemplify two main categories of media processing in MCS, namely *summarizing* and *enriching* information. Typically, summarized content is useful to deal with information overload, while enriching is most suitable for providing better or different access to information.

2.1 Conversational Content

Let us now consider a different dimension for classifying MCS. Conversations can have different purposes and this is typically determined by the type of content that is exchanged. Even very informal conversations are purposeful.[3] It might be strange to classify MCS according to a conversational purpose, but if one considers that certain capabilities can enable different types of interaction, this classification dimension appears logical. Consider for instance, a video-conferencing compared to e-mail exchange. It is apparent that a brainstorming activity is less easy to be performed with e-mail than by using a more synchronous system such as video-conferencing. Moreover, compared to audio alone, video-conferencing has the advantage of allowing participants to interact through more efficient channels such as, for instance, by seeing each other or by means of a shared blackboard.

With respect to this dimension, we mainly distinguish between two broad categories of conversational content:

- *Task-oriented conversations*: have a defined purpose. For instance, people meet in order to solve problems, take decisions, solve conflicts, elaborate ideas, report about some activity, receive approval, seek and transfer knowledge, improve acquaintance, etc.
- *Information sharing conversations*: still have a purpose but it is less defined. Different participants might have different purposes, even conflicting ones. One may call these casual conversations. While it may seem that conversations of this type are more likely to occur within synchronous communication (e.g. telephone calls

[3] By definition, conversation is an informal exchange of information between two or more parties that typically takes places orally. We extend this definition to a broader sense which includes various degrees of formality and the use of different (and, possibly, multiple) modalities.

or instant messaging), they are becoming more and more frequent in social networks or discussion forums. A conversation might start out of a comment about a post and diverge towards unpredictable directions, although keeping the participants interest alive.

The need for differentiating between these two types of conversation arise from the goal of finding efficient techniques for capturing the essential elements of the conversations for further indexing and retrieval. For example, a decision-making conversation will contain utterances that are proposals as well as contributions whose goal is to express agreement and disagreement. These types of contributions are keys to capture the rationale of the conversation when we want, for instance, to build summaries of the discussion. In other words, task-based conversations are likely to contain specific types of contributions whose goals are to advance towards the completion of the task at hand.

As for information-sharing conversations, these are pragmatically[4] less structured. One can still expect to find contributions similar to those of task-based conversations, but it less important to recognize the global purpose of the conversation. This does not mean that information-sharing conversations have a less rich conversational content. In reality, the content is very rich and extremely difficult to analyze due to the intrinsic freedom and unpredictability.

2.2 Conversational Support

We use the term conversational support to refer to the types of interactions that are enabled by MCSs. Conversations can be carried out in several fashions. For instance, *face-to-face meetings* with more than two people can benefit of the co-presence to enable parallel direct channels between any subset of participants for private communications. This feature enables the spontaneous creation of sub-groups in face-to-face meetings. In contrast, these private channels are not available for more than two people in remote conferencing systems. This problem has been addressed in the new collaboration platform, Wave[5], recently created by Google. In Wave it is possible to create a private sub-conversation within a larger conversation. The sub-conversation is only visible to the participants that are members of a selected sub-group of people. Actually, the system is even more flexible because it allows inviting in the sub-conversation people that are not even in the containing conversation. This feature enables the creation of private "hyper sub-groups" that cannot be created in face-to-face meeting without the help of additional technologies[6]. These technologies not only allow overcoming the

---

[4] Pragmatically used in the sense of "pragmatics", the linguistic discipline.

[5] `http://wave.google.com`

[6] One can imagine calling remote participants by phone and including them in the private sub-group conversation. But this requires an additional technology (i.e., the phone) and the remote participants do not have an identical access to the interaction as physically present ones.

interactional limitations of some system, but also help in creating new interactional opportunities.

We now introduce in our classification schema the concept of *conversational affordance* in order to characterize what conversational opportunities are offered by a MCS. We distinguish between two broad categories of conversational affordances:

1. *Natural conversational affordances*. This category includes all those interactional opportunities that are typically available in non-mediated conversational systems and contribute to the feeling of making conversations more natural. For instance, video in a remote-conferencing system makes interaction closer to face-to-face meetings.
2. *Enhanced conversational affordances*. This category includes all the system features that enable interactions that would not be possible in non-mediated systems. For instance, the use of an animated avatar in a cyberworld (e.g. SecondLife) would enhance the interaction through a customized virtual physical identity.

2.3 Information Architecture

By means of the category information architecture we intend to characterize how information is organized and managed in a MCS. Being mostly based on different media, information needs to be *integrated*. For instance, a MCS for video-conferencing that is capable to record both audio-video streams and notes from participants would be better integrated if the notes were time-stamped and aligned with the audio-video stream.

In MCSs information can be captured from several sources and possibly be stored into a computer memory. This can be done in a *centralized* or *distributed* way. For instance, in a video-conferencing system, the audio-video stream captured from remote sources could be recorded in a central server. This would be more difficult to achieve if the system is based on a peer-to-peer technology. In some cases, it is the very nature of the conversation that imposes one or the other of the two approaches. Consider for instance interlinked blog posts. Taken together, they constitute a conversation which is distributed over different Web sites.

Another important feature of MCS's information architecture is the possibility of enriching the media with *metadata*. Metadata describe the different pieces of multimedia content according to a selected ontology. Metadata are important when the content itself cannot be directly indexed or when some relevant information contained in one media is available in another media. As an example, consider a bi-modal meeting capturing system that captures audio streams and interactions with shared documents (e.g. it tracks the text under the mouse pointer). The system is capable of both transcribing the speech from the audio stream and aligning the transcript with information about which part of the document is being currently used by the participants. With this method, the speech transcript is enriched with information that allows improving indexing by resolving verbal references with their referents taken from the documents. For example, if one says "the author" pointing at the document, the system will add the content of the author of the document as metadata for the spoken words. This strategy will allow finding information that would otherwise be impossible to retrieve by indexing only one modality [Lalanne et al. 2005].

A MCS contains several media that are typically tied together by some relationship. The simplest example is temporal alignment of audio, video streams and captions. We will use the notion of *layered media* in a MCS if relationships among media are captured and managed. In other words, media can be combined or mashed-up (in a predefined or customized way) in order to have different views on the conversation. A layered MCS can exploit information from different layers (even in real-time) for enhancing the conversational experience. For example, an augmented reality conversation system could show the notes taken by another participant in real time on a participant's head-mounted displays.

MCSs are supposed to be capable of capturing conversation and making available the content for later use in possibly different applications. It is important that all the layers and metadata of a MCS are *interoperable* with other applications for further processing. Most of the current MCSs are based on non-standard proprietary (often undisclosed) representations. This means that the user is either limited by the MCS capabilities or locked in with the MCS vendor if the data are only interoperable with the same vendor's products. If in contrast MCS data were interoperable, then they could be easily aggregated or mashed-up to obtain enhanced experiences or better access. Interoperability is obtained by adopting industry standard such as XML or MPEG.

Finally, a MCS is *extensible* if new functionalities can be added through plugins or other systems that communicate to the MCS via public API[7]. One notable example of extensible and interoperable MCS is the micro-blogging platform Twitter[8]. Interoperability is obtained through public APIs and it has sparked a whole ecosystem of additional applications beyond simple clients.

2.4 Indexing and Retrieval

Many MCS already provide indexing and retrieval capabilities, especially those systems that include text as one of their constitutive media. However, text-based indexing only enables keyword search and it could be insufficient in conversational information retrieval. Consider the following straightforward example. In indexing the content of Twitter conversations, one would like to retrieve parts of the conversation where there was a high interaction (i.e. high density of exchanges) about a given topic. By indexing the text of each tweet without calculating the

---

[7] API stands for Application Programming Interface.

[8] Twitter is a micro-blogging platform: `www.twitter.com`.

density of the exchanges (e.g. the number of exchanges per minute) we might end up in retrieving uninteresting loose interaction (e.g. replies occurring after a long lapse of time). This example shows that in conversation we need to define additional and different notions of relevance. In this specific case, relevant are only those results that contain terms related to the topic and that temporally occur within a short time interval.

The above is a straightforward example that shows how relevance may be different from its classical notion in information retrieval. Thing can easily become much more complicated in cases where users need to access to information related to the dynamics of the conversation, rather than to the content of the conversation. In these cases we need to take into account the *pragmatic level* of conversation. In other words, we need to understand and provide metadata about what purpose the participant intended to achieve by producing a certain contribution to the conversation. For example, in a decision-making conversation we need to detect if a contribution counts as a proposal, an explanation, an agreement, a disagreement, etc. Only by doing that we would be able to answer questions such as "why was the proposal of buying a new Apple computer rejected?" [Pallotta et al. 2007]. Pragmatic information could also be useful in synchronous MCS during interaction. Consider for instance a system that recognizes the participant's status of attention just by detecting if he/she is or not in front of the screen[9]. This information can be used by the system for several purposes, including starting recording the conversation during the absence.

The *semantic level* is also important, not onls for textual media but for other types of media. In order to achieve semantic descriptions of conversational contributions we need to extract features from each contribution and map them onto semantic categories (e.g. from ontologies). A simple example would be that of MCS where it would be possible to determine the *frame*[10] associated to a given situation. In the context of MCS we could see two applications of this concept. The first applies to the analysis of recorded conversations in order provide metadata for adequate indexing. This can be, for instance, the recognition of call types (e.g. complaints, troubleshooting, and inquiry) in a help-desk call center, for classification purposes.

The second application is during the conversation. The system can be instructed that the ongoing conversation serves a purpose for which a pre-defined structure is identified. During the conversation the MCS can detect whether the conversation is on track with the selected structure and provide a measure of how much the process in converging to the goal set (e.g. taking a decision). For example, again in the help-desk scenario, the system can instruct the operator on what questions to ask the caller or detect if the conversation is becoming too "hot" and advise the operator to cool it down.

---

[9] This can be easily achieved by processing the video stream with a face-detection algorithm.

[10] We adopt the terminology of Frame Semantics [Fillmore 1977], where a situation is characterized by a set of inter-related semantic elements.

2.5 Usability

The last (but not least) dimension of MCS classification is *usability*. What we mean by usability in this chapter has a narrow scope compared to the general notion of usability in computer systems and interfaces[11]. We focus here on a restricted notion of usability that is tailored for the MCS's interfaces. To exemplify it, consider the scenario where someone wants to contribute to a discussion in a public forum, by criticizing the overall contribution of another participant to the whole discussion. Unfortunately the system only allows replying to individual posts of a thread. The alternative would be creating a new thread of the discussion. But still the reply is "relevant" to all the posts of the participant being criticized. To overcome this limitation the users of this MCS need to agree on some workaround by using some ad-hoc notation to express a global comment on a participant (e.g. refer to the participants with their names preceded by "@").

This example motivates us in defining the following criteria that help us in evaluating the MCS's usability:

1. *Unobtrusiveness*. An unobtrusive MCS interface is one that enables the user to use the available interactional affordances without having to switch attention from the conversational activity to the interface controls.
2. *Cognitive Load*. We want to measure the level of understanding required to learn and use available interactional affordances of the MCS. The higher the level, the less natural the interface will be perceived by the user.
3. *Responsiveness*. A responsive MCS interface timely reacts to the user's input without slowing down the interaction. Notice that this criterion does not cover connection lags or delays. It is more about the number of steps (e.g. clicks) needed in order to activate a relevant functionality during the conversation.
4. *Predictability*. This simply means that the system's behavior is predictable. This include, among other things, consistency of the interface throughout the all the MCS available functionalities. This is particularly important when the interface is loaded from a server into the local client. For instance, consider the case in which a Web-based VoIP interface once loaded has added, changed or removed features without any warning to the user.

As we previously mentioned, with the above criteria we do not intend to cover all the possible usability aspect of MCS interface. If needed, other categories can be added to the above list. In Table 1 we summarize the categories used in our MCS classification schema.

---

[11] Being most of the time MCS computer-based systems, standard usability metrics, standards and evaluation techniques also apply.

Table 1 MCS Classification Schema

Conversational content	**Task-oriented conversations**			**Information-sharing conversations**		
Conversational support	**Natural conversational affordances**			**Enhanced conversational affordances**		
Information Architecture	**Media integration**	**Centralized / distributed**	**Metadata**	**Layered Media**	**Interoperable**	**Extensible**
Indexing and Retrieval	**Standard IR**		**Semantic Indexing**		**Pragmatic Indexing**	
Usability	**Unobtrusiveness**	**Cognitive Load**		**Responsiveness**	**Predictability**	

3 Review of Multimedia Conversation Systems

In this section we review some existing MCSs according to the classification schema presented in section 2. We will consider one only representative MCS for each significant combination of features. Similar system will also be mentioned but not reviewed in details. Due to the large number of individuals in the MCS ecosystem, some will unavoidably be left out. It is important to note that our goal is to show how the classification schema is applied to MCSs and to assess its adequacy in classifying this type of systems. As such, it is not meant to be exhaustive. We also advocate that a specific combination of features from the classification schema require a particular technique for indexing and retrieval. The result of the classification exercise will serve as the basis to discuss the adequacy of available indexing and retrieval techniques, and to provide a roadmap for the development of more suitable ones, whenever appropriate.

3.1 Asynchronous MCSs

The first class we consider is that of asynchronous MCSs. In this category we find many mono-modal systems such as e-mail, newsgroups, micro-blogging. Among multi-modal ones we consider blogs, social media and networks.

3.1.1 Comments in Social Media: YouTube and Joost

YouTube is an interesting case because it provides the possibility of replying to a video post with another video. This means that a "video" conversation is in principle possible although most of conversations consist of comments to a video post or to other comments. According to our classification schema, YouTube is an information-sharing system, which in part justifies the lack of advanced conversation structuring devices. Actually, YouTube does not explicitly support conversations although many users engage in conversations in comments. It is very often the case that a comment does not refer to the video itself but to some other user's comments. From the interactional perspective, it offers two natural conversational affordances: the reply link and thumb-up/down button. In contrast, the video reply option can be considered as an enhanced conversational affordance.

From the Information Architecture perspective, YouTube is a centralized system with a basic level of media integration. Available metadata would be tags, but they are not available for tagging the comments. Additionally, adding a tag to a video can only be made by the video owner and it cannot be used as conversational action by other participants to the conversation (e.g. it is not possible to post a comment that add a tag to a video). YouTube offers developers APIs and provides access to comments and means to post text and video comments. Retrieval of comments is obtained by submitting the video ID through the API which returns as results an XML document containing all the related comments. Limited to its basic conversational functionalities the system is interoperable. While YouTube provides tools to overlay different media (e.g. advertisement over the video), it does not provide any means to overlay conversational content over the video (such as a way of navigating through the video replies while watching the video itself). Functionalities of YouTube are not easily extensible as it might be the case of other social media platforms such as Facebook. As we already mentioned, retrieval of conversational content is made through the API and YouTube does not provides any search facility for comments. From the usability perspective, watching video content is the supposedly primary activity of YouTube. Conversations are byproducts of the commenting feature. This means that an analysis of unobtrusiveness is not appropriate in this context. However, we can consider here a similar service offered (and now discontinued) by Joost[12].

Joost is a commercial on-demand video content provider, which in its early beta version[13] offered a desktop client, shown in Figure 1, where conversations could be carried out while watching of a video content. In practice, a widget was overlaid on the video and people watching the same video could chat about it. Conversations were neither recorded nor used as comments of the videos. They only had a temporary validity. While this provided a conversational affordance, both the video viewing and the conversation experiences were interfering with each other. This led us come to the conclusion that the use of conversational affordance during the movie watching experience add a probably too high level of obtrusiveness. Therefore, conversations around video content are better engaged asynchronously and outside the movie watching activity. YouTube and Joost are prototypical of many other video-based social media such as Vimeo, DailyMotion, etc.

---

[12] `http://www.joost.com`

[13] The Joost player is no longer available for download.

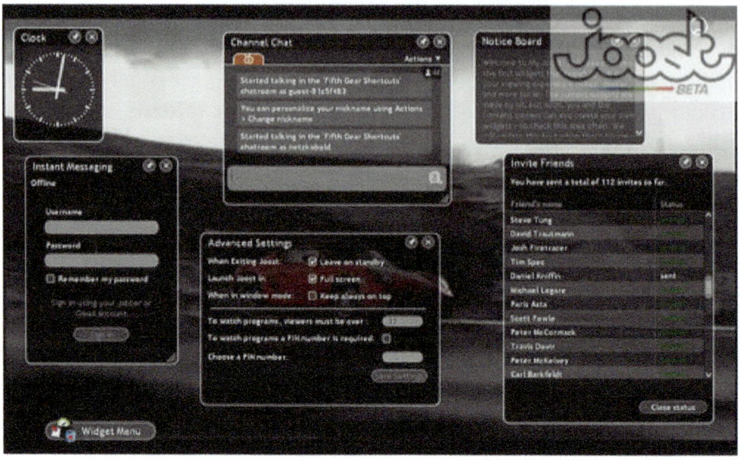

Fig. 1 Joost desktop client with embedded chat

3.1.2 Micro-blogging: Twitter

Another prototypical MCS is the micro-blogging system Twitter. Twitter became very popular because of its innovative way of fostering conversations on the Web. According to our classification schema, Twitter is an information-sharing conversational system which features some very interesting conversational affordances. First, Twitter is an open space where everybody can be heard (read) by everybody connected to the system in nearly real-time. However, Twitter is different than an instant messaging system because replies can be posted at any time and they are persistent. It is therefore an asynchronous system. Twitter provides the following two natural conversational affordances: i) the possibility of replying to one specific user (in private or publicly); ii) the possibility of restricting the scope of the conversation along several dimensions: participants, topic, and location. Twitter provides additional enhanced conversational affordances such as systematically and publicly demonstrating support (or agreement) to a particular message through "retweet" (i.e. quoting a previous message and its author) or to a particular user through addition to lists (i.e. collections of messages from selected users).

It is worth noting that Twitter does not allow any affordance for explicitly replying to individual messages. This limitation eventually affects retrieval because it is not possible to select a message and retrieve, for instance, all the replies to that specific message. This implies that a retrieval system for Twitter which would go beyond simple keyword search would have to heuristically infer the conversational relationship between two messages just from the temporal proximity. This restriction is due to the fact that tweets (Twitter messages) are not messages but author's status updates where one can "mention" other users (i.e. using the "@" prefix). This is a clear case where a technology initially conceived as a mono-directional information sharing tool (i.e. for status sharing) has become a conversational tool.

Twitter is obviously centralized and its support of metadata is restricted to tags (i.e. keywords included in the message and preceded by the symbol "#"[14]). Twitter is interoperable because it provides public APIs, allowing the creation and the retrieval of tweets. Retrieval of tweets can be done by content or by filtering the results on author, friends, mentions (replies), "re-tweets", time and location. An advanced search feature of Twitter is related to sentiment analysis and opinion mining [Pang and Lee 2008]. Twitter provides support for searching tweets with positive and negative attitudes[15]. We consider this as a useful semantic level that enables applications such as online reputation management [Stoke 2008].

Finally, Twitter is not extensible because no plug-ins or links to external applications can be embedded. Related to this aspect, it is interesting to point out that Twitter is fundamentally a social network. This means that the original goal was to build social relationships between users, which is primarily achieved through conversations. Social relationships cannot be maintained without social activities and among those activities conversation is one of the most prominent.

3.1.3 Social Networks: Facebook

Similarly to Twitter, Facebook provides conversational support by allowing the embedded conversation on "walls". A Facebook wall is the open space where conversations happen. In contrast to Twitter there exist as many walls as users and walls are multi-modal (while still asynchronous). Compared to Twitter, Facebook also provides more natural and enhanced conversational affordances. The most basic one is the commenting system, which is however limited to one level of threading (i.e. no comments to comments) and the conversational structure is often simulated by temporal proximity or explicit reference to the content or author of the previous comment. It also includes the approval system (i.e. "likes"). As an enhanced affordance, Facebook provides the ability of including in the conversation contributions that are imported from outside (feeds) or are the result of some actions such as uploading a media, tagging a picture, changing the status, the profile or joining groups, events and causes. However, these actions can only start new conversations and Facebook does not provide any means to qualify these contributions as replies or comments. Another enhanced conversational affordance is the notification system that warns users when a new comment to their post or to a post to which they previously commented has been posted. This affordance helps in tracking multiple conversations when new contributions appear. We consider this affordance as enhanced because in face-to-face conversations the burden of paying attention at how the conversation evolves pertains to the listener.

In Facebook conversations different media can be embedded as part of the main topic (e.g. a page, a link, an event). However, it is not possible to do the same with

[14] The symbol "#" is a convention that has emerged from the users of Twitter. Twitter does not natively provide any specific support for tags.

[15] Testing the sentiment analysis feature, we realized that Twitter returns tweets with the selected attitude based on the use of emoticons.

comments. Facebook is interoperable with other applications since the content of conversations can be created by external applications through feeds and APIs. Facebook is also extensible as it provides a development platform for applications. As it is the case for Twitter and other social networks, Facebook has also become a MCS deviating from its original purpose of being a social network. Facebook provides support for instant messaging to support synchronous conversations for logged-in users. This functionality is not integrated with the Wall or the commenting system but only with the list of friends. It is just an additional functionality that is similar to standard instant messaging technology we will review in the next section on synchronous MCS.

A wealth of metadata is available for Facebook applications developers such as the social graph, profile information, tags for shared media, and comments[16] and this information can be search by means of a the Facebook Query Language (FQL)[17]. However, it does not provide advance support for conversation. One interesting feature that could help in structuring conversations is custom tags[18]. Each element in the Facebook document model (which includes comments) can be tagged with a custom tag within a specific application.

From usability perspective the conversational features of Facebook integrate well with its overall functionality. Conversations with comments and chat are not the primary functionalities but they can be accessed when necessary with minimum effort and with high responsiveness. However, the result of action might not be predictable. Depending on the privacy settings actions can have different visibility. One may not be easily aware of how current privacy settings will affect the visibility of the contributions. For instance, one expects that the comments made to a friend's contribution will be only visible to shared friends. It is not actually the case because Facebook makes visible the comments to all the people who have access to the user's wall and to people who have commented the same contribution, be shared friends or not.

3.1.4 Blogs

Another member of the asynchronous MCS group is the blog technology that was initially conceived as a publishing tool and later became a conversational platform. Blogs are websites whose content is dynamically updated but for which an historical trace is left on the website. Blog content is time-stamped similarly to Twitter (which is a micro-blog whose posts are limited to 140 characters). Blog post can be commented by external viewers. They can contain multiple media and can have references to other blog posts and thus be considered as replies. However, the conversational content is distributed over different sites. This distribution implies that any retrieval technique needs to perform a crawl of the sites where the contributions are stored. This also entails that there may be multiple entry points in the conversations and that one post can contribute to several conversations. In Figure 2, the post A participates in two conversations (B<A<E and A>F<G),

---

[16] http://wiki.developers.facebook.com/index.php/API#Data_Retrieval_Methods
[17] http://wiki.developers.facebook.com/index.php/FQL
[18] http://wiki.developers.facebook.com/index.php/Custom_Tags

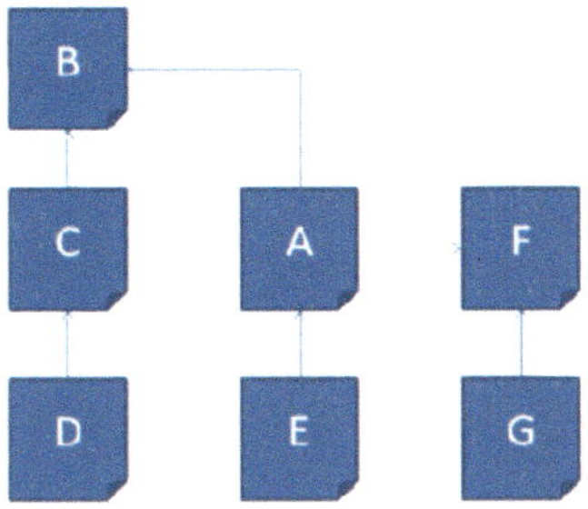

Fig. 2

while the post B has started two independent conversations (B<C<D and B<A<E)[19]. Notice that the same author might have posted multiple posts (e.g. A and D), but also that authors are by default unaware of the replies to their post unless they explicitly search for them. In order to avoid this loss of information, blogging platforms have enabled an acknowledge system that inform the author of a post when the post is referenced in another blog post. This mechanism is referred as *linkback*[20]. This mechanism partially solves the above issues by creating a centralized (and possibly replicated) version of the conversation that contains external blog posts as comments.

Being actually fully fledged websites, blogs offer a wealth of media and can contain as many metadata as possible. As a matter of fact, blogs are hosted in content management systems and the way the media are integrated can be sometime rigid. From the indexing and retrieval perspective, blogs still retain their document connotation. Blog search engines (such as Technorati) indexes blogs on their textual content. Additionally, relevance ranking is modified and restricted to inter-blogs links. In such a way it would be impossible to retrieve all blog posts that commented (linked) a specific post. Moreover, inter-blogs comments are not treated separately and in the best case they are simply excluded from the index. While blogs represent a huge portion of Internet conversations they are not treated in an appropriate way and the conversational structure is very difficult to recover.

From the usability perspective, blogs are not designed for supporting conversation since they are publishing tools. Using them as a conversational device might require advanced skills. However, many blogging platforms provide natural affordances to post comments and easy ways to use trackbacks.

3.1.5 Collaboration tools: Google Wave

We now discuss the last type of asynchronous MCS: *wikis*. A wiki is a Web technology that allows people to collaborate through the sharing of information. Wikis are more precisely collaborative knowledge management systems therefore,

---

[19] We use the notation ">" to indicate the "replies-to" relation: A>B means A replies to B.

[20] Three linkback protocol are commonly in use in different blogging platforms: pingback, trackback and refback.

viewing them as conversational system may be challenging. However, in some cases they provide support for conversation in the form of discussion forums or commenting structure. We use the term wiki to point towards a unique case of asynchronous MCS which can be categorized as a collaborative technology. This is the case of Google Wave, which resembles a combination of e-mail and instant messaging fused into a wiki. To date, Google Wave seems to be the most advanced conversational technology and in fact it has been designed for this purpose from the very beginning. Although Wave is still in its infancy, it promises to instantiate to some extent nearly all the categories of our classification schema. Our goal is to asses if our classification scheme would help us in better understanding this tool as a MCS.

In a nutshell, Wave allows the creation and the updates of conversations. Each participant to the conversation can read, reply and even edit other participant's contributions (in this aspect it resembles a wiki). Contributions are time-stamped and they are visualized as a threaded conversational structure (see Figure 3). The time ordering can be enforced by the interface during the navigation (e.g. one can choose to move through the threads or move along the timeline). Wave has real-time conversational capabilities, but this does not make it a synchronous MCS. The reason is that real-time does not replace asynchronous conversations. In other words, the contributions can be viewed as soon as they are entered but they are persistent and available later in time. This contrasts the instant messaging tools that are synchronous MCS since the contributions have limited persistence (i.e. the session time span).

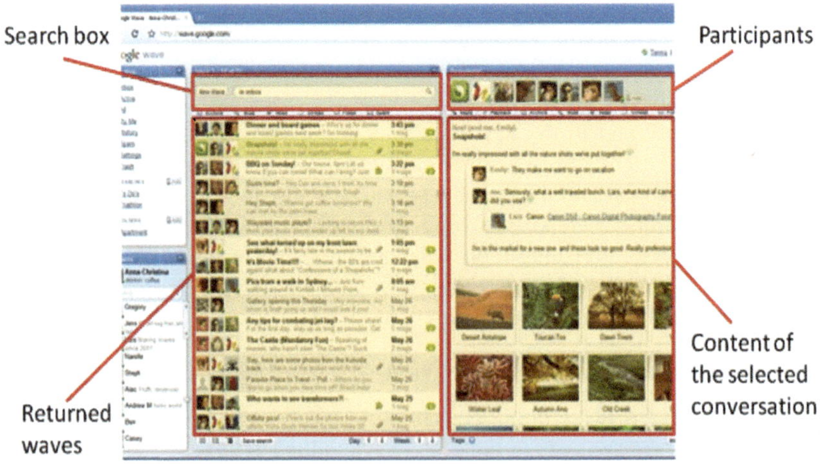

Fig. 3 The Google's Wave Client

While one claimed purpose of Wave is "personal communication and collaboration tool", according to our schema it can be still considered as a general purpose information-sharing MCS since no specific task-oriented process is explicitly supported by the tool. Wave provides many conversational affordances. First of all, conversations can be easily tracked thanks to its appropriate visualization and replies can be inserted exactly where needed, similarly to what we normally do when putting inline replies on messages quotations in e-mail. While this might seem an enhanced affordance, we would rather consider as a natural one because email conversations are part of basic Internet literacy.

Other interesting conversational affordance is the ability to create private sub-conversations that are only visible to invited members. This affordance mimics what might happen during a face-to-face conversation when a group of people decides to start a sub-conversation. In the physical world, this event could cause disruption to the whole conversation and also would prevent the sub-group from following the general conversation. In Wave this undesired effect is mitigated (if not eliminated) due to its asynchronous nature. Parallel sub-discussion can be started and will only be visible to the invited people. If needed, those engaging in the sub-discussion will be able to catch-up the general conversation later. Moreover, if the sub-discussion becomes larger than just a private parenthesis, it can be promoted to a full wave[21] and taken away from the hosting conversation. We would categorize this as an enhanced conversational feature because it improves the quality of conversational experience beyond what could happen in real conversations.

With respect to the information architecture, Wave is a fully distributed system. In fact, the Wave protocol includes the concept of "federation" that allow the connection and synchronization of several Wave servers. The Wave federation protocol, FedOne, is based on XMMP server architecture (formerly known as Jabber). One of the main benefits of being distributed beyond load balancing is security. In fact, local conversations (i.e. those conversation held only by participants within of a corporate intranet) are neither visible nor accessible outside the local Internet domain.

Wave is supposed to provide large support for metadata that can be collected and made available through public APIs. Specifically to the context of MCS, we consider the "annotation" concept in the Wave's document model. Annotations are associated to conversation elements and they can be processed and produced by either the Wave client or Wave extensions called "robots". Annotations can be built-in or customized. This feature makes it possible to overlay additional structure to Wave conversations that can be used to provide both real-time and post-processing services. Among the built-in annotations we find formatting styles and "tags". While formatting styles have a finer granularity (i.e. they may apply to single characters), tag have in contrast the highest scope since they only apply to the whole conversation.[22]

---

[21] Wave is a synonym of conversation.

[22] A finer level of granularity for tags would have been beneficial for some applicative purposes such as the qualification of a contribution (e.g. proposal, explanation, disagreement).

The Wave extension system makes it possible to have layered media that are synchronized with the conversational events. Moreover, the conversations and the embedded media can be re-played later in time. By design, Wave is interoperable and based on open standards. It is obviously extensible by means of server-side (i.e. robots) and client-side (i.e. gadgets) extension. In addition, since it is an open protocol, both Wave servers and clients can be implemented by third parties.

The most interesting aspect of Wave is its indexing and retrieval features. We might expect that the search industry giant Google will offer advance search features especially tailored for retrieving relevant content from conversations. To date, what we can observe using the Google's Wave client is that some basic and advanced search capabilities are available but none of them allows to directly accessing to the conversational structure of waves. In other words, the query returns a list of conversations that match search criteria expressed through operators. The only search operator that looks into the conversational content is the "about:" criterion that corresponds to a standard string search in a text file. No notion of relevance is actually exploited. While tag-based search is possible, the above mentioned restriction of having only conversation-wide tags makes it impossible to search for tagged contributions in the conversation. Fortunately, the extensions system which allows customized annotation enables this level of granularity as well as more advanced search capabilities (e.g. search for waves whose contributions are tagged with selected tags). Needless to say that, for the moment, Google's support for wave search is limited to a very shallow level and does not go into semantic or pragmatic levels.

As for conversational usability, from the unobtrusiveness perspective, Wave shifted from a mode where conversations were attached to some sort of content or media (e.g. a blog post or a video) to a mode where conversation become the main object and media are just part of this main object. In other words, a conversation becomes a multimedia document whose structure corresponds to a conversational structure (made of replies). This shift might create the need for users to reconsider their usual beliefs about the structure of documents and their editing. Wave is designed to support two concurrent purposes: communication and collaborative editing of documents. If editing is the current focus, then the conversation can be distracting and vice-versa. Therefore, we can conclude that wave activities interfere to each other and negatively affect unobtrusiveness.

Related to this double nature of Wave, a cognitive effort seems to be required from users to understand the Wave's concepts and dynamics. For instance, it is not easy to realize that editing a contribution to the conversation does not count as a reply to that contribution, but it only adds an additional author to the contribution. In other words, if one edits a previously authored contribution, the contribution appears to belong to both users and there is no way to figure out who has made which part of it, let alone if some parts have been deleted. However, we believe that this is not a fundamental flaw of the Wave architecture but rather a client limitation. The Wave server is capable of detecting events at a very low-level of granularity which are all recorded. Nearly nothing is lost of the interaction and this can be easily witnessed by the fact that by re-playing the conversations all multiple edits of the same contribution are performed as individual steps. Finally, since we are not evaluating the particular Google's client we might skip the assessment of responsiveness and predictability.

We believe that Google Wave is an important step towards a shift of paradigm where the conversational (and collaborative) dimensions of Internet artifacts are made prominent. Moreover, we believe that this shift is becoming clearer as we look at other technologies which were originally designed for other purposes instead of supporting conversations (e.g., blogs, micro-blogs and social media in general).

3.2 Synchronous MCS

Synchronous MCS are those system that enable real-time conversations. This does not mean that conversations are not recorded; instead, it simply means that temporal co-presence of participants is required to contribute to the conversation. Whenever it becomes possible to contribute to a conversation at different times and still keep the conversation consistent, the system is classified as asynchronous. Synchronous MCS typically require a higher degree of attention from the participants and they are better suited for highly-interactive and for task-based conversations. Like asynchronous MCSs, they can be mono or multi-modal. In our work we will focus on a few representative examples of synchronous MCSs for which we will assess the features along our proposed classification schema.

3.2.1 Instant Messaging

The first type of synchronous MCS is mono-modal and based on textual exchange. It is the case of Instant Messaging (or chat) systems such as Yahoo!, MSN and AOL messenger, IRC, ICQ, Jabber and Google Talk. These are simple MCSs that offer a single conversational affordance based on the metaphor of entering a room in order to join a discussion. Typically, once a conversation is joined, only the subsequent messages of participants are displayed to the user in the client. It is very rare that replies can be addressed to a specific participant let alone to a selected contribution. The thread of the conversation is neither built nor visualized. Often, it is possible to send private messages to one participant but there are no affordances to create sub-groups or sub-chats. Of course, there are many ad-hoc user-created workarounds to overcome the above limitations; however, if more advanced features to support conversational activity are needed, it is better to switch to more comprehensive remote conferencing systems. Needless to say that all these systems in their basic version do not offer any support for producing and managing metadata and they are hardly interoperable or extensible. A notable exception is that of Jabber which turned into a more comprehensive middleware for messaging, XMMP. XMMP became the core foundation of Google Wave we discussed in a previous section. Through XMMP, Wave is also a real-time synchronous conversational system. As we pointed out before, we see synchronicity alone as a limitation rather than an advantage. So, we categorized Wave as being an asynchronous system because it makes persistent all conversations being them synchronous or asynchronous.

3.2.2 Internet Telephony Systems

VoIP systems (e.g. Skype, Apple iChat and SIP-based systems[23]) provide more sophistication than IM systems, not just by the fact that text is replaced by audio (and video in some circumstances), but also by the fact that they provide more advanced conversational affordances. Consider as a prototypical case that of Skype. In Skype, in addition to two party conversations, a conference call capability (i.e. multi-party conversations) is available. Moreover, instant messaging contributions are time-stamped so that the two media are aligned, if recorded. Technically, Skype is a distributed MCS because it is based on a peer-to-peer architecture. However, many metadata such as the list of contacts as well as timestamps of the chat contributions are available to developers through APIs.

Natively, Skype does not offer any recording capability which is instead offered by third party plug-ins at various level of sophistication. This means that Skype is an open platform for developing advanced MCSs. From the conversational affordance point of view, Skype (as many other VoIP systems) adopt the telephone metaphor. This means that conversations are framed as phone calls. This might seem a natural affordance for the average user, but it severely limits the scope of its use as a conversational system. To make this remark clearer, consider a non-standard use of Skype as a tool for supporting mixed remote and face-to-face conversations. In this scenario Skype could support the recording of the meeting and also the creation of parallel channels of conversations between sub-groups through private chats. Apparently, we fall outside the scope of simple phone calls and we might consider this use as a more collaborative tool.

As an interesting trend, however, we notice that many VoIP systems are evolving into full-fledged conferencing systems, e.g., CISCO Webex or Adobe Connect. That is why we intentionally skip for VoIP the assessment of indexing and retrieval as well as the usability. We will review these aspects for the case of a full-fledged remote and face-to-face conferencing system.

3.2.3 Remote Conferencing Systems

We now consider the case of the CISCO WebEx system as another example of synchronous MCS. WebEx extends VoIP to the larger concept of remote conferencing or meeting by providing a reliable infrastructure to stream audio-video data. In terms of conversational capabilities, it is apparent that this type of MCS offers the most advanced affordances that in many cases adequately replace the need of face-to-face meetings. Conversations can be recorded, edited and replayed. WebEx captures: all public chat, all data including annotations, share polls results, notes, presenter video, audio, and third-party audio. WebEx APIs provide access to session's metadata but not all of these metadata are related to the content of the conversations. This makes it difficult to transform recorded conversations into useful information for indexing. Currently, indexing and retrieval of recorded multimedia conversational content for video-conferencing systems has not reached

[23] For list of VoIP client and systems see
`http://en.wikipedia.org/wiki/Comparison_of_VoIP_software`

enough maturity to be considered for commercial applications. A great amount of research has been done in research projects such as the AMI[24], CALO [Tur et al. 2010] and Memetics [Buckingham Shum et al. 2006], addressing some of the challenges that indexing conversational content poses.

In the next section we present an approach to indexing conversational content that could overcome some of the major problems of the standard IR approaches.[25]

4 Indexing and Retrieval of Conversations

In this section we challenge standard IR techniques [Baeza-Yeates and Ribeiro-Nieto 2000] for accessing to textual documents (e.g. Web pages) and we propose a novel approach which appears to be more suitable for indexing and searching conversational content.

Search requirements elicited for conversational content showed that users have a high interest in retrieving information about the dynamics of the conversation (i.e. the processes vs. the topics) [Pallotta et al. 2007]. While topics are equally important, the indexation of topical information alone would not be enough to meet the users' requirements for this type of search because topical and process information is usually retrieved together by users. This is especially true when users are looking for task-based information from conversations such as what decision were made, what tasks were assigned to whom, or how conflicts of opinion were solved (or left unsolved). In order to address this limitation, it has been proposed to enrich the raw conversational content with information related to its conversational structure. One natural choice is argumentative structure [Pallotta et al. 2004]. This choice has been shown to be adequate when the user's goal is to retrieve decision-related information from conversations [Pallotta et al. 2007].

4.1 Argumentative Structure of Meetings

We consider two levels of description, which can be used to characterize features of conversational content that cannot be obtained by simply looking at content-bearing uttered words. The first level is based on the assumption that conversations develop a latent argumentative structure, which is similar to that of scientific texts. Therefore, turns can be classified according to their contribution in exposing a claim and supporting it by an argument. The second level of description is based on a more explicit model of argumentation, which was inspired by the IBIS[26] model [Kunz and Rittel 1979]. In order to compose a multi-dimensional indexing schema, we can add the two features identified above to standard shallow dialogue

[24] `http://www.amiproject.org/`

[25] We intentionally skip the usability assessment here because we believe that this type of assessment deserves a larger discussion.

[26] The Issue Based Information Systems (IBIS) model characterizes the collaborative decision making process in debates by means of asserting formal relations among four argumentative categories: *issues*, *alternatives*, *positions* and *preferences*.

features such as dialogue acts, adjacency pairs [Armstrong et al. 2003; Clark and Popescu-Belis 2004], and topics [Hsueh and Moore 2007]. This schema would allows us to find relevant answers from conversations to complex questions like "*Why did John reject the proposal made by David on microphones issue?*".

4.2 Latent Argumentative Analysis

We describe in this section a classifier based on Latent Argumentative Analysis (LASt). The work described in this section stems from a work on argumentative analysis of scientific articles in medicine and in molecular biology [Ruch et al. 2004]. Indeed, as stated in professional guidelines, articles in experimental sciences tend to respect strict argumentative patterns with at least four categories: PURPOSE, METHODS, RESULTS, and CONCLUSION [Teufel and Moens 1999]. The LASt classifiers were trained on journal corpora using Naive Bayes and Support Vector Machines (with linear kernel) algorithms. The results obtained are similar for both algorithms: on balanced data F-score = 84 %. When applied to a test corpus of conversations (i.e. the ICSI meeting dialogues [Janin et al. 2003]), the LASt system discriminated between argumentative classes as shown in Table 2. The benchmark was obtained from manual annotation by two different annotators. The 70 selected utterances were those where both annotators agreed[27].

Table 2 Confusion matrix for LASt on ICSI data

	C	M	P	R
C	10	3	2	0
M	1	20	0	0
P	5	6	11	0
R	3	4	1	10

The results in Table 2 show that machine learning is effective for this type of shallow argumentative classification. However, better feature selection should be further investigated in order to avoid confusion between RESULT and CONCLUSION. We will see that this level of indexing solves only partially the issues recognized in searching conversational content as illustrated in Section 4.4.1.

4.3 Discourse-Level Argumentative Indexing

A model of conversations was proposed in [Pallotta et al. 2005], which is based on extended IBIS argumentative categories and a simple rhetorical relation, the "replies-to". The argumentative structure defines the different forms of argumentation used by participants in the conversation, as well as its organisation and synchronisation. This model responds to the demand of describing specific task-based conversations, namely

[27] The global agreement measured by the kappa-score was below 0.5.

those that occur when decisions have to be made. We may find this type of conversation as a result of the transcription of remote conferencing systems as well as from the conversations in other MCSs when used for that specific purpose.

When analyzing a conversation, a pure hierarchical structure (e.g. the IBIS structure) is too restrictive. Consider, for instance, an answer that replies to two previous questions in the conversation. In this case, we need a relation that links the answer to both of the questions. This relation is called "*replies_to*", and links a contribution to one or more previous (possibly in time) contributions. The *replies_to* relation induces an *argumentative chain* structure on the conversation which is local to each contribution and which enables visualizing its context. Contributions may have an empty *replies_to* relation. There can be more than one contribution, which replies to the same contribution. Contributions can overlap in time, as for instance in cases where the acceptance of a justification is provided as a backchannel during the presentation of the justification.

We also realized that there is an invariant structure of conversations, which can be obtained by a more general schema by simply varying a parameter. The general structure of a discussion is the follow:

DISCUSS(issue/alternative)
  PROPOSE(solution/idea/alternative/opinion)
    ASK_FOR(explanation/justification)
      PROVIDE(explanation/justification)
        ACCEPT(explanation/justification)
        REJECT(explanation/justification)
    ACCEPT(solution/idea/alternative/opinion)
    REJECT(solution/idea/alternative/opinion)

This structure reflects the IBIS model in terms of conversational structure. The only structural constraints are imposed by the *replies_to* relation, which is graphically represented above by a dependency tree. We require that an argumentative contribution "replies to" the parent argumentative contribution in the tree, as for instance, in:

***replies_to*(ACCEPT(explanation),PROVIDE(explanation)).**

4.4 Three Case Studies

To better understand the intuitions underlying the proposed indexing schemas we present the solutions of four (real) examples from the ICSI meeting corpus, which pose important problems to standard IR techniques for indexing and searching conversational content. The ICSI meeting corpus has been used to perform these experiments since it represent a de-facto standard in conversational corpora and because no corpora exist made of conversations directly extracted from the described MCSs.

4.4.1 First Case

In this case study we describe a situation where filtering the results based on the extracted LASt categories helps in improving precision by eliminating false-positives returned by a standard IR system based on the TF-IDF schema[28]. In answering the query:

"*What are the **decisions** of the meeting?*"

the dataset returned by a standard IR system when applied to an ICSI dialogue contains a false positive:

*"I - I couldn't **decide** which was the right way to do it"*

With the LASt classifier the above utterance is classified as METHODS. By filtering out METHODS utterances, we avoid getting it as a false-positive as we only allow turns classified as CONCLUSION to be returned as "decisions".

If we want to improve the recall and include in the returned answers turns that not contain the term "decision", we may add the LASt labels to indexed terms. In this case, other false-positives are returned by LASt.

1. *"**so** it may be that having three people is very different from having two people or it may not be"*
2. *"**but it seems like that** would be more a case of the control condition compared to uh an experimental condition with more than two"*
3. *"**it would be** an underestimate of the number of overlaps because um i wou- i wouldn't be able to pick it up from the way it was encoded so far"*

The above three utterances are labeled as CONCLUSION by the LASt classifier although in reality, the system should return no answers because *there are no decisions in this meeting, only proposals!*

4.4.2 Second Case

We now make a further step by discussing why argumentative structuring might help in overcoming some inherent difficulties in retrieving relevant passages from conversations. In standard IR, these types of problems remain unsolved even if we enrich the indexed terms with (flat) argumentative indexes (i.e. discourse-level argumentative categories without the *replies_to* relation) as we did for in the first case with LASt categories. To illustrate the problem, let us consider again the dataset returned by a standard IR system when applied to ICSI dialogues with the query:

[28] TF-IDF stands for Term Frequency-Inverted Document Frequency.

"What are the solutions proposed during the discussion about the issue of overlap?"

Looking at the conversation, we find that among all the utterances annotated as DISCUSS(issue) and those annotated as PROPOSE(solution) (and thus represented in the argumentative index), there are several containing the lexical term *overlap* specified in the query. Among the relevant answers returned by a standard IR system for the above query there are the following two turns belonging to the same argumentative chain DISCUSS(issue)-PROPOSE(solution):

1. *"...so basically um as you know uh part of the encoding includes a mark that indicates an* ***overlap****. it's not indicated with um uh tight precision it's just indicated that - okay so it's indicated to - to - so the people know what parts of sp- which - which stretches of speech were in the clear versus being* ***overlapped*** *by others ".*

2. *"...so i used this mark and um and uh uh divided the == i wrote a script which divides things into individual minutes of which we ended up with forty five and a little bit . and uh you know minute zero of course is the first minute up to sixty seconds" .*

The first one is retrieved because it is relevant to the topic "*overlap*" and also annotated as DISCUSS(issue), while the second is returned by the matching the argumentative index: PROPOSE(solution). It is apparent that only the second turn is relevant to the query, since the first one is only related to the second by being the contribution where the issue "overlap" is introduced, and for which the second turn is a proposal of a solution. The system is not able to make any difference between the two turns, which are judged as equally relevant to the query for two different reasons.

In order to solve this problem, in our relevance ranking algorithm we need to take into account to the structured representation in the form of the argumentative sub-chain DISCUSS-PROPOSE in order to filter out the turn that was only used to select the proposal for the right issue (i.e. the overlap) from the answer set.

4.4.3 Third Case

We now consider another example where very simple queries imply non-trivial indexing and ranking strategies to be correctly answered. Consider a simple query like: *Who disagreed on issue T?*

In a standard IR approach, one can imagine to enrich the index by simply attaching an additional "disagreement" term to all contributions included in the argumentative chains induced by the *replies_to* relation of type:

DISCUSS(issue) < PROPOSE(alternative) < REJECT(alternative).

This solution allows us to have indexed both the content of the argument (i.e. terms of the contribution DISCUSS(issue) and the names of the people who have disagreed (i.e. supposing that this information is also indexed for each contribution). However, if we collapse all this information into term indexes, the argumentative structure is lost and we obtain again a false positive: the person who started the discussion of the issue T is considered as the one who disagreed, whereas the system should have returned only the speakers of the PROPOSE and REJECT contribution.

Moreover, if there are several disagreements from different speakers, the speaker corresponding to a PROPOSE(alternative) contribution can be erroneously paired with the speaker of the REJECT(alternative) which is of course outside of the *replies_to* relation. A different approach which solves the above problem would be to answer the query by:

1. gathering all the chains rooted with contributions of type DISCUSS(issue) that are relevant to topic T;
2. selecting the PROPOSE(alternative)-REJECT(alternative) pairs, for each retrieved argumentative chain;
3. returning the speakers of the selected pairs.

4.4.4 Fourth Case

We conclude our case study by considering the conversation excerpt in Figure 4 and the query:

"*Why did John reject the proposal made by David on issue microphones?*"

This query can be answered by putting together the different types of information we have discussed so far. More specifically, the search algorithm will (classically) retrieve all the material dealing with the topic T and filter those contributions tagged as DISCUSS(issue). Then we need to reconstruct the argumentative chains rooted into those contributions and select those that satisfy the constraint of having the PROPOSE(alternative) contribution produced by David and the REJECT(alternative) contribution spoken by John. The query will return the content of the contribution spoken by John and labeled as PROVIDE(justification). This contribution might have been provided spontaneously by the speaker, as in the example in Figure 4, or as a reply to an ASK(justification) contribution linked back to the above REJECT(alternative) contribution.

In this example, the first contribution is labeled with two argumentative acts since both argumentative roles are carried by the same contribution: the PROPOSE(alternative) replies to DISCUSS(issue) since an issue is raised and a proposal is immediately provided for it.

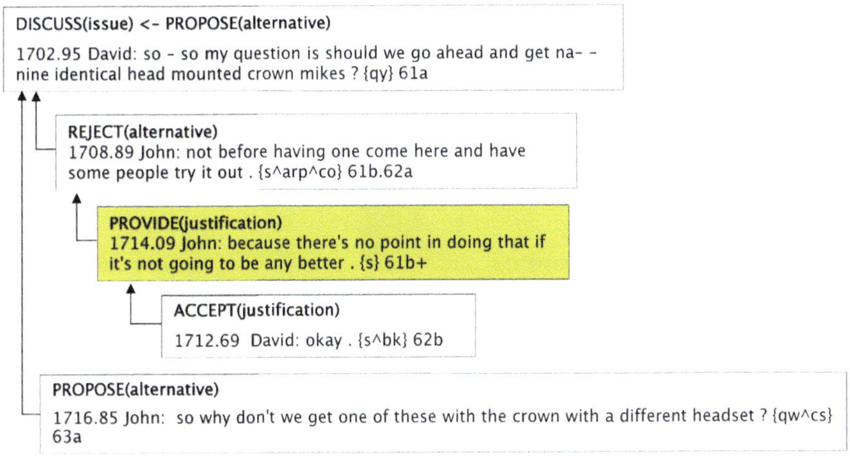

Fig. 4 Argumentative Structuring of an excerpt of the ICSI meeting data

4.5 Discussion

Current term-based indexes (or indexes using term variants such as stems), which are commonly used in IR to provide a basic string normalization process are insufficient to handle argumentative representation level of conversations. Exploiting available additional indexes, such as argumentative (latent and discourse-level) indexes was shown to be a viable alternative that should be taken into account.

Moreover, we might face situations were mono-modal information is not enough to determine the exact dynamics of a conversational process. This is the case when some conversational actions are performed through different channels other than speech or text. These channels can be gestures, strokes on a blackboard, actions on a computer application or any other interaction with artefacts. In some situations, the way to determine the meaning of a conversational clue can also be dependent on the conversational context. For instance, it might happen that after a question like "*Do you have any objection?*" asked by a participant, the following silence might count as an "acceptance" conversational action.

Recent experiments [Pallotta et al. 2009] have shown that the task of identifying argumentative categories in order to build the argumentative structure of conversation is a feasible task obtaining an F-measure of nearly 90%.

This level of analysis is not only useful for building indexes but also for creating derived documents from the conversations. These documents can be of different nature but essentially they are aimed at fulfilling the goal of having condensed versions of the conversations for faster review or easier understanding. To ground our intuitions about the needs of these types of documents we run a survey by asking what type of information would be more appropriate and useful after a meeting. As we expected, a high percentage of participants to the poll have chosen the summary of conversations (see Figure 5). Very few of them would afford reading the transcript or replaying the whole conversation for review.

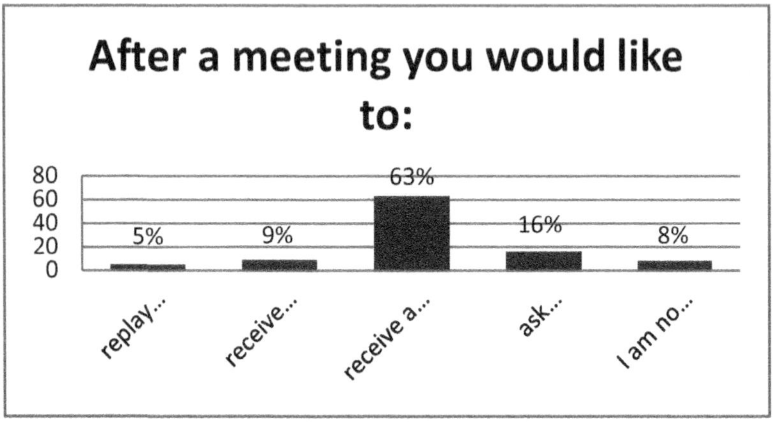

Fig. 5 Survey on after-meeting documents

5 Conclusions

In this chapter, we explored the realm of multimedia conversational systems (MCSs) and presented a framework for classifying them along several dimensions. From the analysis of several existing systems emerged the need of novel indexing and retrieval methods for conversational content. However, the information necessary to build such indexes is not always directly available from the system-produced metadata. Therefore, the conversational content needs to be processed further in order to extract relevant features for automatically building appropriate indexes. We discussed a specific type of indexing techniques which has been shown to be appropriate for indexing task-based conversations, in particular decision-based discussions. We presented a novel indexing schema based on argumentation structure and provided results on the current performance of the core technology used to implement such an indexing schema.

The chapter should be regarded as an effort to improve the awareness of the main drivers to develop IR technologies that are more adequate and more successful when applied to conversational data. MCSs should be designed having in mind that conversations are means to achieve collaboration objectives and the outcomes of conversations are not just their recordings or transcripts. We need to document the conversational process and their outcomes in a more actionable way so that this information can be easily assimilated into knowledge bases.

Conversations contain a wealth of tacit knowledge that is often discarded unless some heavy manual effort of extraction and consolidation is done. New content processing technologies should focus on supporting the knowledge assimilation process from conversational data and improve their access through new types of indexing and information summarization techniques. We strongly believe that this can be achieved by rethinking the way indexes are created and by building more meaningful descriptions of the conversational content. There is obviously no easy way for achieving this goal, but the good news is that many language processing technology have reached a high maturity level and are ready to be exploited for this challenge.

References

[Armstrong et al. 2003] Armstrong, S., Clark, A., Coray, G., Georgescul, M., Pallotta, V., Popescu-Belis, A., Portabella, D., Rajman, M., Starlander, M.: Natural Language Queries on Natural Language Data: a Database of Meeting Dialogues. In: Proceedings of NLDB 2003, Burg/Cottbus, Germany (2003)

[Baeza-Yeates and Ribeiro-Nieto 2000] Baeza-Yates, R., Ribeiro-Nieto, B.: Modern Information Retrieval. Addison-Wesley, Reading (2000)

[Buckingham Shum et al. 2006] Buckingham Shum, S., Slack, R., Daw, M., Juby, B., Rowley, A., Bachler, M., Mancini, C., Michaelides, D., Procter, R., De Roure, D., Chown, T., Hewitt, T.: Memetic: An Infrastructure for Meeting Memory. In: Proceedings of 7th International Conference on the Design of Cooperative Systems, Carry-le-Rouet, France, May 9-12 (2006)

[Clark and Popescu-Belis 2004] Clark, A.S., Popescu-Belis, A.: Multi-level Dialogue Act Tags. In: Proceedings of 5th SIGdial workshop on Discourse and Dialogue, Boston (April 30-May 1, 2004)

[Fillmore 1977] Fillmore, C.J.: Scenes-and-frames semantics, Linguistic Structures Processing. In: Zampolli, A. (ed.) Fundamental Studies in Computer Science, vol. 59, pp. 55–88. North Holland, Amsterdam (1977)

[Hsueh and Moore 2007] Hsueh, P., Moore, J.D.: Combining multiple knowledge sources for dialogue segmentation in multimedia archives. In: Proceedings of the 45th Annual Meeting of the ACL (2007)

[Janin 2003] Janin, A., Baron, D., Edwards, J., Ellis, D., Gelbart, D., Morgan, N., Peskin, B., Pfau, T., Shriberg, E., Stolcke, A., Wooters, C.: The ICSI Meeting Corpus. In: Proceedings of the IEEE International Conference on Acoustics, Speech, and Signal Processing, ICASSP 2003 (2003)

[Kunz and Rittel 1979] Kunz, W., Rittel, H.W.J.: Issues as elements of information systems. Working Paper n. 131, Univer-sität Stuttgart, Institüt für Grundlagen der Planung (May 1979)

[Lalanne et al. 2005] Lalanne, D., Ingold, R., von Rotz, D., Behera, A., Mekhaldi, D., Popescu-Belis, A.: Using Static Documents as Structured and Thematic Interfaces to Multimedia Meeting Archives. In: Bengio, S., Bourlard, H. (eds.) MLMI 2004. LNCS, vol. 3361, pp. 87–100. Springer, Heidelberg (2005)

[Pallotta et al. 2004] Pallotta, V., Ghorbel, H., Ballim, A., Lisowska, A., Marchand-Maillet, S.: Towards Meeting Information Systems. In: Proceeding of 6th International Conference in Enterprise Information Systems ICEIS, Porto, Pourtugal (April 2004b)

[Pallotta et al. 2005] Pallotta, V., Nieksraz, J., Purver, M.: Collaborative and Argumentative Models of Meeting Discussions. In: Proceeding of CMNA 2005 international workshop on Computational Models of Natural Arguments (part of IJCAI 2005), Edimburgh, UK (July 30, 2005)

[Pallotta et al. 2009] Pallotta, V., Delmonte, R., Bistrot, A.: Abstractive Summarization of Voice Communications. In: Proceedings of the LTC 2009 conference on Language Technology and Computers, Poznan, Poland (November 6-8, 2009)

[Pang and Lee 2008] Pang, B., Lee, L.: Opinion Mining and Sentiment Analysis. Foundations and Trends in Information Retrieval 2(1-2), 1–135

[Ruch et al. 2003] Ruch, P., Chichester, C., Cohen, G., Coray, G., Ehrler, F., Ghorbel, H., Müller, H., Pallotta, V.: Report on the TREC 2003 Experiment. In: Genomic Track, TREC (2003)

[Stoke 2008] Stoke, R.: eMarketing: the essential guide to online marketing. Online Reputation Management. Quirk eMarketing, ch. 10 (2008)

[Teng et al. 2008] Teng, Z., Liu, Y., Ren, F.: A multimedia conversation system with application in supervised learning methods and ranking function. International Journal of Innovative Computing, Information and Control 4(6) (June 2008)

[Teng et al. 2009] Teng, Z., Liu, Y., Ren, F.: An Integrated Natural Language Processing Approach for Conversation System. International Journal of Computational Intelligence 5(2) (2009)

[Teufel and Moens 1999] Teufel, S., Moens, M.: Argumentative classification of extracted sentences as a first step towards flexible abstracting. In: Mani, I., Maybury, M. (eds.) Advances in automatic text summarization. MIT Press, Cambridge (1999)

[Tur et al. 2010] Tur, G., Stolcke, A., Voss, L., Peters, S., Hakkani-Tür, D., Dowding, J., Favre, B., Fernández, R., Frampton, M., Frandsen, M., Frederickson, C., Graciarena, M., Kintzing, D., Leveque, K., Mason, S., Niekrasz, J., Purver, M., Riedhammer, K., Shriberg, E., Tien, J., Vergyri, D., Yang, F.: The CALO Meeting Assistant System. To appear in IEEE Transactions on Audio, Speech and Language Processing (2010)

Multimodal Aggregation and Recommendation Technologies Applied to Informative Content Distribution and Retrieval

Alberto Messina and Maurizio Montagnuolo

Abstract. In the modern age, cross-media production is an innovative technique used by the media industry to ensure a positive return on investments while optimising productivity and market coverage. So that, technologies for the seamless fusion of heterogeneous data streams are increasingly considered important, and thus research efforts have started to explore this area. After having aggregated heterogeneous sources, what has to be addressed as the next problem is efficiency in retrieving and using the produced content. Tools for personalised and context-oriented multimedia retrieval are indispensable to access desired content from the aggregated data in a quicker and more useful way. This chapter describes the problems connected with this scenario and proposes an innovative technological framework to solve them, in the area of informative content (news) distribution and retrieval. Extensive experiments prove the effectiveness of this approach in a real-world business context.

1 Introduction

In recent years, the global diffusion of the Internet and the progress in developing Web multimedia applications are enabling the delivering of dynamic heterogeneous content such as news, blogs and audio/video podcasts. As a result of this technological breakthrough, the content of modern Web is characterised by an impressive growing of availability of multimedia data together with a strong tendency towards integration of different media and modalities of interaction. The mainstream paradigm consists in *bringing into the Web* what was thought (and produced) for different media, like e.g. TV content, which is acquired and published on websites and then made available for indexing, tagging, browsing. This instaurates what can be

Alberto Messina · Maurizio Montagnuolo
RAI Centre for Research and Technological Innovation,
C.so Giambone, 68 I10135 Torino, Italy
e-mail: {a.messina,maurizio.montagnuolo}@rai.it

Soro et al. (Eds.): Inform. Retrieval and Mining in Distrib. Environments, SCI 324, pp. 213–232.
springerlink.com

called the *Web Sink Effect*. This effect has rapidly started, recently, to unleash an ineluctable evolution from the original concept of the Web like a resource where to *publish* things produced in various forms outside the Web, to a world where things *are born and live* on the Web. Solutions for intelligent information fusion and organisation are thus becoming indispensable. Here, the challenge lies in the ability of combining and presenting data coming from multiple information sources, i.e. *cross-modal*, and consisting of multiple types of content, i.e. *multi-media*.

This chapter contributes to advance the state of the art in information retrieval and multi-modal information fusion. In particular, the following aspects are addressed: (i) formal definition of a semantic affinity function acting as a kernel able to discover the semantic affinities of heterogeneous information items. Based on this function, semantic dependency graphs among information items are built; (ii) a technique to select representative elements out of the discovered graphs; (iii) development of a fully unsupervised service platform for aggregating, indexing, retrieving and browsing multi-modal news content following the developed theoretical tools.

The chapter is organised as follows. §2 reviews the literature on the reference topics. § 3 presents the theoretical aspects of our research. § 4 describes the architecture developed to test our method in a concrete scenario. § 5 reports about the experiments done on the system to evaluate its performances, and §6 provides final conclusions and some future research directions.

2 Related Work

Modern research efforts in multimedia information retrieval (MIR) have been recently summarised by Lew et al. [11]. Here, one of the key issues pointed out is the lack of a common and accepted test set for researchers conducting experimentation in the field of MIR. This should sound as a serious alarm bell for researchers and practitioners of the field. We believe that this situation is due to a significant lack of precision in the definition of the relevant problems, which leads to the generation of huge research efforts, but only seldom in directions exploitable by the media industry (e.g. broadcasters, publishers, producers) in a straightforward way. We add that in concrete scenarios the accuracy figures obtained by state-of-the-art tools may be not fully satisfactory for an industrial exploitation [12]. The following subsections overviews previous works that are directly relevant to the main topic of this chapter.

2.1 The Role of Semantics in Multimedia Information Retrieval

Several researches have attempted, during the recent years, to give a view on the state of the art of content analysis-based extraction of multimedia semantics. In [8], the author offers a 3-layered view on content indexing including *topics*, *events* and *objects*. However, the distinction made between topics and events is somewhat imprecise and the explanatory examples are arguable. The whole expressed content is a representation of events occurring in places, under a certain contextual topic. Although this view is able to explain a useful part of the semantics conveyed by specific genres of multimedia (e.g. news), it is not sufficient in many other practical cases as movies or documentaries, where further levels of description are needed. Integration between Semantic Web technologies and multimedia retrieval techniques is considered a future frontier [2, 7]. Very recent efforts are concerned with the exploitation of social roles mining in multimedia. In [20], authors perform movie storyline detection by exploiting the relationships of a roles social network. In [17] authors use social networks to perform radio programmes segmentation.

2.2 Information Mashup

Information mash-up is becoming a hot topic in the World Wide Web (WWW) community. A mash-up is a Web application that aggregates content from different data sources to deliver a new, hybrid service that was not originally supported. Much of the current work involves techniques for grouping data from only a single domain and from only a single media, such as RSS items aggregated according to a taxonomy of topics [19]. As an RSS feed usually contains only short descriptions of the referenced items' content, the aggregation process may not be a trivial task. The works in [1, 3] employ either the user's interaction, or external knowledge sources, to improve the aggregation performance. As the nature of the data to be aggregated cannot be in principle shrinked to be merely mono-media, tools to integrate multimedia data, e.g. text, audio and video, from mono-modal information sources were investigated [4, 6]. Other approaches are those employing both cross-modal information channels, like radio, TV and the Internet and multi-media data [21, 23].

2.3 Topic Detection and Tracking

Particularly relevant to our application domain are the works pursued under the NIST Topic Detection and Tracking (TDT) project[1]. TDT aims at automatically locating, linking and accessing topically related information items within heterogeneous, real-time news streams. Intuitively, a *topic* is defined as an aggregation of information items that are *semantically relevant* to a real-world event. As an example, a earthquake could be the event that triggered the topic. Any information item, such as a newscast story or a newspaper article, that talks about the earthquake, or e.g. the rescue attempts, the number of casualties, and so on, is semantically relevant to the topic. The identified tasks of TDT are: News Story Segmentation, First Story Detection, Topic Detection, Linking and Tracking. The first story detection task is aimed at recognising information items linked to topics never seen before. This topic is typically approached by representing each information item as a set of features (e.g., newswire text or closed transcriptions of radio and TV speech). When an incoming item is received, its feature set is matched against those of all the past items according to a similarity measure. If, for each past item, the similarity measure is below a fixed threshold, then the incoming item is marked as new.

[1] `http://www.nist.gov/speech/tests/tdt/` (last accessed: 12th January 2010).

Following the same approach, topic and link detection aims at aggregating and linking individual information items related to the same topic. Finally, topic tracking task aims at keeping track of information items similar to a set of example items.

2.4 News Story Segmentation

The news story segmentation task concerns the ability of automatically detecting semantically coherent parts of the news streams, such as a single news story. The TRECVID[2] initiative had news segmentation among its tasks in 2003 and 2004. The common base of the approaches for automatically segmenting TV newscasts into individual news stories consists in using a combination of visual, audio and speech features. In particular, to solve the *news story* segmentation, the common base of the approaches is constituted by the use of a combination of visual, audio and speech features. The baseline features employed in several cases are visual similarity between shots within a time window and the temporal distance between shots [9]. Other heuristics like similarity of faces appearing in the shots and the detection of the repeated appearance of anchor person shots [18, 22] can be also used to improve the accuracy. The audio channel contribution can be employed to detect boundaries for topic or speaker changes [9, 16, 18]. Text transcriptions are very often used, either by searching similar word appearances in different shots or by detecting text similarities between the shots [18].

3 Innovative Technologies for Multi-modal Data Aggregation

This section illustrates our multi-modal aggregation technology. Multi-modality is the ability for end users to access desired information delivered from multiple data sources (e.g., traditional distribution channels like radio, press and TV, as well as non-traditional channels such as the Internet and mobile data networks) and presented in different media formats, such as acoustic, visual and textual modalities. Our technology is multi-modal in that it supports the generic integration and retrieval of such sources of heterogeneous data.

3.1 Cross-Modal Clustering

In our work multi-modal data aggregation is performed by cross-modal clustering based on the concept of *semantic relevance*, which is inspired by the definition originally proposed in [10].

Definition 1. (Semantic relevance). Let π and β be two information items available to the consumers through information streams $\{I\}_{N1}$ and $\{I\}_{N2}$, respectively. In this context, the *secondary* information item β is semantically relevant to the *primary*

[2] www-nlpir.nist.gov/projects/trecvid/ (last accessed: 15th January 2010).

information item π if the fruition of β by consumers satisfies the consumers expectations about π.

Semantic relevance is modelled by the linking function $R(\pi,\beta)$ that measures how likely the secondary information items are relevant to the information needs expressed by the primary information items. Cross-modal clustering is able to discover these semantic relations in heterogeneous data, thus providing facilities to effectively retrieve desired information in cross-modal, multimedia information streams.

3.2 *Semantic Relevance among Heterogeneous Information Items*

Let $\Pi = \{\pi_i\}_{i=1}^m$ and $\mathscr{B} = \{\beta_j\}_{j=1}^n$ be two sets of information items, for which a distance metric in the space $\mathscr{H} = \Pi \cup \mathscr{B}$ is not defined. Let $R : \Pi \times \mathscr{B} \to [0,1]$ be a linking function such that:

- $R(\pi_i,\beta_j) \to 1$ (tends to 1) if $\beta_j \in \mathscr{B}$ is semantically relevant to $\pi_i \in \Pi$;
- $R(\pi_i,\beta_j) \to 0$ (tends to 0) if $\beta_j \in \mathscr{B}$ is not semantically relevant to $\pi_i \in \Pi$.

Figure 1 illustrates the concept of semantic relevance in the context of Web and TV news. Semantically relevant primary information items (e.g., Web assets π_i) and secondary information items (e.g., TV assets β_i) are merged in a new hybrid space $\mathscr{H}$, thus generating a multi-modal aggregation. The following example better clarifies this procedure.

Example 1. Let us consider a Web journalist who is searching on an online news site details about a crime story happened in a town near his own town, because he has to make report that evening in a local amateur journalists club. After several minutes of searching he is still unable to find a satisfactory article, apart from a brief note from an institutional news agency just reporting a headline and few words. He then decides to shift attention from web to television and start a search on a multimedia repository *using the few words he had found on the agency note* few minutes before as keywords. He soon finds out an extended report of the story broadcasted by one of the regional stations, which *almost satisfies* his needs, and decides to download the clip and to show it that night during the club meeting.

In the above example, the regional television report plays the role of β, while the agency note that of π.

We define the *relevance matrix* as a matrix $\mathbf{R_{el}} = (\mathbf{r_1},\ldots,\mathbf{r_m})^T \in [0,1]^{m,n}$, where

$$\mathbf{r_k} = (R(\pi_k,\beta_1),\ldots,R(\pi_k,\beta_n)), \quad k = 1,\ldots,m \tag{1}$$

Intuitively, the construction of the matrix $\mathbf{R_{el}}$ can be seen as a *space transformation* process, $\mathscr{T} : \Pi \Rightarrow \mathscr{A}, \mathscr{A} \subset R^{|\mathscr{B}|}$ which links the information items from the primary space Π to a transformed space $\mathscr{A}$, through the projection on a secondary space $\mathscr{B}$, which works according to the semantic relevance between objects in such spaces.

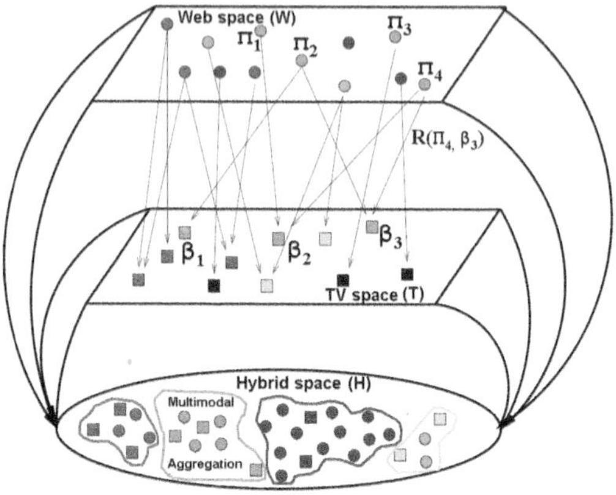

Fig. 1 Illustration of cross-modal clustering

3.3 Hybrid Matching with Semantic Affinity Measurement

Once the matrix $\mathbf{R_{el}}$ has been constructed, the *semantic affinity* between primary information items $\pi_i, i = 1, \ldots, m$ is evaluated by exploiting their projection in the space $\mathscr{B}$. Let $(\pi_a, \pi_b) \in \Pi$ be a couple of primary information items represented by the relevance vectors $(\mathbf{r}_a, \mathbf{r}_b)$. The semantic affinity between π_a and π_b in the secondary space $\mathscr{B}$ is defined as follows:

$$S(\pi_a, \pi_b) = \frac{\langle \mathbf{r_a}, \mathbf{r_b} \rangle}{||\mathbf{r_a}||^2}, \tag{2}$$

where $S(\cdot)$ is the semantic affinity function, $\langle \cdot \rangle$ is the inner product, and $|| \cdot ||$ is the norm induced by the inner product in $[0,1]^n \subset R^n$.

The semantic affinity function is computed for each couple of relevance vectors $(\mathbf{r_a}, \mathbf{r_b})$, $a, b = 1, \ldots, m$. The result is the affinity matrix $\mathbf{A} = (A_{ab}) \in \Re^{m,m}$, where:

$$A_{ab} = \begin{cases} 1, \text{ if } a = b \\ S(\pi_a, \pi_b), \text{ if } a \neq b \text{ and } S(\pi_a, \pi_b) \geq \alpha \\ 0, \text{ otherwise.} \end{cases} \tag{3}$$

3.3.1 Geometrical Interpretation of the Semantic Affinity Function

The semantic affinity measurement $S(\pi_a, \pi_b)$ has a straight forward geometrical interpretation if we consider that it can be written as:

$$S(\pi_a, \pi_b) = \cos(\mathbf{r_a}, \mathbf{r_b}) \frac{||\mathbf{r_b}||}{||\mathbf{r_a}||}, \tag{4}$$

where $\cos(\cdot)$ is the Cosine similarity measurement between vectors $\mathbf{r_a}$ and $\mathbf{r_b}$, as per the definitions of these vectors given in Sect. 3.1. Equation (4) points out that $S(\pi_a, \pi_b)$ is the ratio between the norm of the projection of vector $\mathbf{r_b}$ on $\mathbf{r_a}$ and the norm of vector $\mathbf{r_a}$ itself. This means that $S(\pi_a, \pi_b)$ accounts for the extent to which the vector $\mathbf{r_a}$ is *explained* by vector $\mathbf{r_b}$, or in other terms for the relative error implied by confusing $\mathbf{r_a}$ with the projection of $\mathbf{r_b}$ on $\mathbf{r_a}$, if we take $1 - S(\pi_a, \pi_b)$. This property, which is exemplified in Fig. 2 for a bi-dimensional vector space, allows us to define the following symbolic expressions to represent two particular conditions:

$$\text{iff } S(\pi_a, \pi_b) > \alpha \wedge S(\pi_b, \pi_a) > \alpha \text{ then } Eq(\pi_a, \pi_b) \tag{5}$$

$$\text{iff } S(\pi_a, \pi_b) > \alpha \wedge S(\pi_b, \pi_a) \leq \alpha \text{ then } Ent(\pi_a, \pi_b), \tag{6}$$

where $\cos(\cdot)$ is the Cosine similarity defined in $\mathscr{B}$, and α is a fixed threshold. Equation (5) introduces the *semantic empirical equivalence* relation between π_a and π_b, $Eq(\pi_a, \pi_b)$, while (6) introduces the *semantic empirical entailment* relation from π_a to π_b, $Ent(\pi_a, \pi_b)$. Notice that the latter relationship would not be discovered by using e.g. the plain Cosine similarity measure. Intuitively, the asymmetry given by (2), introduces the possibility to have hierarchies among the aggregated objects, providing also means for a natural procedure to discover representative elements and to have a multi-level granularity of presented information. The disadvantage w.r.t. symmetric functions is the introduction of extra computation, since semantic affinity is an asymmetric measure.

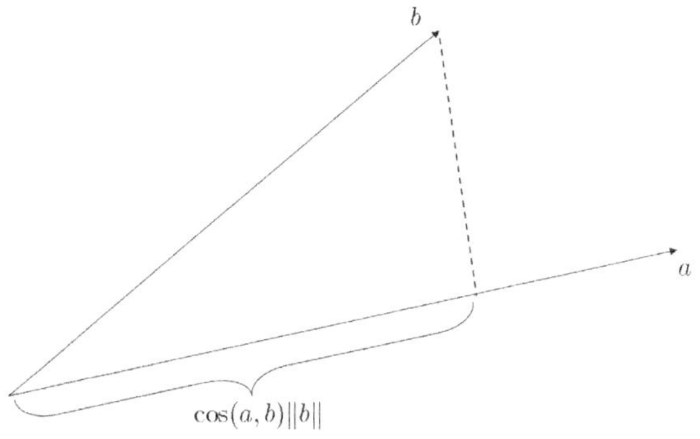

Fig. 2 Geometrical interpretation of the semantic affinity function $S(\pi_a, \pi_b)$

3.3.2 Generalised Semantic Affinity Kernels

Equation (2) is a specific case of a more general family of kernels, which we call Generalised Semantic Affinity Kernels (GSAK). A GSAK of order (m, n), is a function $S_{m,n}(a, b) : R^N \times R^N \Rightarrow R$ of the form:

$$S_{m,n}(a,b) = \frac{\langle a, b \rangle}{\|a\|^m \|b\|^n} \tag{7}$$

where m,n are positive real numbers, and $\langle\cdot\rangle$ is the inner product. GSAKs are a generalisation of the Cosine similarity function. Generalisation is obtained through wiring into the GSAK expression of (7) the dependency on the vectors' norms (i.e., sizes). Therefore, GSAKs express a generic affinity measurement between vectors taking into account both the angle between vectors and the relationship between their size at the same time.

3.4 Induced Partitions

In Sect. 3.3 we introduced the matrix $\mathbf{A}$ as the matrix collecting the semantic affinity values for all couples of vectors in the primary space Π. Once $\mathbf{A}$ is calculated, the primary connectivity graph $G = (V,E)$ is built. Each node of the graph corresponds to a primary information item $\pi_i \in \Pi$. Two nodes $v_a, v_b \in V$ are connected from v_a to v_b if the corresponding element $A_{ab} \in \mathbf{A}$ is greater than α. The α-cut value guarantees that every pair of linked information items has a semantic affinity of at least α in one of the two directions. Figure 3 shows an example of the matrix $\mathbf{A}$ and the corresponding graphs for different values of α[3].

	π_1	π_2	π_3	π_4	π_5	π_6
π_1	1	0.82	0.53	0	0	0
π_2	0.85	1	0	0	0	0
π_3	0	0	1	0.6	0	0
π_4	0	0	0.32	1	0	0
π_5	0	0	0	0	1	0.9
π_6	0	0	0	0	0.28	1

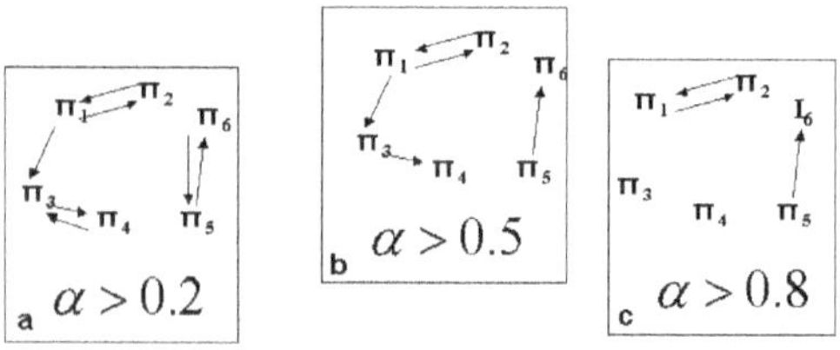

Fig. 3 Example of the affinity matrix between primary information items $\{\pi_i\}_{i=1}^{6}$ and the corresponding connectivity graph for three values of α

Aggregations are generated through the visit of the semantic affinity matrix, i.e. the matrix $\mathbf{A}$ whose elements A_{ij} represent the values of $S(\pi_i, \pi_j)$, by applying the depth first search algorithm on G, and thus finding the set of disconnected subgraphs included in G, i.e. a partition of the graph nodes $\{g_1, g_2, \ldots, g_K\}$ so that $\bigcup_i g_i = G$ and $\forall i \neq j$, $g_j \bigcap g_i = \emptyset$. We can therefore define the induced partition of the primary space Π as:

[3] Remember that since the matrix is asymmetrical, the correspondence with semantic affinity values has to be reconstructed reading rows first, since $A_{ab} = S(\pi_a, \pi_b)$.

Definition 2. Let $\mathbf{A}$ be an affinity matrix for two spaces Π (used as primary) and $\mathscr{B}$ (used as secondary), and let be $G = (V, E)$ its graph representation. Let $\{g_1, g_2, \cdots, g_K\}$ the set of disconnected sub-graphs of G. We define the induced partition $D(\alpha)$ of Π as $D(\alpha) = \{\gamma_1, \ldots, \gamma_{|D(\alpha)|}\}, \forall i \gamma_i \in 2^{\Pi}$, where γ_i is the subset of Π corresponding to the sub-graph g_i of G.

Since the depth first search algorithm is valid for not oriented graphs, we have first to *symmetrise* $\mathbf{A}$, by considering simply connected or not connected each couple of nodes. This can be done using the following criteria:

- **Maximum affinity**: π_i and π_j are connected if $\max\left(S(\pi_i, \pi_j), S(\pi_j, \pi_i)\right) > \alpha$
- **Average affinity**: π_i and π_j are connected if $\frac{1}{2}\left(S(\pi_i, \pi_j) + S(\pi_j, \pi_i)\right) > \alpha$
- **Minimum affinity**: π_i and π_j are connected if $\min\left(S(\pi_i, \pi_j), S(\pi_j, \pi_i)\right) > \alpha$

$D(\alpha)$ is the partition of the primary space Π induced by the space $\mathscr{B}$. For example, from Figure 3(a), it would be, following the maximum affinity criterion:

$$D(0.2) = \{\gamma_1, \gamma_2\} = \{(\pi_1, \pi_2, \pi_3, \pi_4), (\pi_5, \pi_6)\},$$

Each part $\gamma_i \in D(\alpha)$ constitutes a set of semantically related primary information items linked according to their semantic relationships. The parameter α defined in (3) governs the structure of the resulting partition. In fact, by raising ($\alpha \downarrow$) or lowering ($\alpha \uparrow$) this parameter, the condition under which two elements of Π are aggregated is respectively relaxed or restricted.

3.5 *Representative Elements*

For each part $\gamma_i \in D(\alpha)$, its representative element $\bar{\pi}_i$ is chosen so that:

$$\bar{\pi}_i = \arg\max_{\pi_{ij} \in \gamma_i} \sum_{k \neq j} S(\pi_{ik}, \pi_{ij}) \,. \tag{8}$$

Equation (8) means that the representative element is the one whose total semantic affinity measurement is maximised. This criterion is based on the empirical observation that the higher is the total semantic affinity measurement of an element w.r.t. the other elements of an aggregation, the higher is the number of elements in the aggregation that are semantically entailed by it, so that the item content is expected to be the most complete w.r.t. the semantics of the partition, and therefore a higher representativeness is expected to be conveyed by the item itself. From the example of Fig. 3(a), the representative element for the part $\gamma_1 = (\pi_1, \cdots, \pi_4)$ is $\bar{\pi} = \pi_1$.

3.6 *Multi-modal Aggregations*

Let be $\mathscr{H} = \Pi \bigcup \mathscr{B}$, i.e. the union of the primary space Π and secondary space $\mathscr{B}$. Given an induced partition $D(\alpha)$ we can finally build the set of multi-modal aggregations $D(\alpha)^* = \{\gamma_1^*, \ldots, \gamma_{|D(\alpha)|}^*\} \subseteq 2^{\mathscr{H}}$ by retrieving the elements of $\mathscr{B}$ which are semantically relevant to each $\gamma_i \in D(\alpha)$. Letting $K = |D(\alpha)|$, this can be done following two distinct criteria:

$$\forall i: \gamma_i^* = \gamma_i \cup B_i, \quad i = 1,\ldots,K \quad B_i = \cup_{j=1}^{|\gamma_i|} \beta_{ij} \quad \beta_{ij} = \{b \in \mathscr{B} : R(\pi_{ij}, b) > \eta \tag{9}$$

where η is a parametric threshold. Following this criterion, the function of $D(\alpha)^*$ is that of integrating the partition $D(\alpha)$ with *all* the semantically relevant elements of $\mathscr{B}$. Notice that $D(\alpha)^*$ is not in general a partition of $\mathscr{H} = \Pi \cup \mathscr{B}$, because elements of $\mathscr{B}$ may be semantically relevant to elements of Π belonging to different elements of $D(\alpha)$, and because some elements of $\mathscr{B}$ may be not semantically relevant to any element of Π.

$$\forall i: \gamma_i^* = \gamma_i \cup B_i, \quad i = 1,\ldots,K \quad B_i = \cup_{j=1}^{|\gamma_i|} \beta_{ij} = \{b \in \mathscr{B} : \forall k : R(\pi_{ik}, b) > \eta\} \tag{10}$$

where the function of $D(\alpha)^*$ is that of integrating the partition $D(\alpha)$ with the elements of $\mathscr{B}$ that are semantically relevant to *all* elements of γ_i. Notice that neither in this case $D(\alpha)^*$ is a partition of $\mathscr{H} = \Pi \cup \mathscr{B}$.

3.7 Adaptive Relevance Clustering for Multi-modal Aggregation Quality Improvement

In order to improve the accuracy of the discovered aggregations, we use a graph partitioning technique based on a physical metaphor. As each multi-modal aggregation $\gamma^* \in D(\alpha)^*$ is represented by a connected, oriented graph g_s, it can be modelled and segmented using graph drawing procedures. The assumption is that the graph's topology can be used to describe the underlying concepts of γ^*. Cohesive aggregations would be modelled by dense graphs. On the other hand, less cohesive aggregations would be modelled by spread graphs. At the core of this procedure is a divisive hierarchical clustering method that, starting with a complex aggregation γ^*, splits it in k sub-aggregations, and then continues splitting each of them, until it reaches a stable condition in which partition is considered no more convenient.

The adaptive relevance clustering (ARC) procedure is composed of two parts (see Alg. 1) based on the Fruchterman-Reingold graph layout (FRL) algorithm and Gaussian mixture clustering (GMC) respectively. The FRL method is a energy positioning procedure whereby the graph is modelled as a physical system of electric charges and springs. The algorithm starts by setting the operational parameters and the random initial positions of each node. The Coulomb's and Hooke's laws are then applied, pulling the nodes closer together or pushing them further apart. This process is iterated until the system comes to an equilibrium state, i.e. when the system kinetic energy is minimised. Since the reliability of the model is affected by the size of the plane in which the model is situated (because, e.g. the graph is too complex to be represented in the available space, or a local minima was encountered), we repeat the FRL procedure modifying at each step the plane size and the strength of the forces acting on the nodes, until either the number of nodes concentrated on the plane borders is less than a fixed percentage, or the maximum number of iterations is reached. At the equilibrium, the graph node positions (x_i, y_i) are normalised (w.r.t. the plane size) and modelled by a bi-dimensional random variable $Z = (z_1, z_2) \equiv (x, y)$ with range $R_Z \subset [0,1] \times [0,1]$. Assuming that z_1, z_2 are statistically independent and normally distributed, the density function of Z is modelled by a bivariate Gaussian mixture

$$f_Z(z_1, z_2) = \sum_{k=1}^{K} \omega_k \mathcal{N}(Z|k, \bar{\mu}_k, \bar{\sigma}_k) \tag{11}$$

wher K is the number of mixture components and ω_k, $\bar{\mu}_k$, $\bar{\sigma}_k$ denote the weight, the mean vector and the diagonal vector of covariance matrix of the kth Gaussian, respectively. The value of K is set using the following function:

$$K = \left\lceil 1 + \left(\left\lceil \frac{N_Z}{2} - 1 \right\rceil \right) \left(1 - \exp\left(-\frac{Tr(\Sigma_Z)}{2 \parallel \bar{\mu}_Z \parallel^2} \right) \right) \right\rceil \tag{12}$$

where N_Z, $Tr(\Sigma_Z)$ and $\bar{\mu}_Z$ are, respectively, the total number of observations, the covariance matrix trace, and the mean vector of the random variable Z. K takes values in the range 1 to $\lceil N_Z/2 \rceil$. This way, graphs that show higher variance/mean ratio, i.e. that are more spread, will be partitioned using a higher number of clusters, while those with lower ratio, i.e. that are more cohesive, will be partitioned using a lower number. Given the value of K there are three steps in GMC. First, the mixture is initialised by setting the starting positions of the cluster centroids. For this purpose, we use the k-means clustering (KMC) on the elements of Z. Second, the Expectation-Maximization (EM) algorithm is used to estimate the mixture parameters $\bar{\theta}_k = (\omega_k, \bar{\mu}_k, \bar{\sigma}_k), \forall k = 1 \cdots K$. Final, each observation $z_n \in Z$ is assigned to the cluster k for which $\omega_k \mathcal{N}(z_n|k, \bar{\mu}_k, \bar{\sigma}_k)$ is maximised. This process is iterated recursively till the estimation of K given by Eq (12) gives 1.

Two examples of the output of the ARC algorithm are shown in Fig. 4. The root aggregation in Figure 4(a) has two sub-topics. The former concerns the threats of the Talibans against the Afghan Presidential elections. The latter is about the Taliban's attack to the NATO headquarter in Kabul. The ARC algorithm splits the root aggregation into two parts and removes any link between them. The root aggregation is always browsable by selecting the corresponding node of the Hierarchy Tree. Note that in this case, as the attack is a consequence of the threats, the partition 2 points to the partition 1 according to the intuitive explanation of (3). As a second example, the root aggregation in Fig. 4(b) has two uncorrelated topics. This mainly depends on under-segmentation errors occurring in the TV news segmentation process. The root aggregation is split into two parts. The former is about a talk of the Pope. The latter is about the G8 Summit incidents in Genoa.

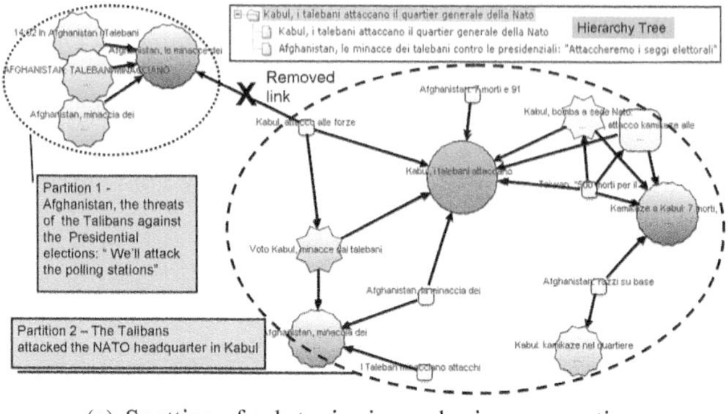

(a) Spotting of sub-topics in a cohesive aggregation.

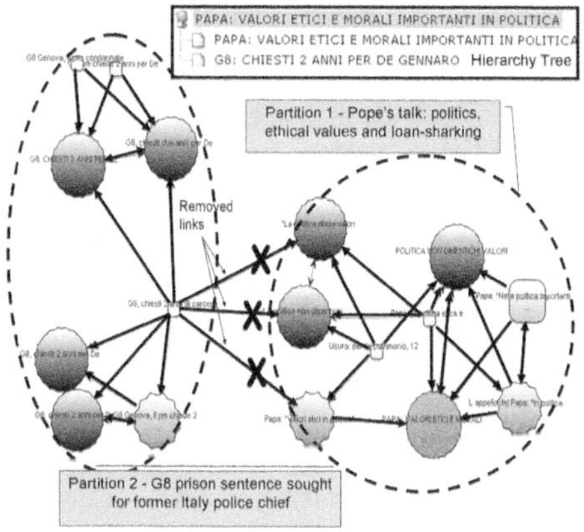

(b) Spotting of uncorrelated topics in an under-segmented aggregation.

Fig. 4 Examples of the ARC algorithm output

4 Integration and Implementation of the Method

This Section overviews the linking elements between the components of the experimental evaluation of the introduced multi-modal aggregation technology, and its formal elements introduced in §3. The block diagram of the developed architecture and a detailed description of its components can be found in [14].

The system is a processing machine having two inputs, i.e. digitised broadcast news streams (DTV) and Web newspapers feeds (RSSF), and one output, i.e. the multi-modal aggregation service (MMAS), that is automatically determined from the semantic aggregation of the input streams. The DTV stream is at first analysed and partitioned into programmes using a visual pattern matching algorithm [13]. On such detected programmes, automatic news boundaries detection is performed. Once detected, transcriptions of the spoken parts in the DTV stream are extracted. Finally, each story is indexed and stored in a permanent documents catalogue. The

RSSF stream consists of RSS feeds from several major online newspapers and press agencies. Additionally, also users weblogs can be treated. The RSSF stream is first parsed to extract the list of the included items. Each RSS item is POS tagged to identify the linguistic elements, like nouns and adjectives, within the RSS item's title and description. The output of this tagging process is used to generate a set of representative queries, which are submitted to the index structure of the TV documents catalogue. For each RSS item, the result of this search operation is a weighted set of TV news stories of decreasing affinity to the target query. On such results, the RSS items are aggregated by the cross-modal clustering method defined in §3.

Algorithm 1. Adaptive Relevance Clustering (ARC).

PART 1: Fruchterman-Reingold layouting (FRL).
Set up the operational parameters $(S_x, S_y, F_a, F_r, perc)$; $epoch \leftarrow 1$
repeat
  Set up initial node positions randomly
  $Z \leftarrow FRL(S_x, S_y, F_a, F_r)$ {Outputs the node positions}
  $F_a \leftarrow aIncr \cdot F_a$; $F_r \leftarrow rIncr \cdot F_r$ {Increment forces}
  $[S_x, S_y] \leftarrow sIncr \cdot [S_x, S_y]$ {Increment plane dimension}
  $epoch \leftarrow epoch + 1$
until $(N_b > perc) \wedge (epoch < maxIter)$
PART 2: Gaussian mixtures clustering (GMC).
Set up number of mixtures (K), EM algoritnm precision (EM_p) and maximum steps (EM_s)
Require: $\forall k = 1 \cdots K$; $\forall d = 1,2$; $\forall n = 1 \cdots N_Z$
$w_k^0 \leftarrow \frac{1}{K}$; $\sigma_{kd}^0 \leftarrow rand()$; $\bar{\mu}_k^0 \leftarrow KMC(Z,K)$; $\theta^0 \leftarrow (w_1^0, \bar{\mu}_1^0, \bar{\sigma}_1^0 \cdots w_k^0, \bar{\mu}_k^0, \bar{\sigma}_k^0 \cdots w_K^0, \bar{\mu}_K^0, \bar{\sigma}_K^0)$

repeat
  E-Step: calculate conditional probabilities at step i
  $\Pr^{(i)}(k|n) \leftarrow \frac{\omega_k^{(i)} \cdot \mathcal{N}\left(z_n | \bar{\mu}_k^{(i)}), \bar{\sigma}_k^{(i)}\right)}{\sum_{k=1}^{K} \omega_k^{(i)} \cdot \mathcal{N}\left(z_n | \bar{\mu}_k^{(i)}), \bar{\sigma}_k^{(i)}\right)}$
  M-Step: update mixture parameters at step $i+1$
  $\theta^{(i+1)} \leftarrow \left(w_1^{(i+1)}, \bar{\mu}_1^{(i+1)}, \bar{\sigma}_1^{(i+1)} \cdots w_K^{(i+1)}, \bar{\mu}_K^{(i+1)}, \bar{\sigma}_K^{(i+1)}\right)$
  $\delta \leftarrow \| \theta^{(i+1)} - \theta^{(i)} \|$; $i \leftarrow i+1$
until $(\delta > EM_p) \wedge (i < EM_s)$
$z_n \leftarrow \arg\max_{\mathbf{k}} \left(\mathcal{N}\left(z_n | k, \theta_k^{(i+1)}\right)\right)$

4.1 MMAS Stream Output Chain

The discovered aggregations are indexed and stored in the multi-modal aggregation index (MAi). For each aggregation, the MAi stores the list of the aggregated RSS items and news stories, as well as a text document including the RSS items titles and description phrases and the news stories transcriptions constituting the aggregation. The MAi is thus the permanent database from which the multimedia news management services (MNMS) are delivered to the users. Currently, the following services are supported: (i) multi-modal news search and retrieval, (ii) topic detection and importance ranking, (iii) multimodal user recommendation, and (iv) interactive graph-based news navigation. The following subsections briefly overviews each of the mentioned services (for additional details refer to [15]).

4.1.1 Multi-modal News Search and Retrieval

The system supports both simple queries (e.g., keywords) as well as more advanced queries (e.g., weighted queries, boolean operators) for searching and retrieving the aggregations. To facilitate the results visualisation, the system provides a browsable Web page showing the ranked results. For each retrieved aggregation, the system lists the basic information, i.e. title, score and update time, and provides the links for the included news stories and newspaper articles. In addition, as the search results are provided in the form of RSS feeds, users can subscribe to the submitted query, and automatically receive a notification when the results page is modified, i.e. when either an already included aggregation is updated or a new one is discovered.

4.1.2 Topic Detection and Importance Ranking

Topic detection is a service through which users can browse the collection of multi-modal aggregations, according to either temporal relevance or threads. This functionality allows users to browse groups of closely related multi-modal aggregations. A group of multi-modal aggregations forms a thread if, for any two multi-modal aggregations H_a and H_b in the group, their content similarity is greater than a threshold value γ. The similarity measure applied during clustering $Sim(H_a, H_b)$ is defined as the average Jaccard similarity between the sets of RSS feed items and TV news stories constituting the two multi-modal aggregations. Each thread is labelled with a title, corresponding to the most recurrent title of the included multi-modal aggregations, and a date. Users can browse the detected threads by either title or date, as well as access the content of the included multi-modal aggregations. The temporal relevance browsing functionality is based on the heuristic that topics corresponding to multi-modal aggregations having longer temporal extent on TV newscasts are more important than those having shorter temporal extent.

4.1.3 Multi-modal User Recommendation

The system provides a recommendation service that helps users find the desired information according to their behaviour and interests. The delivering of the service employes the queries submitted by a user to build a list of related queries. These system-generated queries can be then issued by the user to tune or redirect the search process. The query generation process is based on the assumption that the relevant aggregations for a query q share some terms apart from the original terms used in q. More in detail, the expansion of user queries algorithm works as follows. Let Q be a query submitted by the user u, and $\mathscr{A} = \{\gamma_i^*\}_{i=1}^{|A|}$ be the set of multimodal aggregations retrieved from the MAi for Q. For each aggregation $\gamma_i^* \in \mathscr{A}$, a feature vector $\mathbf{v} = (\mathbf{s}, \mathbf{c}, \mathbf{p})$ is extracted from the analysis of the RSS items' titles and descriptions, the referenced news articles text, and the TV news items' transcribed speech content. The sub-vector $\mathbf{s}$ stores the fraction of word occurrences in the aggregation, according to a reference dictionary. The sub-vector $\mathbf{c}$ stores the normalised (w.r.t. the total number of objects in the aggregation) scores of the categories to which the aggregation belongs, according to the same set defined for the news story categorisation. The sub-vector $\mathbf{p}$ is the set of couples of the proper nouns found by TreeTagger in the RSS items included in the aggregation γ_i^*, and their corresponding frequencies.

The k-means clustering algorithm is run on the set $\mathscr{A}$ using $\mathbf{v}$ as feature vector, until either the desired precision ε is achieved, or the maximum number of epochs N_{iter} is reached. Because of the heterogeneity of the sub-vectors of $\mathbf{v}$, we used the Euclidean distance to compare the sub-vectors in the $\mathbf{s}$ and $\mathbf{c}$ space, and the Jaccard distance to compare the sets in the $\mathbf{p}$ space. Given $\mathbf{v}_a = (\mathbf{s}_a, \mathbf{c}_a, \mathbf{p}_a)$ and $\mathbf{v}_b = (\mathbf{s}_b, \mathbf{c}_b, \mathbf{p}_b)$ two feature vectors, we define a combined distance used by the k-means clustering process:

$$d(\mathbf{v}_a, \mathbf{v}_b) = \frac{1}{3}\left(L2(\mathbf{s}_a, \mathbf{s}_b) + L2(\mathbf{c}_a, \mathbf{c}_b) + \frac{|\mathbf{p}_a \cap \mathbf{p}_b|}{|\mathbf{p}_a \cup \mathbf{p}_b|}\right) \tag{13}$$

Once the clustering process is completed, we select the centroid of the most populated cluster, $C_M = (\mathbf{s}_M, \mathbf{c}_M, \mathbf{p}_M)$ and select the proper nouns $p_1, \ldots, p_K$, $p_i \in \mathbf{p}_M$, such that the corresponding frequencies are greater than a dynamic threshold calculated as the mean of all frequencies in $\mathbf{p}_M$. We then derive the two queries $Q \wedge \{p_1 \ldots p_K\}$ and $Q \vee \{p_1 \ldots p_K\}$. Let us to consider the following example. Suppose a user submits the query *"Donadoni contratto"* (i.e., Donadoni's contract), presumably to find information about the contract of Roberto Donadoni. Let us suppose that the described clustering and selection process discovers the following proper nouns: *Abete*, *Federcalcio* (i.e., football federation), *Lippi* and *Marcello*. Then, the following derived queries would be proposed to the user:

q_1 := `(lippi abete marcello federcalcio)` $\wedge$ `(donadoni contratto)`, i.e. a refinement of Q;
q_2 := `(lippi abete marcello federcalcio)` $\vee$ `(donadoni contratto)`, i.e. an expansion of Q.

4.1.4 Interactive Graph-Based News Navigation

In order to provide interactive visualisation of the generated aggregations for news retrieval, browsing and editing, we developed an advanced Java interface that enables users to track down the structural, spatial (i.e., inter-aggregation) and temporal properties of the selected aggregation and, at the same time, navigate and manage its contents [15]. First, at the top is the spatial-temporal navigation frame, which shows the title of the selected aggregation, the parent-child hierarchy as obtained by applying Algorithm 1, a select menu that users can use to navigate through the aggregations included in the same thread, and a interactive histogram tool displaying the temporal distribution of the Web and TV news included in the browsed aggregation. Second, at the centre is the structural navigation frame that displays the selected aggregation as a browsable graph whose nodes are the included RSS items and whose edges are the affinities among these nodes, as evaluated by (3). The last part of the interface shows, at the bottom, the content (Web articles and TV newscasts) of the selected aggregation.

5 System Evaluations

For evaluation purposes the system was run for several months, collecting about 88,280 online articles and 23,940 news stories, resulting from the segmentation of 3,670 newscast programmes. The online articles were downloaded from 95 RSS feeds supplied by 16 online newspapers and press agencies Web sites. The newscasts were acquired from seven major national channels 24 hours/day and 365 days/year. To accommodate these requirements, the system has been implemented on a distributed multi-CPU architecture. The programme segmentation task takes on average ≈ 3.74 times the programme duration, that is normally a newscast of 30 minutes.

5.1 Evaluation of the TV News Processing System

We tested the DTV processing system by measuring both the news story segmentation and classification efficiency, and the machine processing performance.

5.1.1 News Story Segmentation and Classification Efficiency

Two distinct experiments were performed to test the programme boundary detection accuracy. The first was aimed at identifying 11 different reference clips from a data set of 782 clips randomly acquired from daily television schedules. The second consisted in detecting the starting and ending jingles of seven distinct news programmes (total 14 clips) in real-time, broadcast streams. In the first experiment, the achieved precision and recall were ≈ 0.80 and ≈ 0.87, respectively. In the second experiment the reached precision and recall were, ≈ 1.0 and ≈ 0.90, respectively. The news story segmentation performance was evaluated by measuring precision and recall of segment boundaries compared to manual annotation of story boundaries. Errors in story boundary detection include erroneously splitting a single story into two or more segments (i.e. over-segmentation), and merging two or more consecutive stories into a single segment (i.e. under-segmentation). As under-segmentation effects can be considered as being more penalising than over-segmentation effects, we used an evaluation measurement taking into account starting boundaries and ending boundaries with different weights, as well as considering missing material as having more impact than extra material on the measurement [5]. The system was tested against a test set of 84 programmes obtaining precision and recall of 0.76 and 0.73, respectively. The news story categorisation task was performed using a naive Bayesian classifier. A data set of 25,000 automatic speech transcriptions was collected. Four fifths of the data were used for training, and the remaining data were used for testing, reaching a classification accuracy value of ≈ 0.82.

5.2 Evaluation of the Multi-modal Aggregation System

To test the overall efficiency of the multi-modal aggregation service, we set up a pool of 25 users, taken from the employers of our organisation, who were unaware of the rationales of the system. Each user was asked to perform evaluations through a Web interface showing a random list of aggregations. We set $\alpha = 0.8$ in (3), thus obtaining a total of 4,187 multi-modal aggregations. Each aggregation was then evaluated through the following markers, using a judgement scale from 1 (i.e., disappointment) to 5 (i.e., full satisfaction): (i) global cohesion of the aggregations (C_g); (ii) representativeness of the aggregation title (T_r); (iii) cohesion of the Web (C_w) and TV (C_t) news included in the aggregation. The obtained mean values are: $C_g = 4.1$, $T_r = 4.46$, $C_w = 4.47$ and $C_t = 4.2$.

Similarly, the performance of the adaptive relevance clustering technique was tested measuring the mean global cohesion of 24 root aggregations (C_{gr}), and the mean global cohesion of the set of 77 leaves aggregations derived from them (C_{gl}), obtaining the following results: $C_{gr} = 3.41$ and $C_{gl} = 3.97$. This result indicates the effectiveness of the ARC algorithm to improve the accuracy of the aggregations.

5.3 Evaluation of the Multi-modal Search and Retrieval System

The efficiency of the multi-modal search and retrieval service was evaluated using the mean average precision (MAP). Let $\mathscr{Q} = \{q_k\}_{k=1}^{N}$ be the set of user-generated queries and $\mathscr{H}_k = \{h_{ik}\}_{i=1}^{R_k}$ be the set of retrieved documents for q_k, ranked according their score w.r.t. the submitted query. Average precision (AP) is the average of the precision scores at the ranks where relevant hits (w.r.t. the original query q_k) occur. The mean average precision is the mean of AP over the full set $\mathscr{Q}$:

$$MAP = \frac{1}{N}\sum_{k=1}^{N} AP_k = \frac{1}{N}\sum_{k=1}^{N}\frac{1}{R_k}\sum_{i=1}^{R_k} g_k(i)p_k(i) \,, \tag{14}$$

where $g_k(i)$ is a binary function that returns 1 if h_{ik} is relevant to q_k, and $p_k(i)$ is the precision after i hits of $\mathscr{H}_k$. In our experiments, we evaluated both the MAP for a set of user queries, and the MAP for the queries automatically derived from them.

5.3.1 User-Generated Query Results Quality

In the experiments, users were asked to submit some queries to the system, and then mark each retrieved aggregation as relevant or not to the submitted query. According to TREC specifications, we evaluated 50 queries, achieving a MAP of 0.79. This proves that the proposed approach, namely fusing contributions coming from television and the Web into a single document to be indexed, enables the delivering of an effective search and retrieval service.

5.3.2 Derived Query Generation Efficiency

Analogously to multimodal aggregation efficiency, we used users' ratings to evaluate our query derivation method. Let $\mathcal{Q}^* = \{q_j^*\}$ be the set of derived queries from the set of original queries $\mathcal{Q}$. For each $q_j^* \in \mathcal{Q}^*$, we calculate:

1. Average precision AP_j^* w.r.t. the set of retrieved objects for q_j^*;
2. Relevance degree ρ_j w.r.t. the original query $q \in \mathcal{Q}$ from which q_j^* derives, expressed by a score from 1 (totally unrelated) to 5 (completely related).

The following performance markers can be then defined:

- Mean average precision of the derived queries, α_8
- Average relevance of the derived queries to the original queries:

$$\alpha_9 = \frac{1}{|\mathcal{Q}^*|} \sum_{\rho_j \neq 0} \rho_j \tag{15}$$

- Effectiveness of query derivation w.r.t. the initial topics of interest to the users:

$$\alpha_{10} = \alpha_8 \frac{\alpha_9}{5} \,. \tag{16}$$

We set $k = 64$, $\varepsilon = 0.05$ and $N_{iter} = 20$ for k-means. A set of 140 derived queries produced by the user panel was evaluated, getting the following results: $\alpha_8 = 0.85$, $\alpha_9 = 4.3$ and $\alpha_{10} = 0.73$. The results show that the use of the derived queries improves the number of relevant documents retrieved at the top of the results list. High relevance to the original query is also provided. Thus, the method is helpful in finding new relevant documents for the users who formulated the original query.

6 Conclusions and Future Works

By developing this research we introduced a novel methodology to support the delivery of multi-modal aggregation services based on some novel concepts. The main theoretical innovations of the method are: (i) a semantic affinity function acting as a kernel to discover the semantic affinities of heterogeneous information items, (ii) an asymmetric vector projection model on which semantic dependency graphs among information items are built, and (iii) a technique to select representative elements out of these graphs. To prove the applicability of our technique, we developed a prototype system for aggregating and retrieving online newspaper articles and broadcast news stories. Obtained results are very encouraging and demonstrate the robustness and effectiveness of the proposed method. The development of the research described in this dissertation led to the development of a fully unsupervised platform for content-based collection, retrieval and browsing of hypermedia news content. The main functionalities of this system are: (i) broadcast news stories and Web newspaper articles are fully cross-referenced and indexed to provide a hypermedia retrieval system designed to efficiently and quickly retrieve text, speech and video news information; (ii) TV news reports and newspaper articles are accessible by means of production metadata such as broadcasting channel, date and time, programme title, content category (e.g., politics, current events, sports) and RSS provider name. Additionally, users can search for news information by either full-text retrieval or desired topics; (iii) a functional interface is provided to graphically browse contents and structures of the collected objects; (iv) user evaluations of the effectiveness of the aggregation, indexing and retrieval components of the system confirmed its reliability and validity in a real-world business context [14].

Several future research directions can be foreseen based on the results achieved so far. On a more theoretical side, we foresee at least the following directions: (i) optimisation of the programme segmentation algorithms, taking into account supplemental layers of information (pure textual analysis based on the automatic speech recognition results) and more refined heuristics; (ii) exploitation of the structural information conveyed by the graph-based nature of the discovered multi-modal aggregations, through implementation and evaluation of advanced graph exploration techniques; (iii) refinement of the query construction models; (iv) analysis of methodologies to generate higher-level aggregations (e.g. clustering multi-modal aggregations sharing similar concepts); (v) extension of the cross-modal aggregation method to more than two source streams. From the service perspective we can envisage the following development directions: (i) integration of further information sources such as images and radio data; (ii) construction and publication of scalable and efficient Web sites where to publish the system results; (iii) recommend personalised queries based on user preferences as topics of interest or significant events.

References

1. Ahn, J.W., Brusilovsky, P., Grady, J., He, D., Syn, S.Y.: Open user profiles for adaptive news systems: help or harm? In: Proc. of World Wide Web Conference, pp. 11–20 (2007)
2. Bertini, M., Del Bimbo, A., Torniai, C.: Automatic annotation and semantic retrieval of video sequences using multimedia ontologies. In: Proc. of the 14th annual ACM Intl. Conf. on Multimedia, pp. 679–682 (2006)
3. Das, A.S., Datar, M., Garg, A., Rajaram, S.: Google news personalization: scalable online collaborative filtering. In: Proc. of World Wide Web Conference, pp. 271–280 (2007)
4. Deschacht, K., Moens, M.F.: Finding the Best Picture: Cross-Media Retrieval of Content. In: Macdonald, C., Ounis, I., Plachouras, V., Ruthven, I., White, R.W. (eds.) ECIR 2008. LNCS, vol. 4956, pp. 539–546. Springer, Heidelberg (2008)
5. Di Iulio, M., Messina, A.: Use of Probabilistic Clusters Supports for Broadcast News Segmentation. In: DEXA Workshops, pp. 600–604 (2008)
6. Farrús, M., Ejarque, P., Temko, A., Hernando, J.: Histogram equalization in svm multimodal person verification. In: Intl. Conf. on Advances in Biometrics, pp. 819–827 (2007)
7. Gao, Y., Fan, J.: Incorporating concept ontology to enable probabilistic concept reasoning for multi-level image annotation. In: Proc. of the 8th ACM Intl. Workshop on Multimedia Information Retrieval, pp. 79–88 (2006)
8. Hanjalic, A.: Content-Based Analysis of Digital Video. Kluwer Academic Publishers, Dordrecht (2004)
9. Hoashi, K.: Shot boundary determination on MPEG compressed domain and story segmentation experiments for TRECVID 2004. In: Proc. of TRECVID Workshop 2004 (2004)

10. Kashyap, V., Sheth, A.: Semantic and schematic similarities between database objects: a context-based approach. The VLDB Journal 5(4), 276–304 (1996)
11. Lew, M.S., Sebe, N., Djeraba, C., Jain, R.: Content-based multimedia information retrieval: State of the art and challenges. ACM Transactions on Multimedia Computing, Communications and Applications 2(1), 1–19 (2006)
12. Messina, A., Bailer, W., Schallauer, P., Tablan, V., et al.: Content analysis tools. Deliverable 15.4, PrestoSpace Consortium (February 2007)
13. Messina, A., Borgotallo, R., Dimino, G., Gnota, D.A., Boch, L.: Ants: A complete system for automatic news programme annotation based on multimodal analysis. In: Intl. Workshop on Image Analysis for Multimedia Interactive Services (2008)
14. Messina, A., Montagnuolo, M.: A generalised cross-modal clustering method applied to multimedia news semantic indexing and retrieval. In: Proc. of WWWW Conf. (2009)
15. Montagnuolo, M., Ferri, M., Messina, A.: HMNews: an integrated system for searching and browsing hypermedia news content. In: Hypertext, pp. 83–88 (2009)
16. Quénot, G.M., Mararu, D., Ayache, S., Charhad, M., Besacier, L.: CLIPS-LIS-LSR-LABRI experiments at TRECVID 2004. In: Proc. of TRECVID Workshop 2004 (2004)
17. Vinciarelli, A., Fernandez, F., Favre, S.: Semantic segmentation of radio programs using social network analysis and duration distribution modeling. In: Proc. of IEEE Intl. Conf. of Multimedia and Expo., pp. 779–782 (2007)
18. Volkmer, T., Tahahoghi, S.M.M., Williams, H.E.: RMIT university at TRECVID 2004. In: Proc. of TRECVID Workshop 2004 (2004)
19. Webster, D., Huang, W., Mundy, D., Warren, P.: Context-orientated news filtering for web 2.0 and beyond. In: Proc. of World Wide Web Conference, pp. 1001–1002 (2006)
20. Weng, C.Y., Chu, W.T., Wu, J.L.: Movie analysis based on roles' social network. In: Proc. of IEEE Intl. Conf. of Multimedia and Expo., pp. 1403–1406 (2007)
21. Xu, C., Wang, J., Lu, H., Zhang, Y.: A novel framework for semantic annotation and personalized retrieval of sports video. IEEE Trans. on Multimedia 10(3), 421–436 (2008)
22. Zhai, Y., Chao, X., Zhang, Y., Javed, O., Yilmaz, A., Rafi, F.: University of Central Florida at trecvid 2004. In: Proc. of TRECVID Workshop 2004 (2004)
23. Zhang, Y., Xu, C., Rui, Y., Wang, J., Lu, H.: Semantic event extraction from basketball games using multi-modal analysis. In: IEEE ICME 2007, pp. 2190–2193 (2007)

Using a Network of Scalable Ontologies for Intelligent Indexing and Retrieval of Visual Content

Atta Badii, Chattun Lallah, Meng Zhu, and Michael Crouch

Abstract. There are still major challenges in the area of automatic indexing and retrieval of digital data. The main problem arises from the ever increasing mass of digital media and the lack of efficient methods for indexing and retrieval of such data based on the semantic content rather than keywords. To enable intelligent web interactions or even web filtering, we need to be capable of interpreting the information base in an intelligent manner. Research has been ongoing for several years in the field of ontological engineering with the aim of using ontologies to add knowledge to information. In this chapter we describe the architecture of a system designed to semi-automatically and intelligently index huge repositories of special effects video clips. The indexing is based on the semantic content of the video clips and uses a network of scalable ontologies to represent the semantic content to further enable intelligent retrieval.

1 Introduction

The advent of the Internet and digital media technologies has led to an enormous increase in the production and online availability of digital media assets as well as made the retrieval of particular media objects more challenging. Processing of digital data, such as text, image and video, has achieved great advances during the last few decades. However, as the well known 'semantic gap' [17] still exists between the low-level computational representation and the high-level conceptual understanding of the same information, more intelligent semantic-driven modelling, multi-modal indexing and retrieval for digital data are needed.

The DREAM (Dynamic REtrieval Analysis and semantic metadata Management) project aims at paving the way towards semi-automatic acquisition of

Atta Badii · Chattun Lallah · Meng Zhu · Michael Crouch
Intelligent Media Systems and Services Research Centre (IMSS),
School of Systems Engineering,
University of Reading,
United Kingdom
e-mail: {atta.badii,c.lallah,meng.zhu,m.w.crouch}@reading.ac.uk

Soro et al. (Eds.): Inform. Retrieval and Mining in Distrib. Environments, SCI 324, pp. 233–248.
springerlink.com

knowledge from visual content. This is being undertaken in collaboration with Partners from the UK Film Industry, including Double Negative[1], The Foundry[2] and FilmLight[3]. Double Negative is the test partner who provided the test materials and user requirements and evaluated the system prototype. One of the main challenges for the users in this industry is the storage and management of huge repositories of multimedia data, in particular, video files, and, having to search through distributed repositories to find a particular video shot. For example, when Special Effects Designers need a category of clips containing "fire explosions" which they may wish to use in the making of a new special effect, it is a tedious and time consuming task for them to search for similar video clips which feature specific objects of interest. The first prototype of DREAM has been evaluated in this film post-production application domain and aims to resolve the existing problems in indexing and retrieving video clips.

This chapter presents the DREAM (Dynamic REtrieval Analysis and semantic metadata Management) research project which aims to prepare the way for the semi-automatic acquisition of knowledge from visual content, and thereby addressing the above mentioned problems. The chapter shows how Topic Map Technology [1, 9, 15] has been used to model the semantic knowledge automatically extracted from video clips to enable efficient indexing and retrieval of those clips. The chapter also highlights some related work and presents how semi-automatic labelling and knowledge acquisition are carried out in the integrated framework. The chapter concludes with an analysis of the results of the evaluation of the prototype.

2 State of the Art – Multimedia Information Indexing and Retrieval

From time to time, researchers have sought to map low-level visual primitives to high-level semantic-related conceptual interpretations of a given media content. The objective has been to achieve enhanced image and video understanding which could further benefit in its subsequent indexing and retrieval. Early attempts in this research area have focussed on extracting high-level visual semantics from low-level image content. Typical examples include: discrimination between 'indoor' and 'outdoor' scenes [14, 18], 'city' vs. 'landscape' [10], 'natural' vs. 'manmade' [5], etc. However, the granularity aspect of those research results is considered to be limited due to the fact that only the generic theme of the images can be identified.

Only recently researchers started to develop methods for automatically annotating images at object level. Mori et al [13] proposed a co-occurrence model which formulated the co-occurrency relationships between keywords and sub-blocks of images. The model had been further improved by Duygulu et al [8] using the Brown et al machine translational model [6] with the assumption that

[1] Double Negative, `http://www.dneg.com`
[2] The Foundry, `http://www.the-foundry.co.uk`
[3] FilmLight, `http://www.filmlight.ltd.uk`

image annotation can be considered as a task of translating blobs into a vocabulary of keywords. Zhou and Huang [21] explored how keywords and low-level content features can be unified for image retrieval. Westerveld [20] and Cascia et al [7] sought to combine the visual cues of low-level visual features and textual cues of the collateral texts contained in the HTML documents of on-line newspaper archives with photos, to enhance the understanding of the visual content.

Li et al [19] developed a system for automatic linguistic indexing of pictures using a statistical modelling approach. The 2D multi-resolution Hidden Markov Model is used for profiling categories of images, each corresponding to a concept. A dictionary of concepts is built up, which is subsequently used as the linguistic indexing source. The system can automatically index images by firstly extracting multi-resolution block-based features, then selecting top k categories with the highest log likelihood for the given image to the category, and finally determining a small subset of key terms from the vocabulary of those selected categories stored in the concept dictionary as indexing terms.

The most recent effort in this area focussed on introducing a knowledge representation scheme, such as an ontology, into the process of semantic-based visual content tagging [12, 16]. Athansiadis et al [2] proposed a framework for simultaneous image segmentation and object labelling. The work focused on a semantic analysis of images, which contributes to knowledge-assisted multimedia analysis and bridging the semantic gap. The possible semantic labels, formally represented as fuzzy sets, facilitate the decision making on handling image regions instead of the traditional visual features. Two well known image segmentation algorithms, i.e. watershed and recursive shortest spanning tree, were modified in order to stress the independence of the proposed method from a specific image segmentation approach. Meanwhile, an ontology-based visual context representation and analysis approach, blending global knowledge in interpreting each object locally, had been developed. Fuzzy algebra and ontological taxonomic knowledge representation had been combined in order to formulate the visual contextual information. The contextual knowledge had then been used to re-adjust the labelling of results of the semantic region growing, by means of fine-tuning the membership degrees of the detected concepts.

3 The DREAM Framework

The DREAM framework can be considered as a knowledge-assisted intelligent visual information indexing and retrieval system, which has been particularly tailored to serve the film post-production industry. The main challenge in this research work was to architect an indexing, retrieval and query support framework. The proposed framework exploits content, context and search-purpose knowledge as well as any other domain related knowledge in order to ensure robust and efficient semantic-based multimedia object labelling, indexing and retrieval. Our current framework provides for optional semi-automatic (man-in-the-loop) labelling; thus enabling semantically triggered human intervention to support optimal cooperation in the semi-automatic labelling of what is currently mostly a manual process. This framework is underpinned by a network of scalable ontologies, which grows alongside the ongoing incremental annotation of video content. To support these scalable ontologies, we deployed the Topic Map Technology, which also enables transparent and flexible multi-perspective access to the repository and pertinent knowledge.

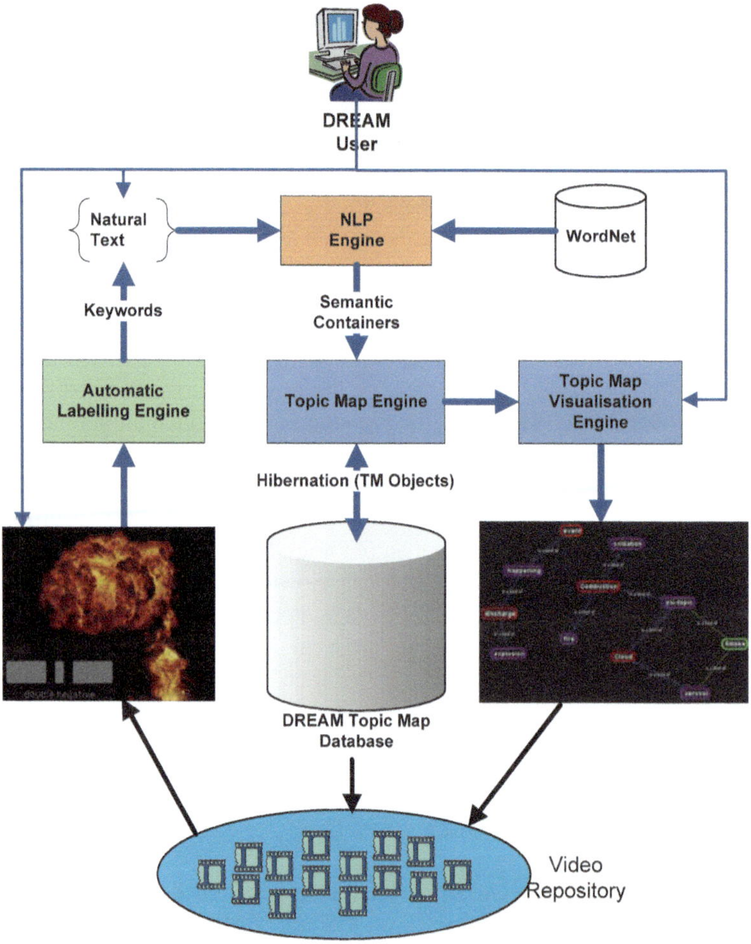

Fig. 1 DREAM Framework for film post-production domain

Figure 1 shows the architecture of the DREAM framework. In the film post-production domain, massive repositories of video clips are held in various locations. As the size of the repository grows, it becomes a nightmare for people trying to find a specific clip for a specific purpose at a specific time. The DREAM framework was developed to help overcome this bottleneck between the large volume of semantically disorganised data and the high-level demand in semantically enhanced efficient and robust indexing and retrieval of such data.

In deploying the DREAM framework, as illustrated in Figure 1, a User is able to query or navigate through the repository of video clips using the Topic Map Visualisation Engine which provides an interface to query the knowledge base or to navigate through the connections between the semantic concepts of the video clips. To support this query and navigate through the knowledge base, a Topic Map Engine has been designed and completed. In order to build the knowledge base, a Collaterally-Cued Automatic Labelling Module has been implemented. This reads in the video clips and extracts the main objects of interest, in terms of semantic keywords, from the clips. The User can then confirm those keywords and/or add more contextualised or domain-specific information so as to make their own viewpoint-specific set of keywords, which is then fed into the NLP Engine. The NLP Engine uses external knowledge such as WordNet to add meaning to the captured information which enhances the situated knowledge element. The situated knowledge element is then in a form that we term as "Semantic Containers", these are passed to the Topic Map Engine, which merges them with the existing knowledge found in the DREAM Topic Map Database. This enables intelligent querying and visualisation of the Semantic Network of concepts which indexes millions of video shots.

4 Knowledge Representation

The DREAM framework deploys Topic Map Technology to incorporate both content and context knowledge e.g. the knowledge of both episodic queries and their overall business context including users' dynamic role-based purposes, and, a priori higher level domain concepts (not just key words) that are so-to-speak "mind-mapped" to suit the users' relevant data-views ("i-schemas") for maximally transparent and flexible multi-perspective access to provide information retrieval that is context-sensitised, concept-oriented and a priori ontology-network-enabled.

Topic Maps, as defined in ISO/IEC 13250 [11], are an international standard for organising and representing information on the Web. A Topic Map can be represented by an XML document in which different element types are used to represent topics, occurrences of topics, and associations (or relationships) between topics. The Topic Map model provides a mechanism for representing large quantities of structured and unstructured information and organising it into "topics". The topics can be interrelated by an association and can have occurrences linked to them. A Topic Map can thus be referred to as a collection of topics and associations. The associations are used to navigate through the information in many different ways. The network can be extended as the size of the collection grows, or it is merged with other topic maps to provide additional paths through the information. The most important feature of the Topic Map is that it explicitly captures additional tacit information. It is this capability that has captured the interest of people working on knowledge management issues, identifying Topic Maps as a mechanism that can help to capture what could be considered "knowledge" from a set of information objects.

The real benefit of integrating Topic Maps within the DREAM Framework is the resulting retrieval gains that in turn confer high-scalability. Normally, the topic maps are linked to each other. It is very beneficial for a user to be able to incrementally develop semantically annotated subnets representing part of the media, and to build this up by linking it with representation of other parts. In this way, various contexts can be applied to the maps as they are linked together. During retrieval, it is natural for users to associate various topics in the way that is best suited to the formulation of the expected domain queries that serve the objectives of the domain process, in other words, the process logic e.g. editing goals that are typically assigned to be completed by the editors. The natural evolution of ideas amongst those engaged in the process may prompt them to re-visit certain associations, extend or refine some and add other new associations. Accordingly the facility for creating new associations amongst the topics exists. Hence, the topic maps are continuously evolving during the entire life of the repository, as new content is added, there are always chances of discovering new associations, both intra and inter content associations.

5 Knowledge Acquisition

This section describes the process of knowledge acquisition through the annotation of video content within the DREAM framework. Variation of the different annotation methods is supported, including automatic annotation and textual annotation, manual annotation, and, visual annotation as shown in Figure 2.

For each new clip, the user decides which annotation method they wish to use. They may choose the automatic annotation process which uses the Automatic Labelling Engine to process the annotation or they may choose the textual annotation process which uses the NLP Engine to process text entered by the user. The output from the automatic annotation process or textual annotation process can further be refined by using the manual annotation and visual annotation, as illustrated in Figure 2. The framework also supports batch processing which allows automatic annotation of a range of video clips using the Automatic Labelling Module. The process of automatic annotation is further described in section 5.1 and in section 5.2 where we cover the process of semi-automatic annotation.

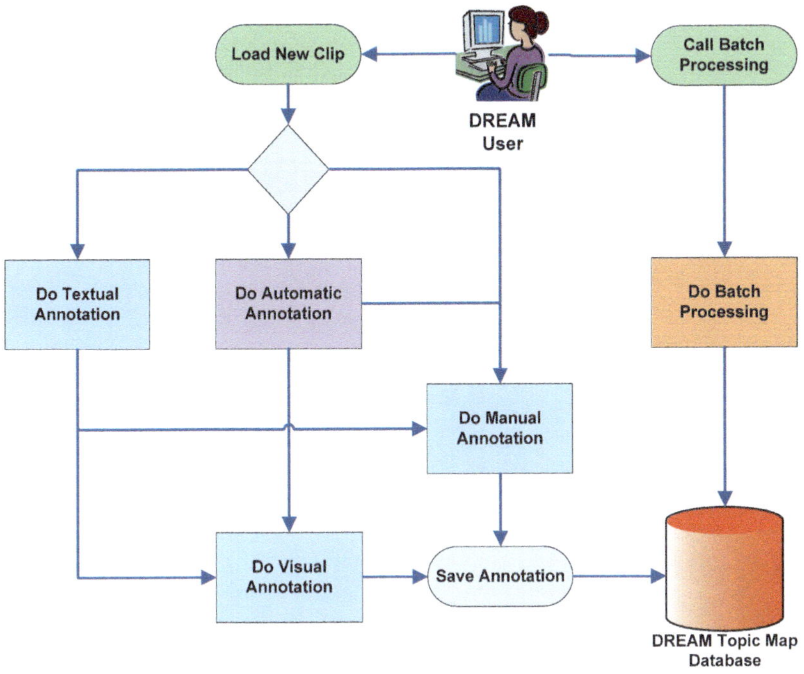

Fig. 2 Video Content Annotation Process in DREAM

5.1 Automatic Video Annotation

The automatic video labelling module [4] is a fundamental component of the DREAM framework. It aims to automatically assign semantic keywords to objects of interest appearing in the video segment. The module had been implemented to label the raw video data in a fully automated manner. The typical user of the system is the video content library manager who will be enabled to use the system to facilitate the labelling and indexing of the video data. With this function, all the objects of interest including moving and still foreground objects will be labelled with linguistic keywords.

Figure 3 shows the typical workflow of the Automatic Labelling Module. The module takes the raw video clips and the associative metadata, i.e. motion vectors and mattes, as input whereby the low-level blob-based visual features, i.e. colour, texture, shape, edge, motion activity, motion trajectory, can be extracted and encoded as feature vectors. It is those visual blobs that were compared against the visual concepts defined in the visual vocabulary of objects. These objects consist of a set of clusters of visual feature vectors of different types of special effect foreground objects such as blood, fire, explosion, smoke, water splashes, rain, clouds etc, to find the best matching visual concepts using the K-Nearest Neighbour algorithm. However, those are all traditional methodologies that would

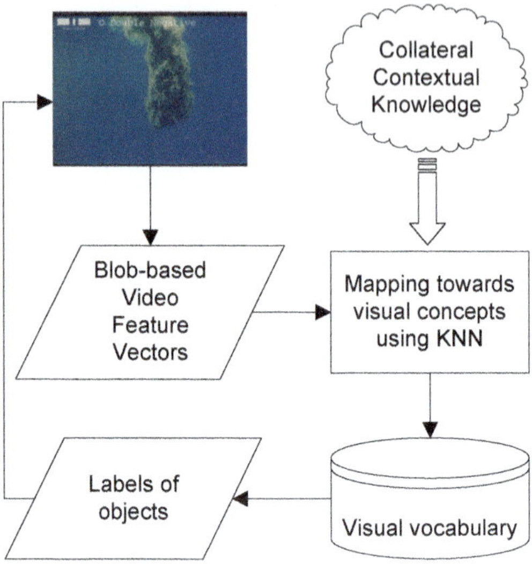

Fig. 3 Workflow of the Automatic Labelling Module

benefit from appropriate semantic enrichment by way of linkages to latent and collateral associations so as to empower the representation of higher-level context related visual concepts. Therefore, we introduced what we refer to as the collateral context knowledge, which was formalised by a probabilistic based visual keyword co-occurrence matrix, to bias (i.e. appropriately converge) the traditional matching process.

Table I shows the labelling accuracy for the DREAM auto-labelling module, including both content and context related labels, based on the training-test ratio of 9:1, 7:3 and 5:5 respectively. Among the 3 different experimental setups, the training-test ratio of 7:3 achieved the superior performance with an average accuracy of 80% followed by 75% for 5:5 and 66% for 9:1. Despite poor performance for several categories, many other categories achieved a very high labelling accuracy percentage; including some at 100%. The main reason for low accuracy results for certain categories was lack of adequate number of samples for training and testing. This was the case in respect of the classes associated with lower accuracy performance as compared with other categories. Therefore, it is possible to conclude that with a benchmarking corpora which can offer a balanced distribution of samples of categorise for training and testing, higher accuracy performance would be obtained for those classes with lower accuracy due to lower number of available samples to train and test.

Table 1 Auto-Labelling accuracy based on the training/test ratio of 9:1, 7:3 and 5:5

Class label	9:1	7:3	5:5	Class label	9:1	7:3	5:5
blood and gore; blood	75%	94%	83%	misc; welding;	100%	100%	100%
blood and gore; gore;	0%	100%	100%	muzzle flash;	0%	100%	100%
bullet hits; sparks;	75%	100%	100%	sparks;	0%	25%	40%
crowds figures;	100%	100%	100%	water;bilge pumps;	100%	100%	100%
explosions fire smoke;explosion;	100%	92%	75%	water; bilge pumps;	100%	100%	80%
explosions fire smoke; fire;	100%	100%	90%	water; boat wakes;	100%	67%	50%
explosions fire smoke; fire burst;	100%	0%	100%	water; bubbles;	0%	0%	0%
explosions fire smoke; flak;	100%	100%	100%	water;cascading water;	0%	0%	50%
explosions fire smoke; sheet fire;	100%	100%	100%	water; drips;	100%	100%	50%
explosions fire smoke; smoke;	86%	90%	90%	water; interesting water surfaces;	0%	50%	67%
explosions fire smoke; steam;	100%	100%	100%	water; rain;	0%	75%	71%
falling paper;	100%	100%	100%	water; rivulets;	0%	100%	0%
lens flares;	100%	86%	83%	water; splashes;	0%	100%	60%
misc; car crash;	0%	0%	0%	weather; clouds;	100%	100%	100%
misc; fecal-matter;	100%	89%	86%	weather; snow;	100%	100%	50%
misc; washing line;	100%	100%	100%	Average	66%	80%	75%

5.2 *Semi-automatic Video Annotation*

In DREAM, we automatically construct Topic Maps for each new video clip. The output from the Automatic Labelling Engine is updated by the User and the resulting unstructured natural text is processed by the NLP Engine, as illustrated in Figure 4. The NLP Engine creates a structured description of the semantic content of the clips. These structured descriptions are termed as Semantic Containers and are further described in [3]. These semantic containers allow the representation of both simple sentences and complex sentences in terms of entities and actions, which are used by the DREAM Topic Map Engine to generate Topic Maps automatically. A semantic container may contain other semantic containers, enabling representation of complex sentences. This enables an entity (topic) to be linked to a group of entities (topics) [3]. As a result, complex sentences are represented by a single semantic container, with a number of "inner" semantic containers detailing the semantic information processed by the NLP Engine. The Topic Map Engine reads those containers and creates a list of topics and associations, which are merged with existing topics and associations in the DREAM Topic Map Database. The new topics are assigned occurrences which index the new clips. The semantic merging of new topics with existing topics creates a network of ontologies which eventually grows each time new clips are indexed.

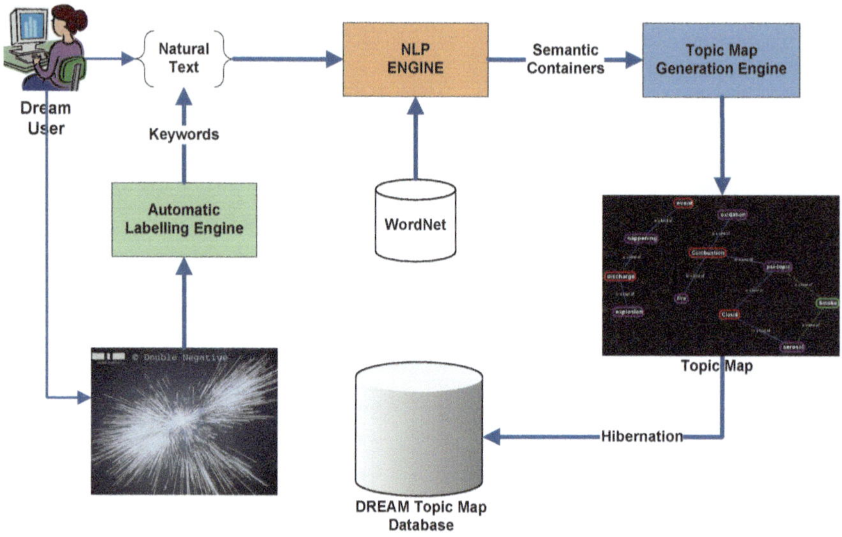

Fig. 4 Semi-automatic labelling process in the DREAM framework

The key to the evolution of the Knowledge Base, as new semantically-defined Topics are added, is the process of adding the synonyms and semantic ancestries, of the concept being added, utilising the WordNet lexical database. This defines the topic in a detailed and robust manner, enabling the linkage to the existing ontology, while ensuring that concepts are stored in the Knowledge Base by their meaning, rather than by word(s) representing the concept. This allows for seamless semantic merging of concepts, and a Knowledge Base that is well-tuned towards retrieval, and visual exploration.

6 Knowledge Retrieval

The Topic Map allows casual browsing of the knowledge base with a richly cross-linked structure over the repository content. Topic occurrences create 'sibling' relationships between repository objects and the video shots. A single resource may be the occurrence of one or more topics, each of which may have many other occurrences. When a user finds/browses to a given resource, this sibling relationship enables them to rapidly determine where the other related resources are to be found. Topic associations create 'lateral' relationships between subjects, the movie concepts – allowing a user to see which other concepts covered by the repository are related to the subject of current interest and to easily browse these concepts. Associative browsing allows an interested data consumer to wander through a repository in a directed manner. A user entering the repository via a query might also find associative browsing useful in increasing the chance of unforeseen discovery of relevant information. A DREAM Topic Map Visualisation, as shown in the Framework diagram [Figure 1] has been implemented to provide such interactive visual browsing of the DREAM Knowledge Base.

Efficient retrieval of desired media is the end goal of the DREAM framework. With the DREAM Knowledge Base built and ever-evolving with newly annotated media being merged into it, the requirement remains for interfaces to be able to achieve this goal. In DREAM, two such interfaces were developed, these being a Visualisation for exploring the Knowledge Base itself, through which media can be retrieved, and a Query Interface that allows directed querying of the Knowledge Base.

6.1 Retrieval Visualisation

The Retrieval Visualisation utilises the DREAM Visualisation Engine to create an interface enabling the user to explore the Knowledge base for concepts, and

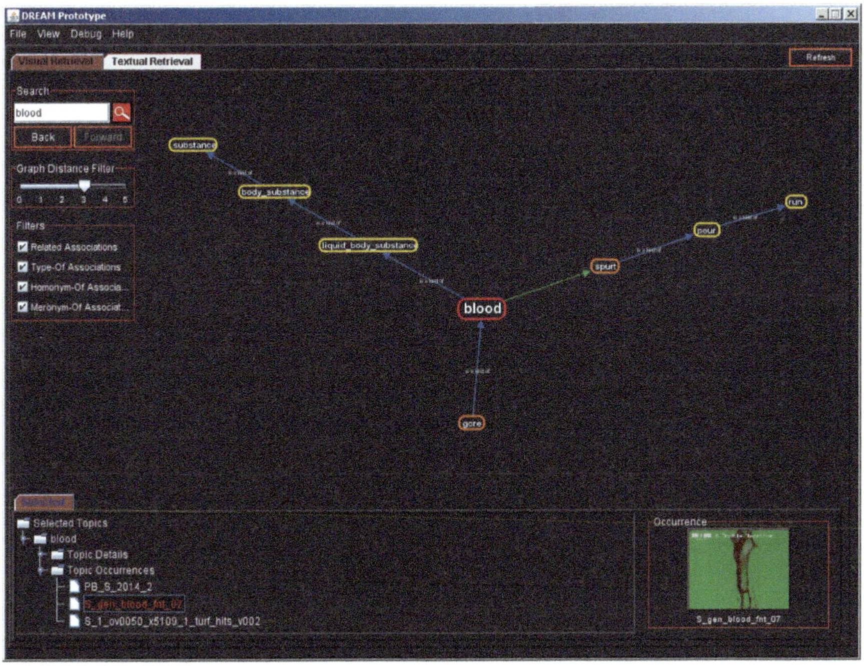

Fig. 5 Retrieval Visualisation Interface

retrieving the Occurrences (Media) attached to them. This operation can range from being rather simple, to allowing for more complicated searching, so that dynamic visual semantic searches become a reality.

The initial entry-point into a search is a simple text search through the Knowledge Base, returning all concepts that match it. This includes searching through the synonyms of each concept, such that any Topic that contains the search string in its synonyms will also be returned, albeit with that specifically matched synonym shown as its display name, for convenience. With the simple search results presented, the user can then choose the path down which they would wish to explore the map further, by clicking on the appropriate Topic that represents the meaning of the word for which they were searching. This process disregards all other search results, and sets the focus upon the selected Topic, whereupon all related Topics are brought into the focus of the visualisation, thus connected to the Topic itself, with the association descriptions labelled on the edge between the nodes.

Colour-coding is used to visually distinguish the different types of association and Topic relevance. Topics with attached Occurrences are coloured orange, whereas empty Topics are coloured Yellow. Currently Selected Topics are enlarged, and coloured Red. Additionally, associations related to Topic definition (typically this 'is a kind of' relationship) are coloured Blue, whereas associations representing a tagged relationship between two Topics (such as "a car crashing into a wall") are coloured Green.

The criteria for the data to be visualised can be modified using a variety of filters, which can be deployed by the user to further direct their search. The first of these filters is a "distance filter", that allows the user to specify the distance from the selected Topic for as far as can be appropriated within the scope of the visualisation of nodes. For example, with a distance of 1, only the immediately associated Topics are shown, whereas, if the distance was 2, all Topics that are within two 'hops', "degrees of separation" from the selected Topic are visualised. This is useful to see an overview of the associations of the Topic as it is situated within the ontological structure, as well as to reduce on-screen Topic clutter, if the selected Topic is one that has a large number of associated Topics. Other filters let the user specify which types of associations they are interested in seeing in the visualisation, depending on the type of search that they may be conducting. For example, showing the homonyms of a Topic would just serve to provide unnecessary clutter, unless the user believes that they may be exploring the wrong route through the semantic associations of a word, thus wishing to have all possible associations depicted on the screen together. Enabling homonym associations then gives the user instant access to jump to a completely different area of the Knowledge Base, with ease, and in a visually relevant manner.

When selecting a Topic, while browsing through the Topic Map, the Occurrences attached to the selected Topic are presented to the user, (as can be seen in Figure 5), in a tree structure mapping out the properties of the selected Topic(s).

The Occurrences are represented here by the clip name, which when clicked displays the key frame for that Clip. The key frames were originally used in the Automatic Labelling process, but also provide an easy visual trigger for user in browsing through.

Additional retrieval techniques can be utilised through this interface, by selecting multiple Topics, with only the common occurrences between selected Topics being returned. This allows for a much finer query into the Knowledge Base, to retrieve very specifically tagged clips. The hierarchical semantic structure of the Knowledge Base is also used to aid the retrieval, by exploring the children of a selected Topic as well, for occurrences. For example, if there was a clip featuring a "car" crashing into a "wall", the user could select the Topics "motor vehicle" and "wall", and the clip would be returned, as "car" is a child of "motor vehicle".

7 User Evaluation of the DREAM Framework

The evaluation of the performance of the DREAM System is a critical task, even more so when the annotations of the video clips are the output of a semi-automatic labelling framework, rather than the result of a concept sorting effort by a team of film post-production domain specialists. The user partner Double Negative examined 145 clips and for each clip the tester at Double Negative gave a score of relevancy from 1 to 5 for each tag as automatically generated by DREAM. Additionally the users commented on the various aspects of the functionality of DREAM and appeared to be able to use its features with ease after some initial training. The usability of the system was ranked as fairly high overall. The users found the visualisation engine with personalise-able colour coding of topics as particularly helpful in navigating the topics of interest from a viewpoint-specific basis. This feature enabled them to avoid cognitive overload when having to consider multi-faceted selection criteria.

Double Negative evaluated the results of processing a sub-set of their library through the DREAM system with the film editing staff (i.e. practitioner users) ranking the accuracy of the Topic Map generated for each clip. As Table II shows, the scores given by the practitioner users were generally very high, but with consistent low scores in certain categories, giving an average overall score of 4.4/5. These scores, along with other evaluation results, indicate the success of the system, but also highlight the areas where performance could be improved. For example, the system had difficulty in identifying clips with very brief exposure of a single indicative feature that constituted the main essence of the clip category for example as in video clips of sparks and flashes whereby the feature to be identified (i.e. the spark or flash) is shown very briefly in the clip.

Table 2 User Evaluation Results

Category Names	Number of Samples	Avg. User Score (Low 1 – 5 High)
Blood and gore; blood;	12	5
Blood and gore; gore;	4	1
Bullet hits; sparks;	4	3.3
crowds figures;	3	3
explosions fire smoke; explosion;	50	4.7
explosions fire smoke; fire;	16	3.6
explosions fire smoke; fire burst;	2	5
explosions fire smoke; smoke;	7	4.4
explosions fire smoke; steam;	5	5
lens flares;	7	4.2
misc; car crash;	1	5
misc; poo;	12	5
misc; washing line;	5	5
misc; welding;	1	1
muzzle flash;	4	1
sparks;	2	1
water; bilge pumps;	2	5
water; boat wakes;	6	5
water; cascading water;	3	5
water; drips;	3	4
water; interesting water surfaces;	5	5
water; rain;	6	5
water; rivulets;	1	5
water; splashes;	3	4.8
water; spray;	3	5
weather; clouds;	4	5
weather; snow;	5	5

8 Conclusion

In this chapter, we have presented the DREAM Framework and discussed how it semi-automatically indexes video clips and creates a network of semantic labels, exploiting the Topic Map Technology to enable efficient retrieval. The framework architecture, which has been presented in the context of film post-production, as a challenging proving ground, has responded well to the high expectations of users in this application domain which demands efficient semantic-cooperative retrieval. We have described how the DREAM framework has also leveraged the advances in NLP to perform the automatic creation and population of Topic Maps within a self-evolving semantic network for any media repositories, by defining the topics (concepts) and relationships between the topics. We also briefly discussed how this framework architecture handles collaborative labelling through its Automatic Labelling Engine. The first DREAM prototype has already been implemented and evaluated by its practitioner users in the film post-production application domain. The results confirm that the DREAM architecture and implementation has proven to be successful. Double Negative have the DREAM system trained with only 400 video clips and deployed within their routine film (post)production processes to achieve a satisfactory performance in labelling and retrieving clips from a repository holding a collection of several thousand clips.

The DREAM paradigm can in future be extended to further domains, including the Web, where any digital media can go through an offline process of (semi)automatic-labelling before being published. The publishing agent will use the semantic labels to create a network of connected concepts, using Topic Maps, and this can be merged with existing concepts on the web space. This will eventually enable more intelligent web interaction, information filtering and retrieval, using semantic concepts as the query parameters rather than keywords. The DREAM Project Team is currently engaged in extending the functionality of the prototype to increase its scalability and ensure a wider uptake across a whole range of application domains particularly in supporting collaborative creative processes in the film, media publishing, advertising, training and educational sectors through our test partners worldwide.

References

[1] Ahmed, K.: Topic maps for Repositories, `http://www.gca.org/papers/xmleurope2000/papers/s29-04.html` (last accessed: January 2010)

[2] Athanasisadis, T., Mylonas, P., Avrithis, Y., Kollias, S.: Semantic image segmentation and object labelling. IEEE Trans. On Circuits and systems for video technology 17(3), 298–312 (2007)

[3] Badii, A., Lallah, C., Kolomiyets, O., Zhu, M., Crouch, M.: Semi-Automatic Annotation and Retrieval of Visual Content Using the Topic Map Technology. In: Proc. of 1st Int. Conf. on Visualization, Imaging and Simulation, Romania, pp. 77–82 (November 2008)

[4] Badii, A., Zhu, M., Lallah, C., Crouch, M.: Semantic-driven Context-aware Visual Information Indexing and Retrieval: Applied in the Film Post-production Domain. In: Proc. IEEE Workshop on Computational Intelligence for Visual Intelligence 2009, US (March 2009)

[5] Bradshaw, B.: Semantic based image retrieval: a probabilistic approach. In: Proc. of the eighth ACM Int. conf. on Multimedia, pp. 167–176 (2000)

[6] Brown, P., Pietra, S.D., Pietra, V.D., Mercer, R.: The mathematics of statistical machine translation: Parameter estimation. Computational Linguistics 19(2), 263–311 (1993)

[7] Cascia, M.L., Sethi, S., Sclaroff, S.: Combining Textual and Visual Cues for Content-Based Image Retrieval on the World Wide Web. In: Proceedings of IEEE Workshop on Content-Based Access of Image and Video Libraries (1998)

[8] Duygulu, P., Barnard, K., Freitas, N., Forsyth, D.: Object recognition as machine translation: Learning a lexicon for a fixed image vocabulary. In: 7th European Conf. on Computer Vision, pp. 97–112 (2002)
[9] Garshol, L.: What are Topic Maps, `http://xml.com/pub/a/2002/09/11/topicmaps.html?page=1` (last accessed: January 2010)
[10] Gorkani, M.M., Picard, R.W.: Texture orientation for sorting photos 'at a glance'. In: Proc. of the IEEE Int. Conf. on Pattern Recognition (October 1994)
[11] ISO/IEC 13250:2000 Document description and processing languages – Topic Maps, International Organisation for Standardization ISO, Geneva (2000)
[12] Maillot, N., Thonnat, M., Boucher, A.: Towards ontology based cognitive vision. Mach. Vis. Appl. 16(1), 33–40 (2004)
[13] Mori, Y., Takahashi, H., Oka, R.: Image-to-word transformation based on dividing and vector quantizing images with words. In: MISRM 1999 First Int. Workshop on Multimedia Intelligent Storage and Retrieval Management (1999)
[14] Paek, S., Sable, C.L., Hatzivassiloglou, V., Jaimes, A., Schiffman, B.H., Chang, S.F., McKeown, K.R.: Integration of visual and text based approaches for the content labelling and classification of Photographs. In: ACM SIGIR 1999 Workshop on Multimedia Indexing and Retrieval, Berkeley, CA (August 19, 1999)
[15] Pepper, S.: The TAO of Topic Maps: finding the way in the age of infoglut, `http://www.ontopia.net/topicmaps/materials/tao.html` (last accessed: January 2010)
[16] Schober, J.P., Hermes, T., Herzog, O.: Content-based image retrieval by ontology-based object recognition. In: Proc. KI 2004 Workshop Appl. Descript. Logics (ADL 2004), Ulm, Germany, pp. 61–67 (September 2004)
[17] Smeulder, A.W.M., Worring, M., Anntini, S., Gupta, A., Jain, R.: Content-Based Image Retrieval at the End of the Early Years. IEEE Transactions on Pattern Analysis and Machine Intelligence 22(12) (December 2000)
[18] Szummer, M., Picard, R.W.: Indoor-outdoor image classification. In: IEEE Int. Workshop on Content-based Access of Image and Video Databases (1998)
[19] Li, J., Wang, J.Z.: Automatic linguistic indexing of pictures by a statistical modelling approach. IEEE Trans. Pattern Analysis and Machine Intelligence 25(9), 1075–1088 (2003)
[20] Westerveld, T.: Image Retrieval: Content Versus Context. In: Proceedings of Content-Based Multimedia Information Access, pp. 276–284 (2000)
[21] Zhou, X.S., Huang, S.T.: Image Retrieval: Feature Primitives, Feature Representation, and Relevance Feedback. In: IEEE Workshop on Content-based Access of Image and Video Libraries (2000)

Integrating Sense Discrimination in a Semantic Information Retrieval System

Pierpaolo Basile, Annalina Caputo, and Giovanni Semeraro

Abstract. This paper proposes an Information Retrieval (IR) system that integrates sense discrimination to overcome the problem of word ambiguity. Word ambiguity is a key problem for systems that have access to textual information. Semantic Vectors are able to divide the usages of a word into different meanings, by discriminating among word meanings on the ground of information available in unannotated corpora. This paper has a twofold goal: the former is to evaluate the effectiveness of an IR system based on Semantic Vectors, the latter is to describe how they have been integrated in a semantic IR framework to build semantic spaces of words and documents. To achieve the first goal, we performed an in vivo evaluation in an IR scenario and we compared the method based on sense discrimination to a method based on Word Sense Disambiguation (WSD). Contrarily to sense discrimination, which aims to discriminate among different meanings not necessarily known a priori, WSD is the task of selecting a sense for a word from a set of predefined possibilities. To accomplish the second goal, we integrated Semantic Vectors in a semantic search engine called SENSE (SEmantic N-levels Search Engine).

1 Background and Motivations

Popular Information Retrieval (IR) systems are based on ranked keyword search [4], in spite of its obvious limits when facing word polysemy (multiple meanings for one word), and synonymy (multiple words having the same meaning). The result is that often IR systems miss relevant documents if they do not contain query keywords verbatim due to synonymy, while irrelevant documents can be retrieved due to polysemy. These problems call for alternative methods that consider not only the lexical level of indexed documents, but also the meaning one.

Pierpaolo Basile · Annalina Caputo · Giovanni Semeraro
Department of Computer Science - University of Bari "Aldo Moro"
Via E. Orabona, 4, 70125 Bari, Italy
e-mail: {basilepp, acaputo, semeraro}@di.uniba.it

Soro et al. (Eds.): Inform. Retrieval and Mining in Distrib. Environments, SCI 324, pp. 249–265.
springerlink.com

Any attempt to work at the meaning level must solve the problem that, differently from words, meanings do not occur in documents and are often hidden behind words. For example, for the query "apple", some users might be interested in documents dealing with "apple" as a "fruit", while others might wish documents related to "Apple computers". Some linguistic processing is needed in order to gain a more effective "interpretation" of information needs behind the user query as well as of words in the document collection. This linguistic processing should hopefully produce insights into the meaning of content in form of machine-readable semantic information.

Words ambiguity is a specific challenge for computational linguistics. Ambiguity means that a word can be interpreted in more than one way, since it has more than one meaning. Usually ambiguity is not a problem for humans, thus it is not perceived as such. Conversely, ambiguity is one of the main problems encountered in the automated analysis and generation of natural languages.

Two main strategies have been proposed to cope with ambiguity:

1. **Word Sense Disambiguation:** the task of selecting a sense for a word from a set of predefined possibilities; usually the so called *sense inventory*[1] comes from a dictionary or thesaurus.
2. **Word Sense Discrimination:** the task of dividing the usages of a word into different meanings, by ignoring any existing *sense inventory*. The goal is to discriminate among word meanings based on information found in unannotated corpora.

The main difference between the two strategies is that disambiguation relies on a sense inventory, while discrimination exploits unannotated corpora.

In the past years, several attempts were proposed to include sense disambiguation and discrimination techniques in IR systems. This is possible because discrimination and disambiguation are not an end in themselves, but rather "intermediate tasks" which contribute to more complex tasks, such as information retrieval. This opens the possibility of an *in vivo* evaluation, where, rather than being evaluated in isolation, results are evaluated in terms of their contribution to the overall performance of a system designed for a particular application (e.g. Information Retrieval).

The goal of this paper is to present an IR system which exploits semantic spaces built on words and documents to overcome the problem of word ambiguity. The starting point is an IR system called SENSE (SEmantic N-levels Search Engine). SENSE is a semantic search engine that maintains multiple indexes of documents by integrating the lexical level represented by keywords with semantic levels. SENSE is used to evaluate how discrimination and disambiguation can help the lexical level. We have performed an evaluation in the context of CLEF 2009 Ad-Hoc Robust WSD task [1] by comparing the original version of SENSE, which implements a WSD algorithm, to a new version that integrates Semantic Vectors to perform Word Sense Discrimination.

The paper is organized as follows: Section 2 presents the IR model involved in the evaluation, which embodies semantic vectors strategies, while Section 3 describes

[1] A *sense inventory* supplies the list of all possible meanings for any word.

the semantic search engine SENSE. Experimental evaluation and reported results are described in Section 4, while a brief discussion about the main work related to our research is in Section 5. Conclusions and future work close the paper.

2 An IR System Based on Semantic Vectors

Semantic Vectors rely on the WordSpace model [23]. This model is based on a vector space in which points are used to represent semantic concepts, such as words or documents. Using this strategy it is possible to build distinct vector spaces for words and documents. These vector spaces can be exploited to develop an IR model as described in the following.

The core idea behind semantic vectors is that words and concepts are represented by points in a mathematical space, and this representation is learned from text in such a way that concepts with similar or related meanings are near to one another in that space (geometric metaphor of meaning). The semantic similarity between concepts can be represented as proximity in a n-dimensional space. Therefore, the main feature of the geometric metaphor of meaning is not that meanings can be represented as locations in a semantic space, but rather that similarity between word meanings can be expressed in spatial terms, as proximity in a high-dimensional space. This straightforwardly suggests an approach to word sense discrimination.

One of the great virtues of semantic vectors is that they make very few language-specific assumptions, since just tokenized text is needed to build semantic spaces to be employed as models of relevance in IR. Even more important is their independence of the quality (and the quantity) of available training material, since they can be built using entirely unsupervised distributional analysis of free text. Indeed, the basis of semantic vectors model is the *distributional hypothesis* [17], according to which the meaning of a word is determined by the set of textual *contexts* in which it appears. As a consequence, in distributional models words can be represented as vectors built over the observable *contexts*. This means that words are semantically related as much as they are represented by similar vectors. For example, if "basketball" and "tennis" occur frequently in the same context, say after "play", they are semantically related or similar according to the distributional hypothesis.

Co-occurrence is defined with respect to a context, for example a window of fixed length, or a document. Co-occurring words can be stored into matrices where the rows represent the terms and the columns represent contexts. Each row corresponds to a vector representation of a word. The strength of semantic association between words can be computed by using cosine similarity. This kind of techniques need to handle the potentially very high dimensionality of vectors. The solution is usually represented by *dimensionality reduction* techniques that allow representing high-dimensional data in a lower-dimensional space without losing information. *Latent Semantic Analysis (LSA)* [20] collects the text data in a words-by-documents co-occurrence matrix, that is then decomposed with singular-value decomposition (SVD) into smaller matrices, by capturing latent semantic structures in the text data. The main drawback of SVD is the scalability.

In our work, we adopt the Semantic Vectors package [29] , which rely on a technique called Random Indexing (RI), introduced by Kanerva [18], for creating semantic vectors. This technique allows to build semantic vectors with no need for (either term-document or term-term) matrix factorization, because vectors are inferred using an incremental strategy. This method allows to solve efficiently the problem of reducing dimensions, which is one of the key features used to uncover the "latent semantic dimensions" of a word distribution.

RI is based on the concept of Random Projection: the idea is that high dimensional vectors chosen randomly are "nearly orthogonal". This yields a result that is comparable to orthogonalization methods, such as Singular Value Decomposition [19], but saving computational resources. Specifically, RI creates semantic vectors in three steps:

1. a context vector is assigned to each document. This vector is sparse, high-dimensional and ternary, which means that its elements can take values in {-1, 0, 1}. A context vector contains a small number of randomly distributed non-zero elements, and the structure of this vector follows the hypothesis behind the concept of Random Projection;
2. context vectors are accumulated by analyzing terms and documents in which terms occur. In particular the semantic vector for any term is computed as the sum of the context vectors for the documents which contain the term;
3. in the same way, the semantic vector for any document is computed as the sum of the semantic vectors (built by step 2) for terms which occur in that document.

For example, let us denote by $A^{n,m}$ a huge term-document matrix with n rows and m columns, and by $R^{m,k}$ a matrix made up of m k-dimensional random vectors, where k can be chosen freely. We define a new matrix denoted by $B^{n,k}$ as follows:

$$A^{n,m} \cdot R^{m,k} = B^{n,k} \quad k << m \tag{1}$$

The new matrix B has the property to preserve the distance between points, that is, if the distance between two any points of A is d, then the distance d_r between the corresponding points in B will satisfy the property that $d = c \cdot d_r$, as shown in Figure 1. A proof of that is reported in [12].

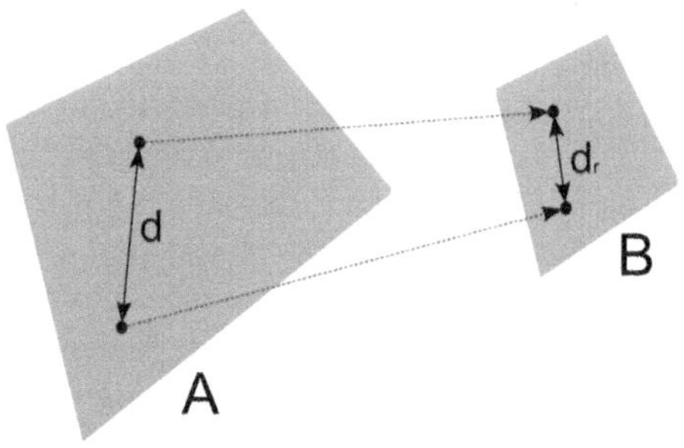

Fig. 1 Random Projection

Therefore, Random Projection allows to build two distinct semantic spaces on terms and documents, which have the same dimension. Vectors in the term (or word) space can be used as queries, while vectors in the document space represent the collection (or search space).

Figure 2 sketches how word and document spaces are obtained using Random Projection. If a term-document matrix plays the role of matrix A in equation 2 then a WordSpace is built, while if a document-term matrix is used as matrix A then a DocumentSpace (DocSpace in Figure 2) is produced. It is important to note that the reduced dimension k is equal in both Word and Document spaces because the semantic vector for any document is computed as the sum of the semantic vectors of the terms that occur in it.

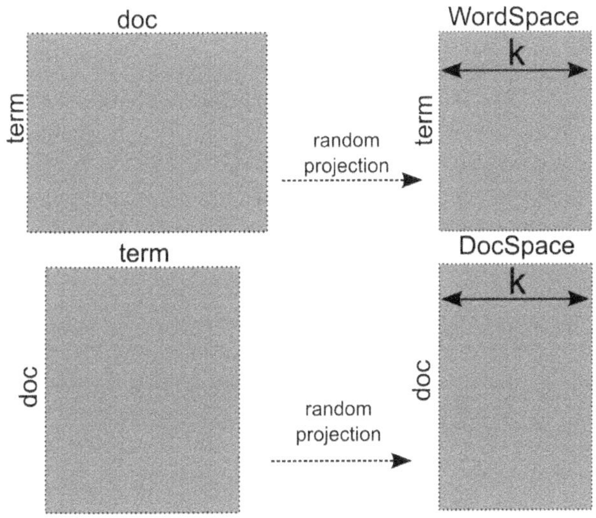

Fig. 2 Word and document spaces

Figure 3 shows a WordSpace with only two dimensions ($k = 2$). If the two dimensions refer respectively to `LEGAL` and `SPORT` contexts, we can note that the vector for the word *soccer* is closer to the `SPORT` context than to the `LEGAL` context. Conversely, the word *law* is closer to the `LEGAL` context. The angle between *soccer* and *law* represents the similarity degree between the two words. It is important to make clear that contexts in a WordSpace have no label (labels in Figure 3 have been added for explanation's sake), thus all we know is that each dimension in the WordSpace is a context, but we cannot have an idea about the specific nature of a context. Similarly, if we consider a DocumentSpace rather than a WordSpace, documents semantically related will be represented close in that space.

The Semantic Vectors package supplies tools for indexing a collection of documents (and their retrieval) by means of the RI strategy. This package relies on

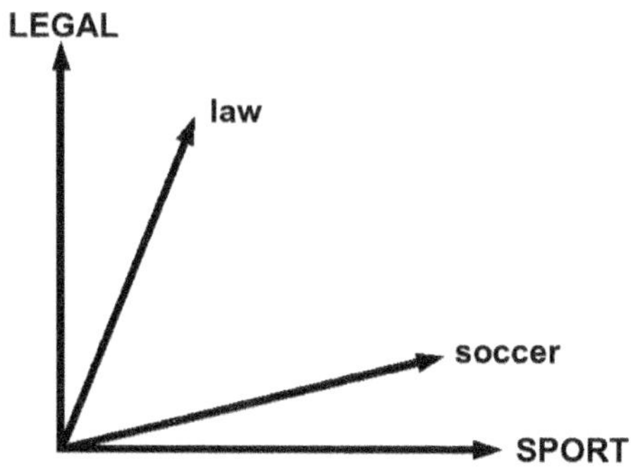

Fig. 3 Word vectors vectors in a 2-dimensional WordSpace

Apache Lucene[2] to create a basic term-document matrix, then it uses the Lucene API to create both a WordSpace and a DocumentSpace from the term-document matrix, using Random Projection to perform dimensionality reduction without matrix factorization. In order to evaluate the effectiveness of the Semantic Vectors model when used to perform the task of sense discrimination, we had to modify the standard Semantic Vectors package by adding some ad-hoc features. Indeed, documents used for the evaluation are structured in two fields, HEADLINE and TEXT, and are not tokenized using the standard text analyzer in Lucene.

Moreover, an important factor to take into account in a semantic space model is the number of contexts, that sets the dimension of the context vectors. Details about the system setup and results of the evaluation are reported in Section 4.

3 Word Sense Disambiguation in a Multi-level IR System

SENSE (SEmantic N-levels Search Engine) is an IR system which relies on Word Sense Disambiguation to build a semantic representation of the document collection. SENSE is based on the N-Levels model [6]. This model tries to overcome the limitations of the approaches based on ranked keyword [4] by introducing *semantic levels*, which integrate (and not simply replace) the lexical level represented by keywords. Semantic levels supply information about word meanings, as described in a reference dictionary or other semantic resources. SENSE is able to manage documents indexed at separate levels (keywords, word meanings, named entities, up to now) as well as to combine keyword search with semantic information coming from the other indexing levels. In particular, for each level:

1. a *local scoring function* is used in order to weigh terms belonging to that level according to their informative power;
2. a *local similarity function* is used in order to compute document relevance by exploiting the above-mentioned scores.

[2] http://lucene.apache.org/

Finally, a *global ranking function* is defined in order to combine document relevance computed at each level. SENSE is thoroughly described in [5]. An overview of its architecture is reported in Section 3.1, while the *global ranking function* used to merge the ranked document lists coming from each indexing level is in Section 3.2.

In SENSE, features at the word meaning level are *synsets* obtained from WordNet [15], a semantic lexicon for the English language. In order to assign synsets to words, a WSD algorithm is needed.

The idea behind the adoption of WSD is that each document is represented, at the meaning level, by the senses conveyed by the words in its content, together with their respective occurrences. Documents are represented by using a synset-based vector space. Consequently, the vocabulary at this level is the set of distinct synsets recognized by the WSD procedure in the collection.

The weight of each synset for a document is computed using the Okapi BM25 model [26,27]. In order to implement BM25 in SENSE, we exploited the technique described in [22]. In particular, we adapted the Okapi BM25 model to cope with semi-structured (or multi-field) document representations.

First of all, in the multi-field representation the weight of each term is computed taking into account the aggregate amount of the term weights for all fields, as follows:

$$weight(t,d) = \sum_{c \in d} \frac{occurs^{d}_{t,c} * boost_c}{((1-b_c) + b_c * \frac{l_c}{avl_c})} \tag{2}$$

where $occurs^{d}_{t,c}$ denotes the number of occurrence of the term t in the field c of the document d, l_c is the field length and avl_c is the average length for the field c. b_c is a constant related to the length of field c, similar to b constant in classical BM25 formula, while $boost_c$ is the boost factor applied to field c.

Then, the similarity between query and document is computed exploiting the accumulated weight for each term t that occurs both in the query q and in the document d.

$$R(q,d) = \sum_{t \in q} idf(t) * \frac{weight(t,d)}{k_1 + weight(t,d)} \tag{3}$$

where k_1 denotes a free parameter, $weight(t,d)$ is computed according to formula 3 and $idf(t)$ denotes the inverse document frequency of t, computed according to the classical BM25 model, as follows:

$$idf(t) = log \frac{N - df(t) + 0.5}{df(t) + 0.5} \tag{4}$$

where N is the number of documents in the collection and $df(t)$ is the number of documents where the term t appears.

3.1 SENSE Architecture

Figure 4 depicts the system architecture and shows the modules involved in the information extraction and retrieval processes.

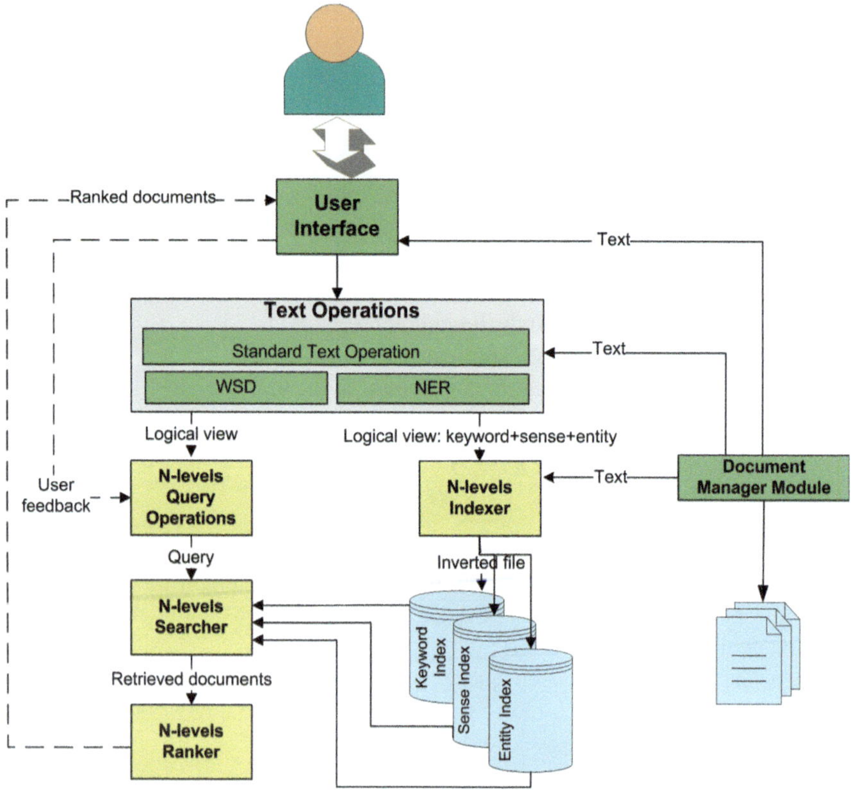

Fig. 4 System Architecture

Some modules are mainly devoted to perform typical Natural Language Processing (NLP) operations, and to manage the interaction with the user. In more detail:

- Document Manager (DM) - It manages document collections to be indexed. It is invoked by User Interface to display the results of a user query.
- Text Operations - It performs basic and more advanced NLP operations. Implemented basic operations are: *Stop words elimination*, *Stemming* (the *Snowball stemmer*[3] has been integrated), *POS-tagging* and *Lemmatization*. For POS-tagging, we implemented a JAVA version of *ACOPOST tagger*[4] which adopts the Trigram Tagger T3 algorithm [7] based on Hidden Markov Models. For

[3] http://snowball.tartarus.org/

[4] http://acopost.sourceforge.net/

lemmatization, we used the WORDNET Default Morphological Processor, as it is included in the WORDNET distribution for English. Besides basic NLP processing, more advanced procedures were designed for constructing semantic indexes ("Sense Index" and "Entity Index" in Figure 4): *Word Sense Disambiguation (WSD) Named Entity Recognition (NER)*.
- USER INTERFACE - It provides the query interface, which is not just a textbox where keywords can be typed in, since it allows users to issue queries involving semantic levels, too.

The core of the N-Levels indexing and retrieval processes is performed by the following modules:

- N-LEVELS INDEXER - It creates and manages as many inverted indexes as the number of levels used into the N-levels model. While the TEXT OPERATIONS component provides the features (recognized in the text) corresponding to the different levels, the N-LEVELS INDEXER computes the local scoring functions defined for assigning weights to features.
- N-LEVELS QUERY OPERATIONS - It reformulates user needs so that the query can be executed over the appropriate inverted index.
- N-LEVELS SEARCHER - It retrieves, for each level identified by TEXT OPERATIONS, the set of documents matching the query. It implements the local similarity functions defined in the model.
- N-LEVELS RANKER - It arranges documents retrieved by the SEARCHER into a unique list to be shown to the user. For each level involved into the search task, it ranks documents according to the local similarity function, and then merges all the local lists into a single list by using the global ranking function.

The core components of the architecture are implemented by using the Lucene API. Lucene is a full-featured text search engine library that implements the vector space model. In order to implement the N-levels model, we developed an extension of the Lucene API, the N-LEVELS LUCENE CORE, to meet all the requirements of the proposed model.

3.2 Global Ranking

A global ranking function is defined in order to combine document relevance computed at each level. Given a query q, each local similarity function produces a local ranked list of relevant documents. All the local lists must be merged in order to return a single ranked list to the user. The global ranking function is devoted to this task.

Algorithms for merging ranked lists are widely used by meta-search engines, which send user requests to several search engines and aggregate results into a single list [11, 14]. Our strategy for defining the *global ranking function* is thus inspired by prior work on meta-search engines.

Formally, let us denote by:

- U the universe, that is the set containing all the distinct documents in the local lists;
- τ_j={ $x_1 \geq x_2 \geq \ldots \geq x_n$ } the j-th local list, $j = 1,\ldots,N$, defined as an ordered set S of documents, $S \subseteq U$, where $\geq$ is the ranking criterion defined by the j-th local similarity function;
- $\tau_j(x_i)$ a function that returns the position of x_i in the list τ_j;
- $s^{\tau_j}(x_i)$ a function that returns the score of x_i in τ_j;
- $w^{\tau_j}(x_i)$ a function that returns the weight of x_i in τ_j.

Since local similarity functions may produce scores varying in different ranges, the aggregation of lists in a single one requires two steps: the first one produces the N normalized lists and the second one merges the N lists in a single one $\hat{\tau}$.

In SENSE, we consider the normalization strategy based on scores [21]. Score normalization strategies compute $w^{\tau_j}(x_i)$ by using $s^{\tau_j}(x_i)$. In particular we adopt the Z-Score Normalization defined by the equation:

$$w^{\tau_j}(x_i) = \frac{s^{\tau_j}(x_i) - \mu_{s^{\tau_j}}}{\sigma_{s^{\tau_j}}} \tag{5}$$

Z-Score Normalization works on the average of the scores in τ_j, $\mu_{s^{\tau_j}}$, and their variance $\sigma_{s^{\tau_j}}$. Given N local lists τ_j, the goal of the rank aggregation method is to produce a new list $\hat{\tau}$, containing all documents in τ_j, ordered according to a *rank aggregation function* ψ that combines the normalized weights of local lists in a (hopefully) better ranking.

Let R be the set of all local lists, $R = \{\tau_1,\ldots,\tau_N\}$, $hits(x_i,R)$=| $\{\tau_j \in R : x_i \in \tau_j\}$ |. In SENSE, we adopt the CombSUM rank aggregation method [25]: the score of document x_i in the global list is computed by summing all the normalized scores for x_i:

$$\psi(x_i) = \sum_{\tau_j \in R} w^{\tau_j}(x_i) \tag{6}$$

Moreover, we introduce a weight for each list (level). The score of document x_i in the global list is computed similarly to CombSUM, except for the introduction of a boost factor α_j for each local list, in order to amplify (or reduce) the weight of x_i in each local list:

$$\psi(x_i) = \sum_{\tau_j \in R} w^{\tau_j}(x_i) * \alpha_j \tag{7}$$

where α_j underlines the importance of a local list in the global ranking, i.e. the importance of a level in the SENSE system.

4 Evaluation

The goal of the evaluation is to establish how Semantic Vectors influence the retrieval performance. The system has been evaluated into the context of an IR task. We adopted the dataset used for CLEF 2009 Ad-Hoc Robust WSD task [1]. Task organizers made available document collections from the news domain, and topics, which have been automatically tagged with word meanings (synsets) from WordNet using several state-of-the-art disambiguation systems. Considering our goal, we exploited only the monolingual part of the task.

The dataset comprises corpora from "Los Angeles Times" and "Glasgow Herald", amounting to 169,477 documents, 160 test topics and 150 training topics. The annotations concerning word meanings were automatically added by WSD systems from two leading research laboratories, UBC [2] and NUS [9]. Both systems returned word senses from the English WordNet version 1.6. We used only the senses provided by NUS. Each term in the document is annotated by its senses with their respective scores, as assigned by the automatic WSD system.

In order to compare the IR system based on Semantic Vectors to other systems which cope with word ambiguity by means of methods based on WSD, we considered results produced by SENSE as a baseline. All SENSE components involved in the experiments are implemented in Java using the version 2.3.2 of Lucene API. Experiments were run on an Intel Core 2 Quad processor at 2.6 GHz, operating in 64 bit mode, running Linux (UBUNTU 9.04), with 4 GB of main memory. Table 1 reports the different settings of BM25 parameters for the keyword level ($HEADLINE_k$, $TEXT_k$) and the meaning/sense level ($HEADLINE_s$, $TEXT_s$) of SENSE.

Table 1 BM25 parameters used in SENSE

Level	**Field**	k_1	N	avl_c	b_c	$boost_c$
KEYWORD	$HEADLINE_k$	3.25	166,726	7.96	0.70	2.00
	$TEXT_k$	3.25	166,726	295.05	0.70	1.00
MEANING	$HEADLINE_s$	3.50	166,726	5.94	0.70	2.00
	$TEXT_s$	3.50	166,726	230.54	0.70	1.00

In CLEF, queries are represented by topics, which are structured statements representing information needs. Each topic typically consists of three parts: a brief TITLE statement, a one-sentence DESCRIPTION, and a more complex "NARRATIVE" specifying the criteria for assessing relevance. All topics are available with and without tags concerning meanings. Topics in English are disambiguated by both UBC and NUS systems.

We adopted as baseline the system which exploits only keywords during the indexing, identified by *KEYWORD*. Regarding disambiguation we used the SENSE system adopting two strategies: the former, called *MEANING*, exploits only word meanings, the latter, called *KEYWORD+MEANING*, uses two levels of document representation: keywords and word meanings combined.

The query for the *KEYWORD* system was built using word stems in TITLE and DESCRIPTION fields of the topics. All query terms are joined adopting the OR boolean clause. Regarding the *MEANING* system each word in TITLE and DESCRIPTION fields is expanded using the synsets in WordNet provided by the WSD algorithm. In SENSE, the query at *KEYWORD* level was built using word stems in the fields TITLE and DESCRIPTION of the topics. All query terms were joined adopting the OR boolean operator. The terms in the TITLE field were boosted using a factor 8. Boosting factor gave more importance to the terms in the fields TITLE. Moreover, the query at *MEANING* level, was built using the synset with the highest score provided by the WSD algorithm for each token. Synset boosting values are: 8 for TITLE and 2 for DESCRIPTION.

The query for the SENSE system is built combining the strategies adopted for the *KEYWORD* and the *MEANING* systems. In particular, the two levels are combined using a factor of 0.8 for keyword and 0.2 for meaning. Stop words were removed in all runs from both the index and the topics. In particular, we built a different stop words list for topics in order to remove non informative words such as *find*, *reports*, *describe*, that occur with high frequency in topics and are poorly discriminating.

For a fair comparison, we used the very same index built for the *KEYWORD* system to infer semantic vectors (using the Semantic Vectors package, as described in Section 2). It was necessary to tune two parameters in Semantic Vectors: the number of dimensions (the number of contexts) and the frequency[5] threshold (T_f). The last value is used to discard terms that have a frequency below T_f. After a tuning step, we set the number of dimension to 2000 and T_f to 10. Tuning were performed using training topics provided by the CLEF organizers.

Queries for the Semantic Vectors model were built using several combinations of topic fields. Table 2 reports the results of the experiments using Semantic Vectors and different combinations of topic fields. Boldface highlights the best result.

Table 2 Results of the performed experiments: Semantic Vectors

Topic fields	**MAP**
TITLE	0.0892
TITLE+DESCRIPTION	**0.2141**
TITLE+DESCRIPTION+NARRATIVE	0.2041

To compare the systems we used: the Mean Average Precision (MAP), due to its good stability and discrimination capabilities. Given the Average Precision [8], that is the mean of the precision scores obtained after retrieving each relevant document, the MAP is computed as the sample mean of the Average Precision scores over all topics. Zero precision is assigned to unretrieved relevant documents.

Table 3 reports the results of each system involved into the experiment. The column *Imp.* shows the improvement with respect to (wrt) the baseline *KEYWORD*. The system SV_{best} refers to the best result obtained by Semantic Vectors (reported in boldface in Table 2).

[5] Here, frequency is intended as the number of occurrences of a term.

Table 3 Results of the performed experiments: comparison between Semantic Vectors and SENSE levels

System	MAP	Imp.
KEYWORD	0.3962	-
MEANING	0.2930	-26.04%
KEYWORD+MEANING	0.4222	+6.56%
SV_{best}	0.2141	-45.96%

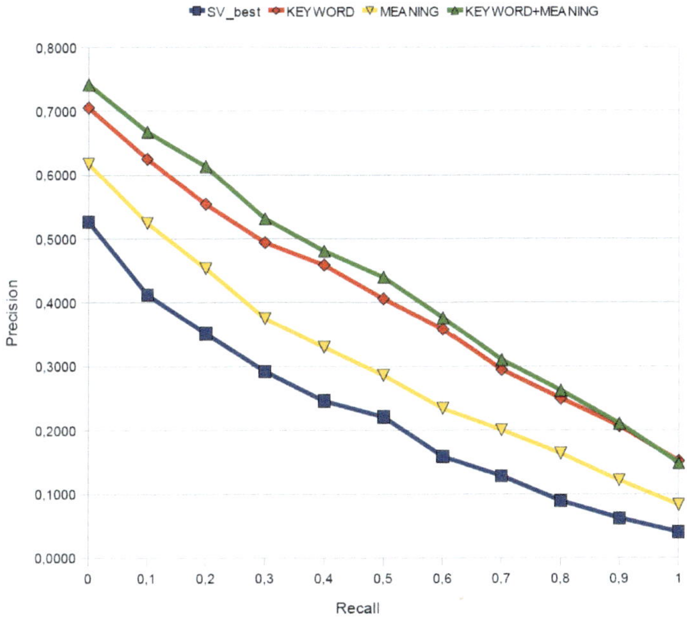

Fig. 5 Precision vs Recall: comparison between Semantic Vectors and SENSE levels

The main result of the evaluation is that *MEANING* works better than SV_{best}; in other words disambiguation outperforms discrimination. Another important observation is that the combination of keywords and word meanings obtains the best result. It is important to note that SV_{best} obtains a performance below the KEYWORD system, about the 46% under the baseline. The keyword level in SENSE uses a modified version of Apache Lucene, which implements Okapi BM25 model [22]. Figure 5 shows the Precision vs Recall plot for each system reported in Table 3.

In the previous experiments we compared the performance of the Semantic Vectors-based IR system to SENSE. In the following, we describe a new kind of experiment in which we integrated the Semantic Vector as a new level in SENSE. The idea was to combine the results produced by Semantic Vectors with the results produced by both the keyword level and the word meaning one. Table 4 shows that

Table 4 Results of the experiments: combination of Semantic Vectors with other levels in SENSE

System	MAP	Imp.
SV+KEYWORD	0.4150	+4.74%
SV+MEANING	0.3238	-18.27%
SV+KEYWORD+MEANING	0.4216	+6.41%

the combination of the keyword level with Semantic Vectors outperforms the keyword level alone.

Moreover, the combination of Semantic Vectors with word meaning level achieves an iteresting result: the combination is able to outperform the word meaning level alone.

Finally, the combination of Semantic Vectors with *KEYWORD+MEANING* obtains the best MAP with an increase of more than 6% wrt *KEYWORD*.

Analyzing results query by query, we discovered that for some queries the Semantic Vectors-based IR system achieves an high improvement wrt keyword search. This happen mainly when few relevant documents exist for a query. For example, query "10.2452/155-AH" has only three relevant documents. Both keyword and Semantic Vectors are able to retrieve all relevant documents for that query, but keyword achieves 0.1484 MAP, while for Semantic Vectors MAP grows to 0.7051. This means that Semantic Vectors are more accurate than keywords when few relevant documents exist for a query.

5 Related Work

The main motivation for focusing our attention on the evaluation of disambiguation or discrimination systems is the idea that word ambiguity resolution can improve the performance of IR systems.

Many strategies have been used to incorporate semantic information coming from electronic dictionaries into search paradigms.

Query expansion with WordNet has shown the potential of improving recall, as it allows to refine the notion of relevance in order to deem documents as relevant even when the query string does not appear verbatim in the document [28]. On the other hand, semantic similarity measures have the potential to redefine the similarity between a document and a user query [10]. However, computing the degree of relevance of a document with respect to a query means computing the similarity among all the synsets of the document and all the synsets of the user query, thus the matching process could have very high computational costs.

In [16] the authors performed a shift of representation from a lexical space, where each dimension is represented by a term, towards a semantic space, where each dimension is represented by a concept expressed using WordNet synsets. Then, they applied the Vector Space Model to WordNet synsets. The realization of the semantic tf-idf model was rather simple, because it was sufficient to index the documents or the user-query by using strings representing synsets. The retrieval phase is similar to the classic tf-idf model, with the only difference that matching is carried out between synsets.

Concerning the discrimination methods, in [13] some experiments in an IR context adopting LSI technique are reported. In particular this method performs better than canonical vector space when queries and relevant documents do not share many words. In this case LSI takes advantage of the implicit higher-order structure in the association of terms with documents ("semantic structure") in order to improve the detection of relevant documents on the ground of terms found in queries.

In order to show that WordSpace model is an approach to ambiguity resolution that turns out to be effective in IR, we summarize the experiment presented in [24]. This experiment evaluates sense-based retrieval, a modification of the standard vector-space model in IR. In word-based retrieval, documents and queries are represented as vectors in a multidimensional space in which each dimension corresponds to a word. In sense-based retrieval, documents and queries are also represented in a multidimensional space, but its dimensions are senses, not words. The evaluation shows that sense-based retrieval improved average precision by 7.4% when compared to word-based retrieval.

Regarding the evaluation of WSD systems in the context of IR, it is important to cite SemEval-2007 task 1 [3]. This task is an application-driven one, where the application is a given cross-lingual IR system. Participants disambiguate text by assigning WordNet synsets, then the system has to do the expansion to other languages, the indexing of the expanded documents and the retrieval for all the languages in batch. The retrieval results are taken as a measure for the effectiveness of the disambiguation. CLEF 2009 Ad-hoc Robust WSD task [1] is inspired to SemEval-2007 task 1.

6 Conclusions and Future Work

We have evaluated Semantic Vectors exploiting an IR scenario. The proposed IR system relies on semantic vectors to induce a WordSpace model exploited during the retrieval process. Moreover, we compared the proposed IR system to a system which exploits WSD. The main outcome of this comparison is that disambiguation works better than discrimination. This is a counterintuitive result: indeed it should be obvious that discrimination is better than disambiguation, since the former is able to infer the usages of a word directly from documents, while disambiguation works on a fixed distinction of word meanings encoded into a sense inventory, such as WordNet.

It is important to note that the dataset used for the evaluation depends on the method adopted to compute documents relevance, in this case the pooling techniques. This means that the results submitted by the groups participating in the previous ad hoc tasks are used to form a pool of documents for each topic by collecting the highly ranked documents. What we want to underline here is that generally the systems taken into account rely on keywords. This can produce relevance judgements that do not take into account evidence provided by other features, such as word meanings or context vectors. Moreover, distributional semantics methods, such as Semantic Vectors, do not provide a formal description of why two terms or documents are similar. The semantic associations derived by Semantic Vectors are similar to how human estimates similarity between terms or documents. It is not clear if current evaluation methods are able to detect these cognitive aspects typical of human thinking. More investigation on the strategy adopted for the evaluation is needed. As future work we intend to exploit several discrimination methods, such as Latent Semantic Indexing and Hyperspace Analogue to Language.

References

1. Agirre, E., Di Nunzio, G.M., Mandl, T., Otegi, A.: CLEF 2009 Ad Hoc Track Overview: Robust WSD Task. In: Working notes for the CLEF 2009 Workshop (2009), http://clef-campaign.org/2009/working_notes/agirre-robustWSDtask-paperCLEF2009.pdf
2. Agirre, E., de Lacalle, O.L.: BC-ALM: Combining k-NN with SVD for WSD. In: Proceedings of the 4th International Workshop on Semantic Evaluations (SemEval 2007), Prague, Czech Republic, pp. 341–325 (2007)
3. Agirre, E., Magnini, B., de Lacalle, O.L., Otegi, A., Rigau, G., Vossen, P.: SemEval-2007 Task 1: Evaluating WSD on Cross-Language Information Retrieval. In: Proceedings of the 4th International Workshop on Semantic Evaluations (SemEval 2007), Prague, Czech Republic, pp. 7–12. ACL (2007)
4. Baeza-Yates, R., Ribeiro-Neto, B.: Modern Information Retrieval. Addison-Wesley, Reading (1999)
5. Basile, P., Caputo, A., de Gemmis, M., Gentile, A.L., Lops, P., Semeraro, G.: Improving Ranked Keyword Search with SENSE: SEmantic N-levels Search Engine. Communications of SIWN (formerly: System and Information Sciences Notes) 5, 39–45 (2008)
6. Basile, P., Caputo, A., Gentile, A.L., Degemmis, M., Lops, P., Semeraro, G.: Enhancing Semantic Search using N-Levels Document Representation. In: Bloehdorn, S., Grobelnik, M., Mika, P., Thanh Tran, D. (eds.) Proceedings of the Workshop on Semantic Search (SemSearch 2008) at the 5th European Semantic Web Conference (ESWC 2008), CEUR Workshop Proceedings, Tenerife, Spain, vol. 334, pp. 29–43 (June 2, 2008), CEUR-WS.org
7. Brants, T.: TnT–a statistical part-of-speech tagger. In: Proceedings of the 6th conference on Applied Natural Language Processing, pp. 224–231 (2000)
8. Buckley, C., Voorhees, E.M.: Evaluating evaluation measure stability. In: SIGIR 2000: Proceedings of the 23rd annual international ACM SIGIR conference on Research and development in information retrieval, pp. 33–40. ACM, New York (2000), http://doi.acm.org/10.1145/345508.345543
9. Chan, Y.S., Ng, H.T., Zhong, Z.: NUS-PT: Exploiting Parallel Texts for Word Sense Disambiguation in the English All-Words Tasks. In: Proceedings of the 4th International Workshop on Semantic Evaluations (SemEval 2007), Prague, Czech Republic, pp. 253–256 (2007)
10. Corley, C., Mihalcea, R.: Measuring the semantic similarity of texts. In: Proceedings of the ACL Workshop on Empirical Modeling of Semantic Equivalence and Entailment, pp. 13–18. Association for Computational Linguistics, Ann Arbor (2005), http://www.aclweb.org/anthology/W/W05/W05-1203

11. Croft, W.: Combining approaches to information retrieval. Advances in information retrieval 7, 1–36 (2000)
12. Dasgupta, S., Gupta, A.: An elementary proof of the Johnson-Lindenstrauss lemma. Tech. rep., Technical Report TR-99-006, International Computer Science Institute, Berkeley, California, USA (1999)
13. Deerwester, S., Dumais, S.T., Furnas, G.W., Landauer, T.K., Harshman, R.: Indexing by latent semantic analysis. Journal of the American Society for Information Science 41, 391–407 (1990)
14. Farah, M., Vanderpooten, D.: An outranking approach for rank aggregation in information retrieval. In: Kraaij, W., de Vries, A.P., Clarke, C.L.A., Fuhr, N., Kando, N. (eds.) SIGIR, pp. 591–598. ACM, New York (2007)
15. Fellbaum, C.: WordNet: An Electronic Lexical Database. MIT Press, Cambridge (1998)
16. Gonzalo, J., Verdejo, F., Chugur, I., Cigarran, J.: Indexing with WordNet synsets can improve text retrieval. In: Proceedings of the COLING/ACL, pp. 38–44 (1998)
17. Harris, Z.: Mathematical Structures of Language. Interscience, New York (1968)
18. Kanerva, P.: Sparse Distributed Memory. MIT Press, Cambridge (1988)
19. Landauer, T., Dumais, S.: A solution to Plato's problem: The latent semantic analysis theory of acquisition, induction, and representation of knowledge. Psychological review 104(2), 211–240 (1997)
20. Landauer, T.K., Dumais, S.T.: A Solution to Plato's Problem: The Latent Semantic Analysis Theory of Acquisition, Induction, and Representation of Knowledge. Psychological Review 104(2), 211–240 (1997)
21. Manmatha, R., Rath, T., Feng, F.: Modeling score distributions for combining the outputs of search engines. In: Proceedings of the 24th annual international ACM SIGIR conference on Research and development in information retrieval, pp. 267–275. ACM, New York (2001)
22. Robertson, S., Zaragoza, H., Taylor, M.: Simple BM25 extension to multiple weighted fields. In: CIKM 2004, Proceedings of the 13th ACM international conference on Information and Knowledge Management, pp. 42–49. ACM, New York (2004), `http://doi.acm.org/10.1145/1031171.1031181`
23. Sahlgren, M.: The Word-Space Model: Using distributional analysis to represent syntagmatic and paradigmatic relations between words in high-dimensional vector spaces. Ph.D. thesis, Stockholm: Stockholm University, Faculty of Humanities, Department of Linguistics (2006)
24. Schütze, H., Pedersen, J.O.: Information retrieval based on word senses. In: Proceedings of the 4th Annual Symposium on Document Analysis and Information Retrieval, pp. 161–175 (1995)
25. Shaw, J.A., Fox, E.A.: Combination of multiple searches. In: Proceedings of the 2nd Text REtrieval Conference (TREC-2), pp. 243–252. NIST, Special Publication (1994)
26. Sparck Jones, K., Walker, S., Robertson, S.E.: A probabilistic model of information retrieval: development and comparative experiments - part 1. Information Processing and Management 36(6), 779–808 (2000)
27. Sparck Jones, K., Walker, S., Robertson, S.E.: A probabilistic model of information retrieval: development and comparative experiments - part 2. Information Processing and Management 36(6), 809–840 (2000)
28. Voorhees, E.M.: WordNet: An Electronic Lexical Database, chap. Using WordNet for text retrieval, pp. 285–304. The MIT Press, Cambridge (1998)
29. Widdows, D., Ferraro, K.: Semantic Vectors: A Scalable Open Source Package and Online Technology Management Application. In: Proceedings of the 6th International Conference on Language Resources and Evaluation, LREC 2008 (2008)

Information Processing in Smart Grids and Consumption Dynamics

Mikhail Simonov, Riccardo Zich, and Marco Mussetta

Abstract. This work suggests an effective approach for information management in smart power grids based on the introduction of a suitable theory of digital energy. It shows a possible way to effectively manage energy dynamics in real life systems in real time. Power grids hold real time information flows already, but the control systems currently adopted use other information sources. We discuss the use of the information and semantic technologies in order to balance the loads in storage-less electric energy domain, and the changes brought by Future Internet and its entities.

Keywords: information retrieval, distributed information processing, smart grid, digitized energy, electric web, electric engineering, load management.

1 Introduction

Human being consumes daily a combination of different kinds of energy for basic living activities, mobility, and for the achievement of social goals. All these activities are dynamic processes, accompanied by energy consumption dynamics, happening in real time at different scales, both locally and globally. Electric energy is generated in a centralized or a distributed way, transmitted to the electrical loads by power grids, and traded as a commodity. In storage-less grids it is shipped and

---

Mikhail Simonov · Riccardo Zich · Marco Mussetta
Politecnico di Milano, Dipartimento di Energia,
Piazza L. Da Vinci 32, 20133 Milano, Italy
e-mail: simonov@ismb.it,
{mikhail.simonov,marco.mussetta,riccardo.zich}@polimi.it

Mikhail Simonov
ISMB, Via P.C. Boggio 61, 10138
Torino, Italy

Marco Mussetta
Politecnico di Torino, Dipartimento di Elettronica,
Corso Duca degli Abruzzi 24, 10129 Torino, Italy

Soro et al. (Eds.): Inform. Retrieval and Mining in Distrib. Environments, SCI 324, pp. 267–286.
springerlink.com

immediately consumed or dispersed [1]. The modern electric energy network has a distributed grid structure [2]. In particular, a *smart grid* is a power grid that includes an intelligent monitoring system that keeps track of all electricity flowing in the system and automatically optimizes the operation of the interconnected elements of the power system. Smart grids use advanced information-based technologies to increase efficiency, reliability and flexibility, integrating alternative sources of electricity such as solar and wind [3]. This real life physical entity has been originally designed following the top down approach because of business motivations. Several traditional stakeholders – all real life physical entities - exist in the electric energy domain: Producers, Transmission system operators, Distributors, Consumers, and other value chain stakeholders. When the load on a system reach the maximum capacity, growth rate or threshold, the network operators must either find additional supplies of energy or ways to curtail the load, because an unsuccessful attempt in doing this within the time allowed will bring the system in an unstable condition, in which the frequency deviates form the original value and blackouts can happen. Real time load monitoring, control, and demand management are used by human or automated operators assisted by Supervisory Control And Data Acquisition (SCADA) systems to balance the loads ensuring the needed power quality [4]. If the actual conditions differ from the predicted patterns, the operator can intervene applying some load shed [5].

When a load exceeds the generating capacity, it unavoidably produces a system overload, causing a frequency drop. The under-frequency measurement is used because the speed of rotation of alternating current generator rotors is directly related to the frequency of the voltages they generate. Monitoring the power grid against the under-frequency phenomena in real-time [6], system protection schemes automatically dropping loads to avoid a total collapse of the system [7] might be set. Several researchers have contributed in the load control and under-frequency load shedding and some recommendations exist already [8]. When the protection scheme is triggered and initiated, load is shed a block at a time isolating power delivery to predetermined areas until the power system stabilizes. The frequencies at which load blocks are shed might be different [9]: for example 49.1 and 48.5 Hz. There might be several blocks/units of load, big enough to save the system. The above process is actuated with some time delay – from six to ten cycles of drop of 0.1 seconds, because selected feeders are assigned to these frequencies on a 10 outage rotational basis. Supervisory control handles the restoration: as the frequency recovers, the load is restored also in small blocks to maintain stability.

In this work we discuss the use of information management in energy domain permitting anticipatory knowledge elicitation.

2 Smart Power Grid as a Distributed Environment

Energy production, transmission, distribution, and consumption are dynamic processes existing in real 4D time-space, requiring dedicated *information processing and management* (Fig. 1). Traditionally the energy-related processes happening in real life were considered as continuous entities, using analogue energy metering devices. Rapid progress in electronic measurement technology of Alternating Current (AC) waveforms has opened the way for a variety of measurement application techniques [10] including those described in the EN 50-160. One possibility is for example to assume the sampling rate every 200 ms or every 10 cycles for 50 Hz system. Frequency assessment using different equations happens in each n-seconds window: 10 seconds contain 50 frequency samples measured. However the above process is reactive because under-frequency is the *consequence* of an unobserved cause happened. In fact, the cause-effect latency overlaps with the decision making time.

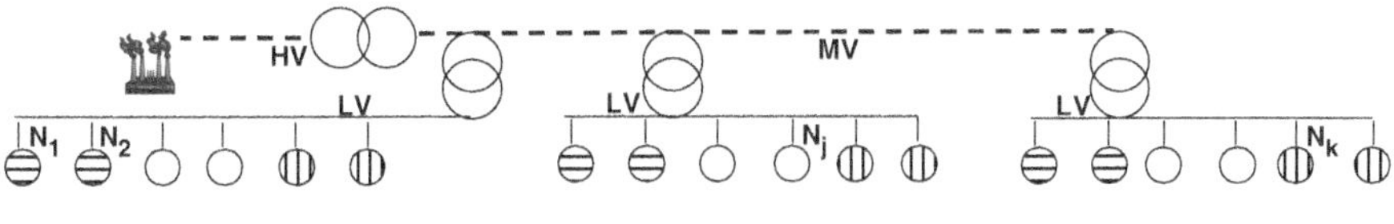

Fig. 1 Power grid

Real time electric energy sampling enables the management of the smart grid in a totally different way, because it makes available the continuous information flow carrying the events, and all the semantic elements related to energy distribution dynamics. This introduces a new set of applications relying on the *distributed information processing* and it requires some new theory. Numeric calculations in digital Internet of Things world digitizing real life objects and operating them from this new conceptualization (Fig. 2) complements old sophisticated math techniques, currently used in energy assessment.

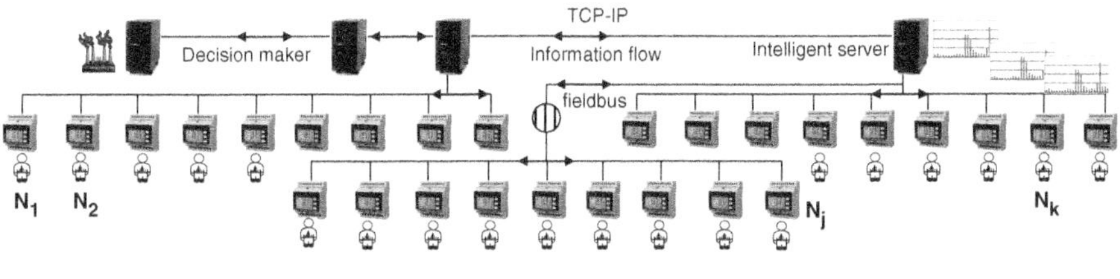

Fig. 2 Smart grid seen as a distributed information space

Nowadays, in fact, electric energy management on wired line is complemented by some wireless SCADA used in Oil, Gas and Water industries as the only economically justified solution when wired communications to the remote site is too expensive or time-consuming. Energy generation has been made highly distributed: traditional top down power grids now accept incoming flows from the *distributed generation*. Accompanied by new methods accounting the needed electric energy flows and new techniques ensuring the total local energy consumption [11], new ICT applications do information mining and intelligent data processing. Therefore, a smart power grid (Fig. 2) is a distributed system-of-system environment in which a continuous information flow exists, and it feeds efficient load management tools. The powerline currently support the local data communication, thus linking the same voltage segment fieldbus, but an additional technical solution is needed to overcome the physical limitations imposed by voltage transformers, which separate High Voltage (HV), Medium Voltage (MV), and Low Voltage (LV) segments. Nowadays, the upper segments are inter-linked now using GSM, UMTS, TCP-IP, satellite links, and Future Internet (Fig. 3).

We can represent the sub-topology behaviour by a virtual node (N_i,R_i) hiding the overall complexity, but requiring the distributed information processing in each LV segment which the elicited knowledge exchange between segments.

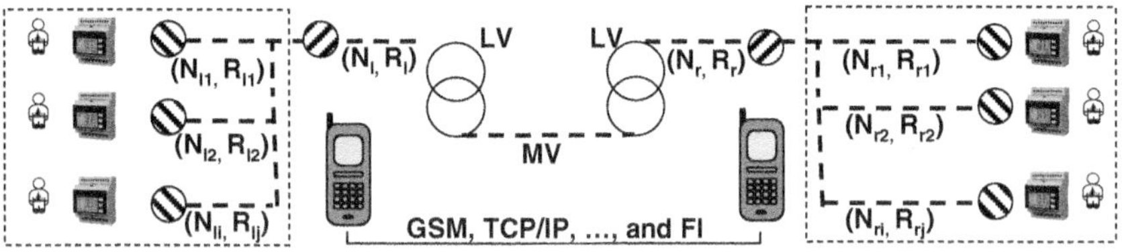

Fig. 3 Low Voltage (LV) segments of power grid and their interlinking

The virtual node behaviour is modelled keeping observable events happening inside segments [11]. Physical energy digitization happens on low computational power smart meters [12], and currently the real time data on a large scale is not yet available. The collection of smart meters [13] from a given power grid (Fig. 1) becomes the set of representative nodes qualifying the distributed environment (Fig. 2). Nodes communicate using standard powerline protocols. The Telegestore project [14] uses two implementations, one of which is based on Lontalk - ANSI/CEA 709.1 – extension, while the second is based on the "Integrated System for data Transmission on Electricity Distribution network" (SITRED) proprietary stack. A possible alternative is described in [15]: the digitized energy becomes a *real-time flow of exchanged events.* This abstraction exemplifies the information exchange in particular distributed environments. It challenges concrete industrial and commercial interests because makes Legacy business cheaper. The load information is not offered to other stakeholders and not tagged semantically loosing the customer side happening events. The information flow elaboration capability of power grids (Fig. 4) is the main task in which ICT contribute in.

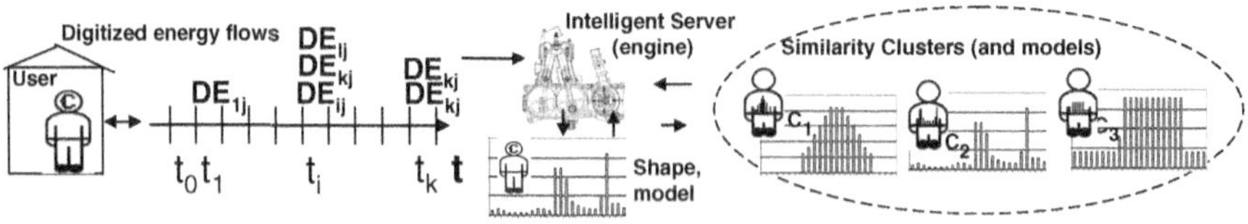

Fig. 4 Event flow and its elaboration in smart power grid

Modern power grids are huge systems-of-systems populated by millions of nodes - changing their production/consumption dynamics in real time - driven mostly by the unpredictable human behaviour. Network condition, in which a full set of the system bus voltages, currents, and frequencies cannot be directly measured or calculated from the real-time synchronized measurements, makes them practically unobservable [16]. Any power grid featuring a large number of interacting components, whose aggregate activity is non-linear and typically exhibits self-organization under selective pressures [17] is a non-linear, non-deterministic, dynamic self-organizing and error-tolerant *complex system*, preserving several basic functions under errors and failures, but requiring the regulation [18]. Power grid's macroscopic behaviour is studied statically, overlooking the individual events and their evolutionary dynamics. An extremely large numbers of different interacting elements make *real-time knowledge management* computationally too expensive. Information about the consumption dynamics might be very useful to achieve the desired "smartness".

3 Complexity, Scalability, and Phasors

Automated metering, with low computational power, originates a certain data amount. Additional telecommunication channels extend LV segments and cross-link the grid's components, but scalability issue limits the remote processing. GSM/GPRS/UMTS communication is relatively expensive, while available broadband capabilities are still limited. Currently the smart metering provides a limited number of data because of the business scenario, cost-efficiency considerations, and scalability limits. Meters can release better resolution giving much more detailed load shapes and even characterization of the events/peaks; however this would increase dramatically the data volume and processing requirements. Utilities adopt an hourly or some 4-6 samples per hour [14] suiting their main business. Ten minutes sampling can show some energy drops, while several relevant events pass unobserved at hourly rate (Fig. 5). Real time sampling wraps the time, makes the relevant events observable. The forecast for neighbourhood about clouds can be formulated on 1 seconds sampling (Fig. 6).

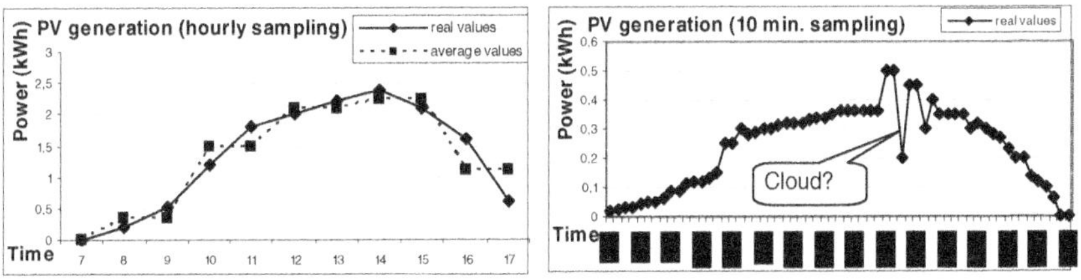

Fig. 5 Photovoltaic production at hourly and 10 min. sampling

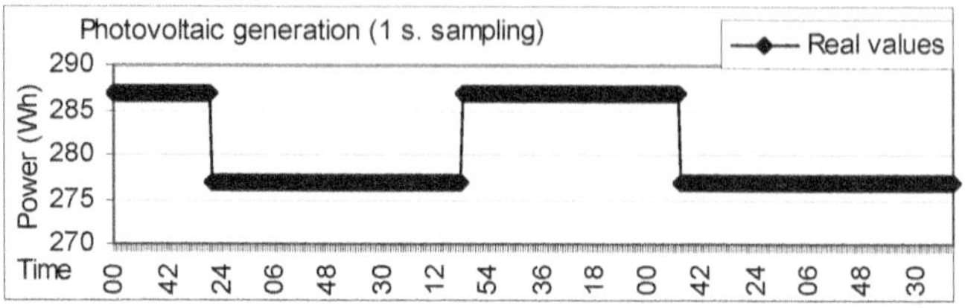

Fig. 6 Photovoltaic production at 1 second sampling

A sampling with a higher resolution – e.g. up to 0.1 s – would constrain the decision making time, and it is needed for load shedding purposes. We can calculate data volumes assuming the plain dataset by counting 9 words (from, to, id, time, voltage, power and some other measurements). Complementing it by some semantic elements (RDF triples), we increase the message length, while in certain implementations an *envelope* might grow up to 1 Kbytes. To simplify, we set to 20 millions the number of grid nodes corresponding to the maximum number of smart meters digitizing and exchanging the energy. In the table (Fig. 7) we observe the data growth caused by increased sampling rates: data volumes generated at high sampling rates overcome the scalability limits, requiring specific distributed information processing techniques to process them remotely by some intelligent agents.

nodes	sampling	data packets (daily)	Tbytes (raw)	Tbytes (1K)
	hourly	480.000,00	0,00	0,48
20.000.000	15 min	1.920.000,00	0,02	1,92
	1 min	28.800.000,00	0,26	28,80
	1 sec	1.728.000.000,00	15,55	1.728,00
	0,2 sec	8.640.000.000,00	77,76	8.640,00
	0,02 sec	86.400.000.000,00	777,60	86.400,00

Fig. 7 Data volumes generated at different rates

The data grow rate, limited storage, cost-effectiveness considerations, the feedback vs. SCADA systems, and the maximum scalability are main challenges calling for the distributed data processing techniques in the Energy domain. A possible ICT architectural viewpoint applied to Energy domain is available in [19].

Real time sensors, called Phasor Measurement Units (PMU), are distributed throughout the power grid in order to monitor the power quality and, in some cases, to trigger automatically some dynamics, for the capability to represent the waveforms of alternating current in real-time [20]. Research suggests that automated management of power systems might be set up using large number of PMU distributed over the grids [21]. Synchronized Phasor Measurements provide sequences of voltage and current measurements synchronized to within a microsecond, thanks to Global Positioning System (GPS) and the sampled data processing techniques from computer relaying applications. Recent blackouts on power systems have stimulated the wide-scale deployment of PMU, providing the most direct access to the state of the power system at any given instant through the sequences measured and enabling many applications, including crisis management [22]. Positive-sequence voltages of a network constitute the state vector of a power system, of fundamental importance till now; however the method does permit only the *a-posteriori analysis*. The Phasor representation is only possible for a pure sinusoid, but real waveform is often corrupted because of the noise. A single frequency component of the signal should be extracted first using Fourier Transformations and then represented by a Phasor. In sampled data systems this becomes the Discrete or Fast Fourier Transform (DFT and FFT respectively), requiring certain *computational power* in *real time*. The Phasor definition remains valid for a time span or *data window*, until signal remains unchanged, but in practice only a portion of time span holds a valid Phasor representation. Since Fourier transform is a function of frequency, the Phasor measurement inherits from the under-/over- frequency observation in the grids dealing with the real time sampled data, remaining reactive, revealing the impact on the power system of the already happened real life events (Fig. 8). Phasor estimates only from within the data window. When a fault occurs, the only Phasor belonging entirely to the pre- or post-fault periods represent a meaningful system state, discarding several data series. Indeed, the Phasor concept is related to a steady-state, but real power systems are never in a steady state. Voltage and current signals have constantly changing fundamental frequency due to load changes, generation imbalances, interactions between real power demand on the network, inertias of large generators, and the operation of automatic speed controls with which most generators are equipped. When faults and other switching events take place, there are very rapid changes in voltage and current waveforms. Some degrees of imbalance due to unbalanced loads and un-transposed transmission lines are common. Estimates of such unbalances (negative- and zero-sequence) range between 0 and 10% of the positive-sequence component. As the power system is rarely at the nominal frequency, some error terms can be tolerated.

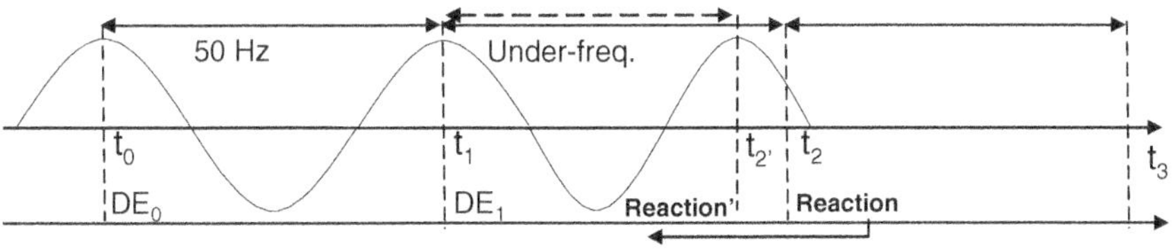

Fig. 8 Power quality measurements vs. anticipatory triggering

Authors look for the *computationally simpler* alternative anticipating energy events to reveal the *causes* instead of *the effects*. The anticipatory knowledge about the events to happen, those preceding under-frequency, would permit better decision making. Since the Phasor transmission datagrams are *digital information exchanged over the networks*, here we develop further the *Digital Energy* concept, already introduced in [11], in order to make fieldbus observable, inter-connected and manageable.

4 Energy Digitization for Electricity Web

The Web of communicating electric devices is not only an abstract distributed environment, but an important candidate for e-business in a liberalized electricity market, made possible thanks to the information objects representing the real asset (energy) and making available its behaviour, e.g. the digital information representing the real energy in the virtual Internet of Things world [23]. To make it happen we need the elementary communicating objects and their relationships, contributing in more complex design patterns handled for business purposes. Intelligent Servers, retrieving and processing the load data, patterns and shapes in order to understand the individual profiles, work over the digitized data sent through the fieldbus (Fig. 2). These information objects exist in the "remote" (LV) segments of the power grids, and their living space is constrained by the scalability limits and the time constraining the decision-making. Load shapes can be processed using data mining techniques, while the changing dynamics' analysis calls for data warehousing. Transient objects carrying instantaneous measurements should be stored somewhere to become accessible and manageable by distributed algorithms. After collecting a sufficiently big number of load shapes, coming form different users, we can run the similarity clustering algorithms [24, 25, 26] to determine the collective profiles, permitting event triggering by standardized policies.

Let assume that an Intelligent Server (as shown in Fig. 2) has obtained individual load profiles, performed the needed calculations, and produced generalizations, representing a possible segmentation. We store models centrally and make them available upon requests, since remote units need them for pattern matching, comparing the individual load shapes with the respective clusters. The centralized data repository should be kept up to date and be scalable to serve many real time requests coming from the remote nodes, while the distributed storage might rely on Peer-to-Peer. The initially elaborated models require periodic updates.

The *energy digitization* is a process performed in the discrete time-space by electronic energy meters. The physical electric energy – notably continuous real world entity - originates some scalar values about the energy production and/or consumption. It becomes the information injected and streamed in the communication networks, typically over powerline communication. The above mentioned information flow becomes the *digital artefact* accompanying and representing the real life electric energy. Being scalar, it will show the additive properties, making much simpler the computations.

Definition 1. *Digitized* or *Digital Energy* (*DE* hereafter) is the *real time dynamic flow of information* exchanged between energy consumers and other stakeholders electronically using any ICT broadband data transmission channels, typically powerline protocols and internet networks (1). Digital Energy *data elements* - information events - are originated by digital metering devices and instantly injected in the communication grids thanks to broadcasting offered by protocols. Digital Energy is a non-material object (*digital thing*) existing in the digital, information space. Digital Energy might be observed and interpreted. It might originate automatically triggered reactions in digital space, those sent back to actuators in the real time-space. The sequence of DE items appears as load shapes.

Smart power grid with DE is dual, because of the co-presence of the physical phenomena of the real energy and the new information sphere representing and synchronously accompanying it. The real energy is directional, while the DE is not: it is propagated through the digital space because of broadcasting (fieldbus protocols specs.) and it might be listen. The synchronization issue is considered separately since the communication network ensures the contemporaneous processing of all broadcasted messages reflecting the current time frame.

$$DE = \{ DE_1, DE_2, DE_3, \ldots, DE_i, \ldots\} \tag{1}$$

Definition 2. *Digital Energy* is measured in scalar units or tokens. DE elements happening along the timeline (time arrow) originate the payload, e.g. real time informative sequence. Each DE element carries some information with well-defined semantics. One digital unit represents one unit of the real energy consumed or produced on the node of the topology, e.g. 1 DE Watt-hour = 1 Watt-hour of real energy. We define the DE corresponding to the real energy directed *towards* the node (energy consumption) as positive, and the *outgoing* one (energy production) as negative, for our convenience.

Definition 3. *Digital Energy* is relative. It depends on the real energy, node and time. Each DE element carries three attributes: DE value, timestamp characterizing the time it was generated, and identification of the originating node (2). It carries at least three semantic meanings in the ICT sphere: one about the real energy snapshot, one about the timing, and one about the node/topology. DE might be represented in RDF or any other convenient way.

$$DE_i = \{DE.Unit_i, DE.Time_i, DE.NodeId_i\} \tag{2}$$

Definition 4. The amount of *Digital Energy* is the arithmetic sum of the units computed for the same node in a given time interval (3). DE is precise because the digital space is discrete. It can replace complex Fourier calculations of real energy by the arithmetic in digital space: DE fully satisfies the fundamental physical laws because directly linked with the real energy. DE is additive, showing the commutative, associative, and distributive fundamental properties.

$$DE = \sum_{i=1,m} DE_i\ (n, t) \tag{3}$$

Definition 5. The resolution of Digital Energy is the constant time interval happening between each two DE elements (4) coming from the same node.

Assuming the digital time space being isometric, all the DE_i elements happen with the same frequency (resolution). However it is also possible to vary individual frequencies (resolutions), leading to the non-isometric digital time spaces. Another possible variation might be the omission of the sub-sequent DE identical to the previous. Both options show also some disadvantages. To simplify, here we can assume an isometric digital space. The resolution of DE determines the time-wrapping properties. A suitably selected resolution makes observable the real life human-related events, which cause energy variations.

$$\text{DE\_Resolution} = t_i - t_{i-1} \tag{4}$$

Definition 6. *Digital Energy* is characterized by intensity, or the information flow density (from within the grid). We define the reference intensity 1 accompanying 1/50 or 1/60 seconds resolution, e.g. one triplet DE packet sent every 0.02 seconds in a 50 Hz power grid. For example 1 minute resolution determines the flow intensity being 50, e.g. 50 samples per second. Each DE_i is an event happening inside the information sphere with a given frequency. The DE intensity is the speed of DE_i events. The Digital Energy is coordinated because naturally ordered by the time. The intensity choice should be justified by the goals and the economic considerations. 15 minutes might be sufficient for the billing purposes, while the anticipatory control would require very frequent sampling.

Definition 7. We call the repetitive minimal value of DE manifested during the observation period as *zero-point*. In households it corresponds to the minimal consumption *caused by* appliances in stand-by. The real energy of online customers never equals to zero because of some appliances in stand-by, except black-out or fault conditions. Zero-point permits configuring consumers and producers in one energy neutral entity or micro grid accordingly the heuristics about their profiles.

Definition 8. *Digital Energy Event* (DE event) is the variation between two adjacent DE_i items (5) reflecting significant increase or decrease of the real energy flows. To reduce the information exchange originated by nodes in a concrete implementation it is possible to consider DE events only. The numeric value showing the significance ε is not set a-priori, but it needs to be tuned. To trigger all appliances, ε should be lower than their minimal thresholds (e.g. 5 W LED-lamp).

$$DE_i\,,\ DE_{i+1},\ DE_{i+2} \mid \{\ (DE_{i+1}.\text{Unit} - DE_i.\text{Unit}) > \varepsilon\ ;\ DE_{i+2}.\text{Unit} \neq 0\} \tag{5}$$

Definition 9. *Digital Energy Pattern* (DE pattern) is the sequence of DE events with the same initial and final events or states (6). In addition we define as "meaningful DE pattern" the DE pattern explaining the non-ambiguous sequence of underlying real life energy dynamic events.

$$\{DE_i\,,\ DE_{i+1}\,,\ DE_{i+2}\,,\ \ldots\,,\ DE_{i+k}\} \mid DE_i.\text{Unit} = DE_{i+k}.\text{Unit} \tag{6}$$

Fig. 9 shows a complete "residential" pattern started at 120 W level describing one possible cycle. In real life the corresponding scenario might be described semantically as "going from the living room to the kitchen to take a drink from the fridge" because of the on-off sequence 100 W (corridor lighting) and 60 W (kitchen lighting), possibly validated by the fridge-related events. Time series with different initial and final events, showing incomplete patterns, are ambiguous. The co-presence of different appliances with the same nominal energy consumption brings ambiguity. Moreover, Fig. 9 shows also an 800 W load gradually introduced, likely characterizing a Fuzzy-controlled, eco-featured, wash-machine, starting to wash. It might be used as the anticipatory knowledge about the whole cycle, heating included. This information might be useful for the power use analysis.

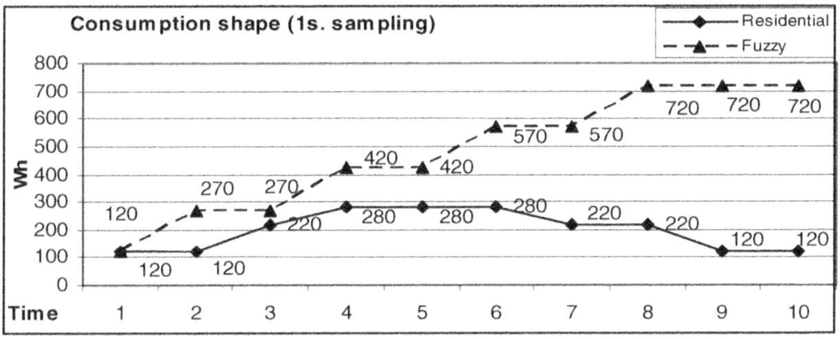

Fig. 9 Load shape at 1 second sampling

Definition 10. The *load shape* (7) is the sequence of DE_i originated by a given node during a certain period of time (24 hours, 1 week, 1 month, 1 year). Load shape is an aggregation of transient elements DE_i made persistent in real time in any convenient way, typically by listening and writing to any storage.

$$\text{Load Shape (t)} = \{DE.Item_1, \ldots, DE.Item_i, \ldots, DE.Item_t\};\ \text{i}= 1,\text{t} \qquad (7)$$

DE enables distributed information processing and knowledge elicitation in smart power grids. The collection of the load shapes describes the short-term consumption dynamics for a given grid. The data-warehouse of the load shapes collected during the long observation periods permits the analysis of the trends and evolutionary dynamics. To explain semantically the long lasting phenomena happening in power grids, additional semantics are required. More existing topologies aggregated in one information space conceptualize a new Internet of Things entity. Compared with the traditional power quality analysis based on the under-frequency monitoring, the digitized energy information is synchronized with the human-related events happening in real life and it might bring the semantic characterizations of the related events, thus being capable to anticipate the under-frequency in power grids (Fig. 8). Real time smart metering injects DE events in the network with the timing permitting timely data retrieval, elaboration and real time decision-making, satisfying the anticipatory control constraints. The computational complexity of DE event processing is much cheaper than the DFT or FFT calculations [27], but it is paid by the intensity of data flows, volumes, complexity and scalability factors. Smart meters act on the time-interval or time-of-use basis relying on the real-time sensors, power outage notification, and power quality monitoring, going much beyond the Automated Meter Reading (AMR), because real-time digital energy has time wrapping feature.

5 Digital Energy in Future Energy Web

The daily AMR data provides total energy information. The hourly sampling supports the variable tariffs but lacks appliance-related events. On Fig. 5, an energy generation drop, invisible in an hourly sampling, has been discovered. Observing Fig. 9 we see patterns that can be described semantically. Real-time DE information makes observable events not managed by current state-of-the-art systems. The elicited knowledge like "cloud passing through", "washing" or "going from the living room to the kitchen" is valuable, considering past attempts by Ariston and Merloni Elettrodomestici [28, 29]. Frequent sampling captures events *caused by* human actors giving some time to undertake controlling action before under-frequency happens. To characterize electric energy consumption by the social meaning, two instruments might cooperatively analyze the user behaviour: one real time digital meter and knowledge server governing smart appliances or an intelligent TV media server, with re-profiling feature. Real-time sampled DE enables understanding of semantics in daily life. Like an explosion process, unobserved at 25 fps but made visible by high-speed 3.000 fps video-camera, the high-rate sampled DE gives very precise details for advanced reasoning about the demand. The information overflow can be solved transforming the information in some knowledge and keeping the elicited semantics at the user site or somewhere else in the distributed topology - between the customer and the utility – with a possible option of the digital energy broker acting in trust on behalf of the end user (Fig. 10).

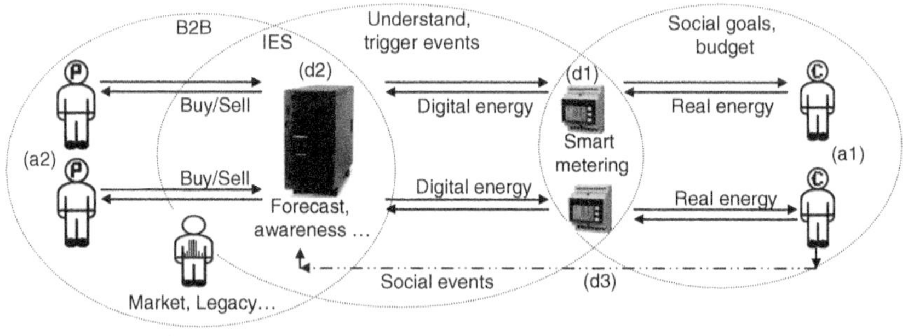

Fig. 10 Intelligent Server

New digital meters were initially introduced to make billing cheaper. They support new business models, variable tariffs and other options, introducing the exchange of information batches between producers and consumers. Therefore, new global liberalized electricity market needs the e-business of energy. It requires an efficient real-time ICT support of energy transactions in the same way it works for other commodities. An energy producer needs an efficient tool accounting and estimating the energy to trade, while an energy consumer needs a tool enabling to choose the most convenient energy provider matching its current and future needs. An online *e-Energy* trading is conceptually similar to online trading at financial markets; however the main difference is the real time self-estimation of the energy availability/needs, thanks to DE data. This requires background calculations about the available/needed energy and about Service Level Agreements to fulfil, mainly because of the uncertainty on both demand and renewable energy production sites: small stakeholders does not have business information systems keeping statistical data.

Here, we propose a real time smart metering system based on architecture shown on Fig. 10. It originates the digital energy flow (d1) elaborated by intelligent energy server (d2). The digitized energy brings real-time information about the individual consumption dynamics and forms the load shape collection stored at (d2) server, handling data-warehouse. Similarity clustering algorithms are run to deliver the segmentation and replace the individual behaviour with the cluster's one. The system might be more precise making available the semantics about the social events (d3). The 1st task is the energy profile estimation giving the future energy consumption, a pre-condition for buying some energy online. The 2nd task is the online energy availability analysis to choose the best option, e.g. the cheapest or the "green" energy. The 3rd task is the online commercial transaction completion.

Current information processing research contributes in decision-making over static load shapes, paying also some attention on the time-related and context-related real time semantic application classes [30]. The evolution of Data Mining, Semantic Web and Knowledge technologies analyze the *inner information* of the enterprises, while many external entities exist in power grids with the consequence that valuable information is not understood or used. Context-awareness in ubiquitous computing has importance since context-aware devices make assumptions about the user's current situation [31]. The context might be seen as a union of a series of assertions, while the time and the context are restrictions to the assertions. Time instance is a number in the time explanation domain, and it is a point in the unlimited linear sequence. Because the knowledge in our domain is strongly time related and context related, the formal knowledge representation based on description logic can be used to construct an automated knowledge system. In electric domain the system is observed using measurements along the timeline, facing the real-time information over-flood [32], but should remain interoperable [33].

The past power generation was aligned with communication infrastructures. Electric energy was delivered top-down from HV large scale production grids to the final consumers through the LV grids, while the communication infrastructures controlling information flows were used by SCADA. In the *distributed power generation* bi-directional power flows and related uncertainty introduced by renewable sources of energy (wind, clouds) has to be accounted. Some data models, interfaces and architectures [34] for running the grid and automating the substations are developed, with a possible integration based on semantic technologies

[35] and extending the IEC TC 57 implementations towards Semantic Web services. We need the distribution information processing, SOA dispatching flows and managing dynamic nodes, and Semantic Web services handling various types of information. A crucial e-business requirement is the availability of a well-defined ontology [36] to enable and deploy Semantic Electric Web services.

In the future smart cities we will move from storage-less power grids with static nodes to the dynamic ones. Electric Vehicle (EV) adding storage capacity to the power grids [37] will play a particular role. An EV consumes some energy from the grid during a certain period, but might drop the connection completing or abandoning the transaction. EV offers some temporary storage because of the batteries and bi-directional energy flows. Hybrid EV is similar to a photovoltaic plant injecting some renewable energy into the grid [38] in daytime, but maybe consuming energy in night-time. A pluggable into electric grid EV changes the role and behaviour of the nodes, becoming Prosumer (Fig. 11).

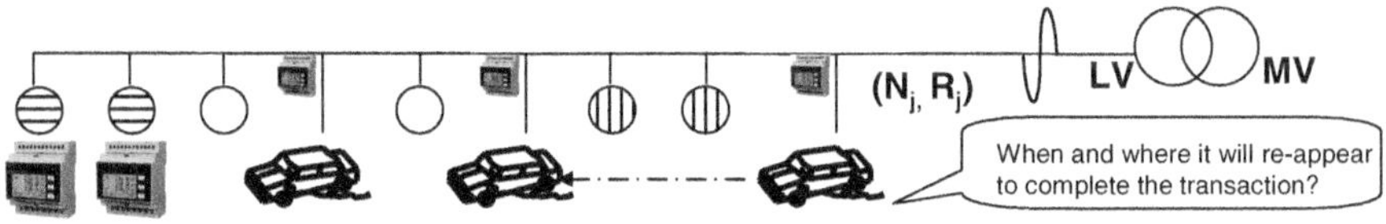

Fig. 11 Electric Vehicle as Internet of Things entity

An EV bus extends the power grid adding a new mobile node, enabling also the geographically distributed commercial transactions, initiated at a given node and completed at a different – not known a-priori – node, which is a significant complication of the business and the respective modelling. The geographic position of EV node might vary, showing the typical "intermittent" behaviour of an Internet of Things entity: the temporary presence at a given location, the uncertain duration of the presence and the energy flows, the uncertain power when plugged, the "off-line" periods, and the re-appearance in a different place to complete the pending operation. The payment of bi-directional energy flows might be done also under different economic conditions: the variability of the tariffs along the time and the compensation through different market operators.

Several new research questions come with this scenario: (1) EV entity presence duration calculation, (2) likelihood to complete the transaction at given node, (3) EV re-appearance in a new location estimation, (4) minimal battery capacity preservation, (5) future flows anticipation, and so on.

The anticipatory knowledge about the expected user behaviour in energy terms might trigger the decision making tools operating from within smart power grids, enabling also the better clustering of the virtual communities [13]. Speaking about the anticipatory declaration of the future mobility plans or patterns, it becomes possible to handle proactively the future load optimization strategies, also in the grids with storage. It makes possible to exploit the storage capacity made available by parked EV, the additional energy generation injected by hybrid EV attached to smart grid, the price difference between the peak and off-peak periods, and the mobility of the storage elements dynamically changing the smart power grids. The complexity, the new broadband connectivity, and the real time automated e-business operations are new components.

6 Use of Anticipatory Knowledge

Power grids are digitalized and very detailed data on the operating status can be collected using DE and elaborated using modern information retrieval. As result, more sophisticated *real time monitoring and diagnosis* become feasible. However, conventional fixed procedural monitoring programs are inflexible and cannot effectively resolve various subtle fault situations. Authors use Knowledge Technology (KT) contributing in powerful monitoring takes [39] since it can perform inference based on the information on the system structure, measured values and protection relay status. The *energy consumption dynamics assessment* is based on the precise behavioural information collectable at run-time from remote nodes. At design stage the maximal power is accountable, but real life raises the energy consumption towards or even beyond the initially designed upper limits. The anticipatory control relies on data warehousing. DE-based behaviour estimation needs few elementary data, does not require complex computations, and suits real-time constraints. Although the full list of electrical appliances at the user site can be obtained at certain stage, their energy consumption characteristics are never constant along the time because of the new technological items added or replaced. It is practically unreasonable to keep updated above lists, suggesting the periodic load shape analysis discovering changed energy consumption patterns. This leads to new portfolio segmentation challenging similarity clustering algorithms use.

We consider an initial batch elaborating available datasets for possible similarity clustering by an intelligent server [40]. The resulting modelling should be made available publicly. The initial modelling and customer segmentation make possible the classification of the current shapes. Any adherence or deviation shows the persisting or changing profiles (because of the new assets maybe) respectively. Representing the individual users by their DE load shapes, we capture the human-related events, show the use of the appliances, and generalize. We keep the consumption dynamics but hide the complexity, because limiting the information flows. Much more ambitious goal is to obtain the *semantic interpretation about the energy consumption dynamics* to preserve the reliability of the complex system, because requiring some automated semantic tags about real life events, accompanied by significant energy consumption variations, which is not trivial. To solve the unpredictability and consider the human factor some direct sensing using intelligent electric appliances and in-home sensors are needed, but constraining to privacy keeping.

Let move from the use of the data mining techniques operating with the snapshots to the *analysis of the historical series of data*. Energy consumption dynamics depend on the very slow external processes - technology evolution, urbanization, climatic changes, market globalization - challenging the data warehousing use. The energy need is constantly growing, but its geographical distribution might vary. The comparison between the current real time snapshot and the cluster's model tells about the current node's use. It might show how many persons stay at residential home, the patients being visited in a hospital environment, the passenger flow intensity in an airport, and an increased or decreased industrial production. Data warehousing being applied to these historical series characterizes the evolution of the power grid and estimates the expected trends.

The real life shows a set of correlated events - originated by actors happening in the time and space – characterized by specific properties which can be fixed using predicates. We assess spatial and temporal relations linking events happening in real life and detected by sensors or digital meters, contrary to other applications ignoring the above difference. We distinguish between an infinite cyclic time with the event chain repeated an infinite number of times, and a limited cyclic time where the number of repetitions is finite. The analysis of the load shapes collected during several years might show some cycles. The repetitiveness of the patterns discovered during the long periods of time is an interesting knowledge enabling new application classes, for example dynamic portfolio restructuring thanks to the cluster analysis and new triggering scheme.

Energy consumers typically buy energy up to the maximal power thresholds kept static along the years. To make them dynamic, we need new real-time tools estimating the energy needs, calling for shape's warehousing. User can be motivated to monitor peak energy dynamics adopting sophisticated business scenario stimulating anticipatory knowledge about expected loads. Prosumers can match individual energy dynamics with the commercial offerings in real time, choosing automatically the optimum. Individual data warehousing predicts the expected future behaviour, while the comparison of the individual DE shapes with the known clusters (Fig. 4) might increase the precision because relying on the validated elaborations. It is helpful when penalties for any over-consumption exist.

The developed tool enables automatic real-time energy trading on liberalised energy markets, where price changes in real time, because expected quantities are calculated using advanced data management algorithms. It permits choosing from the different energy partners in real time, discriminating between renewable or non-renewable stakeholders, but requires some exact knowledge about the energy dynamics. Assuming some distributed photovoltaic plants generating and trading energy in real time, the cloud variability impacting on the production should be accounted before contracting the power sells. A reference application [41] comprises some weather sensors, forecast, and the local algorithms producing short-term trading decisions with some *a-posteriori* validation. The use of the *anticipatory knowledge* and its sharing in power grids enables the collective optimization of the energy resources, showcasing the use of the Future Internet technologies.

The use of the digital energy is based on the real-time processing of smart metering datagrams. Depending on the protocol used, the information is formatted in different ways, but carries same energy consumption dynamics. The difference DE_i - DE_{i-1} shows an event happening in time interval $[t_{i-1}, t_i]$, made available at t_i. Remote node listen individual energy dynamics, but can integrate the group ones, because reasoning over the time series. After some delay for knowledge processing, the system makes the forecast valid for "next" time frame $[t_i, t_{i+1}]$.

The traditional approach based on under-frequency is reactive, because detecting at t_{i+1} any "unhappy" event happened at t_i. System can react currently not earlier than t_{i+2}. Instead, DE-based load management becomes proactive thanks to the evident time gain $[t_{i+1}, t_{i+2}]$: the decision over the digital energy happens at t_{i+1} instead of t_{i+2}. Specialized DE class declaring only changes in consumption dynamics simplifies the networked event processing because triggering anticipated frequency variations. Users can also declare in advance their expected consumptions to permit the anticipatory network load balance and/or price discounts. In DE scenario, the network at any time t_{i-1} is considered stable and balanced. The sum of DE messages, processed at t_i, indicates an immediate local unbalance, which is going to originate the under-frequency. Running the load forecast in parallel drafts also the most likely anticipation of the situation at t_{i+1}.

We have implemented a single Intelligent Energy Server for Consumption Dynamics (IESCD) interoperable in real-time with one off-the-shelf energy analyser (FEMTO D4 device made by Electrex [42]). One experiment was set up in order to showcase the time-wrapping properties of the energy digitisation, varying the DE resolution. The main use of the software is the acquisition of the awareness about the real-time energy consumption dynamics, the triggering of and the reaction to DE events, the consolidation and the local storage of the resulting load shapes, plus an attempt to annotate semantically relevant events, adding some pattern-analysis. Delphi-written IESCD prototype ver.1, of which the initial IESCD Human Machine Interface (HMI) is shown on Fig. 12, runs under Windows OS.

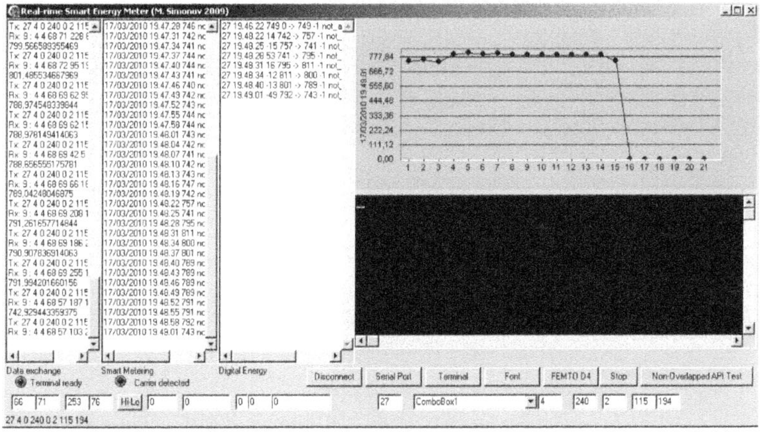

Fig. 12 Intelligent Energy Server for Consumption Dynamics (IESCD)

The experimental residential data acquisition is Netbook-based because alimented from the battery in order to exclude the additional energy consumption dynamics. The communication ComPort Library is an open-source project [43]. Library implementations, including CRC calculation and IEEE 754 data type mapping are proprietary. The main application class contains the event-driven annotation feature, and the timer-controlled event triggering routine, enabling the pattern-matching. Data handling module permits exporting datasets in human readable textual representation, keeping the eventually available annotation. The main difference of the proposed solution, compared with the current smart metering schemes, is the much lower data amount exchanged. In fact the DE event corresponds to the 1st derivate value exceeding the monitoring threshold, so varying this value we have reduced the dataflow up to 68 samples out of 6800 in certain periods, keeping the network observable.

7 Conclusions

In this chapter, we considered in detail a real life distributed environment – smart power grids - and discussed the application of the modern Information Retrieval, Data Mining and Knowledge Management to the available from the environment itself information set, outlining the industrial or commercial potential they might have. An attempt to contribute with a theory in the electric energy digitization helps to abstract the real energy with the digital entity from Internet of Things in order to apply the above-mentioned techniques to energy domain. Information Retrieval in LV segments of power grids is the first issue to face in order to create the local intelligence through the autonomous agents representing the lowest layers and interlinking the upper ones. The information mining over the static snapshots of load shapes permits energy assessing in real life, while the data-warehousing shows the energy consumption dynamics. Some Use Cases were selected to illustrate the concrete application classes, justifying the effort dedicated to the energy analysis.

Future work will be related to the experimentation with higher sampling rates, an attempt to remove some scalability limits, and the algorithmic solutions supporting new Future Internet dynamic entities.

Acknowledgments. This work is developed under the late Ph.D. research programme, drawing upon the contributions of many people from both ICT and Electric Engineering scientific community.

References

1. Burke, J.J.: Power Distribution Engineering. Marcel Dekker, Inc., New York (1994)
2. Fan, J., Borlase, S.: The evolution of distribution, Power and Energy Magazine. IEEE 7(2), 63–68 (2009)
3. European Commission, European SmartGrids Technology Plaform: Vision for Europe's Electricity Networks of the Future, EUR 22040, EC (2006)
4. Qiang, Z., Danyan, C.: Design and Implementation of Distribution Network SCADA System Based on J2EE Framework. In: International Forum on Information Technology and Applications, IFITA 2009, vol. 1, pp. 633–636 (2009)
5. Hirodontis, S., Li, H., Crossley, P.A.: Load shedding in a distribution network. In: Intl. Conference on Sustainable Power Generation and Supply (SUPERGEN 2009), pp. 1–6 (2009)
6. Phadke, A.G., De Moraes, R.M.: The Wide World of Wide-area Measurement. IEEE Power and Energy Magazine 6(5), 52–65 (2008)

7. Thorp, J.S., Phadke, A.G.: Protecting power systems in the post-restructuring era. IEEE Computer Applications in Power 12(1), 33–37 (1999)
8. Thorp, J.S., Phadke, A.G.: IEEE Guide for the Application of Protective Relays Used for Abnormal Frequency Load Shedding and Restoration, IEEE Std C37.117-2007 (2007)
9. Delfino, B., Massucco, S., Morini, A., Scalera, P., Silvestro, F.: Implementation and comparison of different under frequency load-shedding schemes. IEEE Power Engineering Society Summer Meeting 1, 307–312 (2001)
10. Delfino, B., Massucco, S., Morini, A., Scalera, P., Silvestro, F.: IEEE Recommended Practice for Monitoring Electric Power Quality, IEEE Std 1159-2009 (Revision of IEEE Std 1159-1995) (2009)
11. Simonov, M., Mussetta, M., Zich, R.E.: Digital energy: clustering microgrids for social networking. International Journal of Virtual Communities and Social Networking 1(3), 75–93 (2009)
12. Derbel, F.: Trends in smart metering. In: 6th International Multi-Conference on Systems, Signals and Devices, SSD 2009, pp. 1–4 (2009)
13. Singh, T., Vara, P.K.: Smart Metering the Clouds. In: IEEE Int. Workshops on Enabling Technologies: Infrastructures for Collaborative Enterprises, WETICE 2009, pp. 66–71 (2009)
14. Rogai, S.: Telegestore Project Progresses And Results. In: IEEE International Symposium on Power Line Communications and Its Applications, ISPLC 2007, p. 1 (2007)
15. Rusitschka, S., Gerdes, C., Eger, K.: A low-cost alternative to smart metering infrastructure based on peer-to-peer technologies. In: 6th International Conference on the European Energy Market, EEM 2009, pp. 1–6 (2009)
16. De La Ree, J., Liu, Y., Mili, L., Phadke, A.G., DaSilva, L.: Catastrophic Failures in Power Systems: Causes, Analyses, and Countermeasures. Proc. of the IEEE 93(5), 956–964 (2005)
17. Rocha Luis, M.: Complex Systems Modeling: Using Metaphors From Nature in Simulation and Scientific Models. In: BITS: Computer and Communications News. Computing, Information, and Communications Division, Los Alamos National Laboratory (1999)
18. Xinqing, L., Tsoukalas, L.H., Uhrig, R.E.: A neurofuzzy approach for the anticipatory control of complex systems. In: 5th IEEE Int.Conf. on Fuzzy Sys. proc., vol. 1, pp. 587–593 (1996)
19. Tolbert, L.M., Qi, H., Peng, F.Z.: Scalable multi-agent system for real-time electric power management. IEEE Power Eng. Society Summer Meeting 3, 1676–1679 (2001)
20. Adamiak, M.G., Apostolov, A.P., Begovic, M.M., Henville, C.F., Martin, K.E., Michel, G.L., Phadke, A.G., Thorp, J.S.: Wide Area Protection - Technology and Infrastructures. IEEE Trans. on Power Delivery 21(2), 601–609 (2006)
21. Phadke, A.G., Thorp, J.S.: Synchronized Phasor Measurements and Their Applications. Springer, N. Y. (2008)
22. Heydt, G.T., Liu, C.C., Phadke, A.G., Vittal, V.: Solution for the crisis in electric power supply. IEEE Computer Applications in Power 14(3), 22–30 (2001)
23. Tsoukalas, L.H., Gao, R.: From smart grids to an energy internet: Assumptions, architectures and requirements. In: Third International Conference on Electric Utility Deregulation and Restructuring and Power Technologies, DRPT 2008, pp. 94–98 (2008)
24. Liu, G., Wei, H.L., Wang, X., Peng, W.: Research on Data Interoperability Based on Clustering Analysis in Data Grid. In: International Conference on Interoperability for Enterprise Software and Applications, IESA 2009, pp. 97–103 (2009)

25. Coppola, M., Pesciullesi, P., Ravazzolo, R., Zoccolo, C.: A Parallel Knowledge Discovery System for Customer Profiling. In: Danelutto, M., Vanneschi, M., Laforenza, D. (eds.) Euro-Par 2004. LNCS, vol. 3149, pp. 381–390. Springer, Heidelberg (2004)
26. Horowitz, S.H., Phadke, A.G.: Blackouts and relaying considerations - Relaying philosophies and the future of relay systems. IEEE Power and Energy Magazine 4(5), 60–67 (2006)
27. Phadke, A.G., Kasztenny, B.: Synchronized Phasor and Frequency Measurement Under Transient Conditions. IEEE Trans. on Power Delivery 24(1), 89–95 (2009)
28. Aisa, V., Falcioni, P., Pracchi, P.: Connecting white goods to a home network at a very low cost. In: International Appliance Manufacturing, pp. 85–91 (2004)
29. Adams, C.E.: Home Area Network Technologies. BT Technology Journal 20(2), 53–72 (2002)
30. Ma, L., Zhang, Q., Wang, K., Li, X., Wang, H.: Semantic Load Shedding over Real-Time Data Streams. In: International Symposium on Computational Intelligence and Design, ISCID 2008, vol. 1, pp. 465–468 (2008)
31. Dey, A.K., Anind, K.: Understanding and Using Context. Personal Ubiquitous Computing 5(1), 4–7 (2001)
32. Ma, L., Li, X., Wang, Y., Wang, H.A.: An Approach to Handle Overload in Real-Time Data Stream Management System. In: Fifth International Conference on Fuzzy Systems and Knowledge Discovery, FSKD 2008, vol. 3, pp. 3–8 (2008)
33. Uslar, M.: Semantic Interoperability within the Power Systems Domain. In: Proc. of the ACM Conference on Information and Knowledge Management, CIKM 2005, pp. 39–45 (2005)
34. Robinson, G.: Key Standards for Utility Enterprise Application Integration (EAI). In: Proc. of the Distributech 2002, Miami, Pennwell (2002)
35. Uslar, M., Rohjans, S., Schulte, S., Steinmetz, R.: Building the Semantic Utility with Standards and Semantic Web Services. In: Meersman, R., Tari, Z., Herrero, P. (eds.) OTM-WS 2008. LNCS, vol. 5333, pp. 1026–1035. Springer, Heidelberg (2008)
36. Schulte, S., Eckert, J., Repp, N., Steinmetz, R.: An Approach to Evaluate and Enhance the Retrieval of Semantic Web Services. In: 5th ICSSSM Conf. 2008, pp. 1–7 (2008)
37. Brooks, A.: Integration of electric drive vehicles with the power grid-a new application for vehicle batteries. In: The 17th Annual Battery Conf. on Appl. and Advances, p. 239 (2002)
38. Pecas-Lopes, J.A., Rocha-Almeida, P.M., Soares, F.J.: Using vehicle-to-grid to maximize the integration of intermittent renewable energy resources in islanded electric grids. In: International Conference on Clean Electrical Power, pp. 290–295 (2009)
39. Park, C.E., Lee, Y.H., Kim, D.J., Lee, S.C., Moon, U.C.: An Inference Technique based on Semantic Primitives for the Development of Intelligent Load Distribution Systems. In: Int. Conf. on Power System Technology, pp. 1–5 (2006)
40. Sammartino, L., Simonov, M., Soroldoni, M., Tettamanzi, A.: GAMUT: A system for customer modeling based on evolutionary algorithms. In: GECCO 2000, p. 758 (2000)
41. Simonov, M., Mussetta, M., Pirisi, A., Grimaccia, F., Caputo, D., Zich, R.: Real time energy management in smart cities by Future Internet. In: Tselentis, G., et al. (eds.) Towards the Future Internet – A European Research Perspective, IOS Press, Amsterdam (2010)
42. Electrex, Femto D4 RS485 Data Sheet (2009), `http://www.electrex.it/pdf_catalogo/eng/Data_sheet_Femto.pdf`
43. Crnila, D.: ComPort Library, `http://sourceforge.net/projects/comport`

Author Index

GPSR Compliance

The European Union's (EU) General Product Safety Regulation (GPSR) is a set of rules that requires consumer products to be safe and our obligations to ensure this.

If you have any concerns about our products, you can contact us on ProductSafety@springernature.com

In case Publisher is established outside the EU, the EU authorized representative is:

Springer Nature Customer Service Center GmbH
Europaplatz 3
69115 Heidelberg, Germany

Batch number: 10755466

Printed by Printforce, the Netherlands